ADVANCES IN SONOCHEMISTRY

Volume 4 • 1996

ADVANCES IN SONOCHEMISTRY

Editor: TIMOTHY J. MASON
School of Natural and Environmental Sciences
Coventry University
Coventry, England

VOLUME 4 • 1996

JAI PRESS INC.

Greenwich, Connecticut *London, England*

55 Old Post Road, No. 2
Greenwich, Connecticut 06836

JAI PRESS LTD.
38 Tavistock Street
Covent Garden
London WC2E 7PB
England

ISBN: 1-55938-793-9

Manufactured in the United States of America

DEDICATION

This volume is dedicated to the memory of Jacques Berlan who died April 1995. He was a good friend and colleague of mine and will be sorely missed by the international sonochemistry community.

CONTENTS

LIST OF CONTRIBUTORS

C. Bao	Department of Chemistry Yunnan University Kunming, China
*Jacques Berlan**	ENSIGC Toulouse, France
Georgy I. Eskin	All Russia Institute of Light Alloys Moscow, Russia
R. Feng	Acoustic Institute Nanjing University Nanjing, China
John Homer	Department of Chemical Engineering and Applied Chemistry Aston University Birmingham, England
Timothy J. Mason	School of Natural and Environmental Sciences Coventry University Coventry, England
Stuart A. Palfreyman	Department of Chemical Engineering and Applied Chemistry Aston University Birmingham, England
Larysa Paniwnyk	School of Natural and Environmental Sciences Coventry University Coventry, England

*Deceased

Sukhvinder S. Phull	School of Natural and Environmental Sciences Coventry University Coventry, England
David J. Walton	School of Natural and Environmental Sciences Coventry University Coventry, England
J. Yin	Department of Chemistry Yunnan University Kunming, China
Y. Zhao	Department of Chemistry Yunnan University Kunming, China

PREFACE

The general acceptance of sonochemistry and the exciting results obtained using the technique have resulted in the formation of large numbers of new (and established) research groups and industrial interest in the possibilities of scale-up. As a result sonochemists are now facing two problems linked with power measurement. The first is that the results obtained in one laboratory are not necessarily identical to those from another (due to a great variety of different makes and configurations of ultrasonic apparatus currently in use). The second relates to the fact that sonochemical results obtained in small reaction volumes are not always easily scaled up for use in large reactors. There is a common origin for these problems. At present there is no universally accepted absolute method of quantifying the amount of ultrasonic energy used to perform a particular chemical transformation and great efforts are being made to establish protocols. Central to the solution of this problem is the development of probes which will act as dosimeters for ultrasonic irradiation. Some are based on direct measurement of electrical or acoustic power, some on chemical reactions influenced by cavitation, while others are based on the determination of mass transfer through electrochemical measurements. In the first chapter of this volume the editor and (the late) Jacques Berlan have attempted to put together the various dosimetry methodologies available to sonochemists in order to show the range of options available. It is hoped that before the next volume goes to press a decision will have been taken within the sonochemical community as to which methodologies are the most appropriate for sonochemistry.

The U.K. and other European countries have both seen an increased interest in the uses of ultrasound in metallurgy. Several companies are exploring the possibilities of the use of power ultrasound not only for grain refinement and degassing during casting but also in welding and surface treatment. In chapter 2 Georgy Eskin expands on the information which appeared in Volume 2 of this series in a chapter entitled, "The Action of Ultrasound on Solidifying Metals," by Oleg Abramov. The current chapter concentrates on light metal casting technology.

In recent years there have been two major conferences in China devoted to acoustics: in Peking in 1992 the Fourteenth International Congress on Acoustics and in 1994 a conference celebrating the fortieth anniversary of the Acoustics Institute in Nanjing. At both of these there were presentations involving sonochemistry. Much of the material was published in the Chinese literature and so this volume presented an ideal opportunity to gather together recent work in China, thereby making it more readily accessible to the international sonochemistry community. One of the authors of chapter 3, Professor Feng, was involved in both conferences, while Zhao Yiyun spent some time at Coventry University working in the Centre of Excellence in Sonochemistry.

There is very little information available regarding a new subject that embraces the combined use of nuclear magnetic resonance spectroscopy (NMR) and ultrasound which seems to be providing some fascinating information on molecular structure. In chapter 4 one of the originators of such studies, John Homer, has focused attention on this topic and particularly on his own work on the development of sonically induced narrowing of the NMR spectra of solids (SINNMR). This promises to provide a rapid and reasonably inexpensive method for the investigation of the NMR of solids.

Despite the fact that high-frequency, low-power ultrasound has been employed within the food industry for many years in a diagnostic mode, the potential for the use of power ultrasound is far less well known. In fact the very features which make sonochemistry a useful tool in chemical processing have their analogies in food industry and these are explored in chapter 5 by the editor and Larysa Paniwnyk. Ultrasound has been found to aid oxidation/reduction processes, e.g. to enhance the flavor of wines and spirits. It can also be employed to inhibit some enzyme activity and to destroy microorganisms both of which can reduce food spoilage. The mechanical effects of ultrasound applied to improve emulsification, defoaming, degassing, filtration, and drying are all important to the food industry. Also highlighted are two additional technologies which are still awaiting commercial exploitation: the extraction of food products from vegetable material, and the remarkable improvements possible in food freezing.

The simultaneous application of ultrasonic irradiation to an electrochemical reaction which has been termed sonoelectrochemistry has been shown to produce a variety of benefits in almost any electrochemical process. These include: enhanced chemical yield in electrosynthesis and the control of product distribution; improved electrochemical efficiency in terms of power consumption, improved mixing, and diffusion in the cell; minimization of electrode fouling; accelerated degassing; and often a reduction in the amount of process-enhancing additives required. In a major chapter devoted to this topic, Suki Phull and Dave Walton have attempted to cover the majority of applications of ultrasound in electrochemistry including: electrochemical synthesis, electroanalytical chemistry, battery technology, electrocrystallization, electroinitiated polymerization, and electroplating.

Timothy J. Mason
Series Editor

DOSIMETRY FOR POWER ULTRASOUND AND SONOCHEMISTRY

Jacques Berlan and Timothy J. Mason

OUTLINE

Advances in Sonochemistry
Volume 4, pages 1–73

ISBN: 1-55938-793-9

1. INTRODUCTION

The last few decades have seen enormous developments in the uses of ultrasound for the provision of versatile tools for testing, detecting, imaging, chemical processing, and research [1–15]. In sonochemistry progress has been particularly rapid over the last few years and many of the chemists and engineers involved in this field are attempting to quantify the effects of power ultrasound on chemical systems. Such studies necessarily involve obtaining optimum and reproducible effects from ultrasonic irradiation. Inextricably tied in with this approach must be the need for accurate acoustic power measurement, i.e. dosimetry. In this chapter we will attempt to give an overview of the current state of knowledge in this field of research.

In order to make any progress in quantifying sonochemistry three conditions must be fulfilled:

- the nature and characteristics of the ultrasonic device should be known,
- the properties of the medium to be sonicated should be appreciated,
- the coupling between the sonic device and the treated material should be optimized.

Each of these will have a profound effect in determining the character of the propagating acoustic field, which can be defined in terms of: its wave length (l), frequency (f), sound velocity (c), particle displacement (d), velocity (v), sound intensity (I), transmitted acoustic power (W_t), and other derived quantities [1,3,16,17]. Some of these depend on the ultrasonic device itself (e.g. frequency, amplitude); other conditions are mainly dependent on the irradiated medium (e.g. sound velocity, wavelength), but some of them strongly depend on the coupling efficiency between the sound emitter and the medium (e.g. particle velocity, sound intensity and transmitted power).

A very important point occurs in the transmission of acoustic power into a liquid which is termed the cavitation threshold. When very low power ultrasound is passed through a liquid and the power is gradually increased, a point is reached at which the intensity of sonication is sufficient to cause cavitation in the fluid. It is only at powers above the cavitation threshold that the majority of sonochemical effects occur because only then can the great energies associated with cavitational collapse be released into the fluid. In the medical profession, where the use of ultrasonic scanning techniques is widespread, keeping scanning intensities below the cavitation threshold is of vital importance. As soon as the irradiation power used in the medical scan rises above this critical value, cavitation is induced and, as a consequence, unwanted even possibly hazardous chemical reactions may occur in the body. Thus, for both chemical and medical reasons there is a considerable drive towards the determination of the exact point at which cavitation occurs in liquid media, particularly in aqueous systems. Historically, therefore, the determination of the cavitation threshold was one of the major drives in dosimetry.

Once the cavitation threshold is passed the determination of sound intensity and sonic power are of crucial importance since the use of too high an ultrasonic power often results in deleterious effects in the treated material (e.g. excessive heating, or degradation of material due to the breaking of chemical bonds), Thus in medical imaging, nerve blockage has been reported at 1–5 MHz in 10 minutes when the intensity was 2 W cm^{-2} [16]. In the same way, solvent degradation is likely to occur above the cavitation threshold, and this could result in unwanted side reactions [2]. On the other hand none of the anticipated sonochemical or mechanical effects will occur if the acoustic energy input is too small. Thus very poor results in terms of polymer degradation [6,7] or particle disruption [18] are obtained at low power.

Accurate power monitoring is also needed for two other reasons:

- it is well known that over a long period of use transducer efficiency decreases, and this will naturally result in poorer sonochemical effects,
- obtaining reproducible results requires strictly monitored conditions, and more particularly the necessity for identical power input.

For all of these reasons, many methods have been devised to measure ultrasonic power [16–23]. It is extremely important that we define exactly what type of power is to be measured, e.g. power associated with the transducer system, with physical effects generated in the liquid, or with chemical effects induced by cavitation.

2. POWER MEASUREMENTS: A MANY-SIDED PROBLEM

2.1 Basic Definitions

If W is the acoustic energy density in W cm^{-3} and I is the quantity of energy in W cm^{-2} propagated each second through the unit surface area,

$$W = I/c \tag{1}$$

where c is the speed of the sound through the medium of propagation [2,23].

The vibration, or motion amplitude of an imaginary particle in such an acoustic field is related to I according to,

$$I = \tfrac{1}{2}\rho\, c\, A^2\, \omega^2 \tag{2}$$

where ρc (the product of the medium density and the speed of the sound, respectively) is the acoustic impedance of the medium, A is the maximum elongation, and ω the pulsation ($\omega = 2\pi f$ where f is the frequency).

The acoustic pressure amplitude is:

$$P = (2\,\rho\, c\, I)^{1/2} \tag{3}$$

The velocity amplitude V of an imaginary particle is also related to I through,

$$I = \tfrac{1}{2}\rho\, c\, V^2 \tag{4}$$

and the acceleration amplitude is given by:

$$\Gamma = \omega\, V = \omega\, (2I/\rho\, c)^{1/2} \tag{5}$$

Several others parameters can be used [1–3,16,17] which are related to the main field characteristics and a knowledge of these is of crucial importance to understand and monitor ultrasound effects. The methods used for power measurements and testing of transducers will rely on the measurement of one of these characteristics by direct or indirect methods.

Each application will require different transducer properties to get the expected effect. For example, ultrasonic cleaning requires a broad uniform sound field, drills and welders require high amplitude motions, ultrasonic flowmeters utilize narrow uniform sound beams with stable amplitudes, and so on. Thus it is unlikely that one dosimetry method can be used for all types of transducers and sonic devices, a feature previously highlighted by Welkowitz [22].

2.2 General Considerations

An ultrasonic device [1,3–7,12,24] is generally composed of (a) a generator, producing a high frequency current; (b) a transducer, converting this current into a

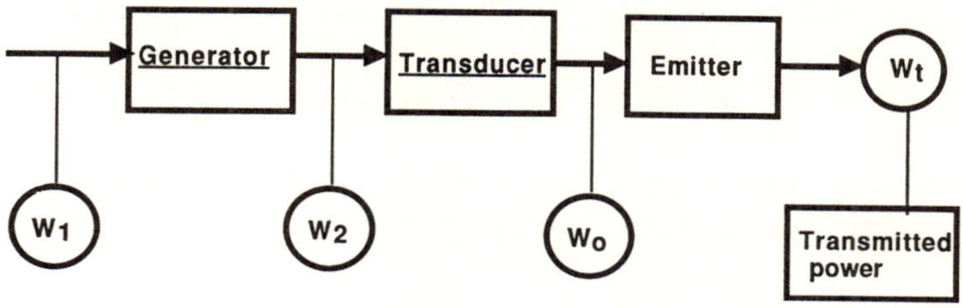

Scheme 1. Positions for the measurements of ultrasonic power in a typical sonication system.

mechanical vibration at the same frequency; and (c) an emitter to transmit this vibration into the treated medium.

Power measurements can be made at three different levels (Scheme 1). It is possible to measure the total power consumption of the system, that is the input power to the generator (W_1), or the input power to the transducer (W_2), or the transmitted power which enters the treated medium (W_t).

These three quantities are not totally independent. The W_2/W_1 ratio depends on the generator and gives its energetic efficiency. The W_t/W_2 ratio depends on the efficiency of the coupling between the emitter and the sonicated medium. Due to different acoustic impedance, not all of the energy output from the transducer is transmitted to the medium, and part is reflected at the emitter/medium interface [1] and this is degraded into heat in the transducer. This will depend of course on the nature of the irradiated medium (density, viscosity, gas content,...) and on the experimental conditions (temperature, external pressure), but it will also depend on the mass and almost certainly on the geometry of the reactor.

W_1 and W_2 can easily be measured with a wattmeter or with an oscilloscope by determining the applied voltage (V), the current intensity (I) and the phase shift ($\cos\phi$). Determination of the phase shift at the entry of the transducer may provide interesting information [25,26]. However the determination of these quantities is very seldom reported in patents or research papers.

Total energy consumption, W_1, is very important in power ultrasound, since it can determine the viability of an industrial process, but it is much less important in ultrasonic scanning.

Energy input to the transducer, W_2, is critical for reproducibility, and is probably the easiest parameter to handle. As already mentioned, it depends at least partly on "matching" of the acoustic source with the irradiated medium. This can be clearly illustrated experimentally [26]. Using the device illustrated in Figure 1, a cup-horn (Sodeva; working frequency 20 kHz driven by a generator from Sonics and Materials) was filled with different liquids. The liquid height H was varied from 8 to 20 cm and the consumed power at the generator, W_1 was maintained constant at a low level (2 to 10 watts). Voltage V, current intensity I, and phase shift cos Ý at the transducer were measured with an oscilloscope; this gives the input power to

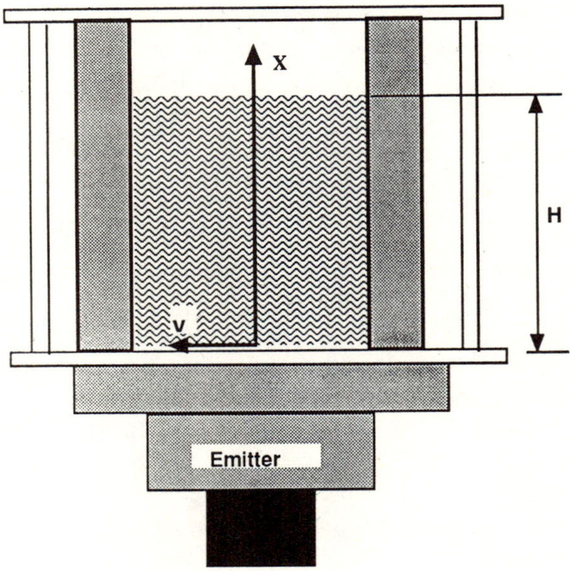

Figure 1. Sodeva "cup-horn" device.

the transducer, W_2, which was plotted against the liquid height. A typical result, obtained using 0.5 M aqueous NaOH, is given in Figure 2.

It can be seen that W_2 depends on the liquid height. This means that it depends on the acoustic load and on matching this with the emitter. Most interestingly, it passes through maximum values, and the distance between any two maxima gives the half-wave length of the sound in this medium. This experiment was repeated with several liquids as shown in Table 1 and the measured wavelength was compared with literature data.

This observation is quite interesting as it provides a very easy route to the wavelength measurement in complex media (e.g. chemical reactions) the value of which cannot be found in the literature. It also clearly illustrates the interest of sonochemists in equipment which includes automatic matching, such as the Undatim Sonoreactor. Such systems are based on the search for a maximum ultrasound input through automated tracking of the frequency around the nominal vibration frequency of the probe.

The third power, transmitted power (W_t), is also very important, but its measurement is much less straightforward. Theoretically, W_0, the output power from the transducer, can be calculated [1], and similar equations have been established to calculate W_t or the sound intensity I in $\mathrm{Wcm^{-2}}$ [1,27], such as,

$$I = \frac{4e_{11}^2 V_{\mathrm{eff}}^2 \times 10^{-7}}{T^2 Z 300^2} \qquad (6)$$

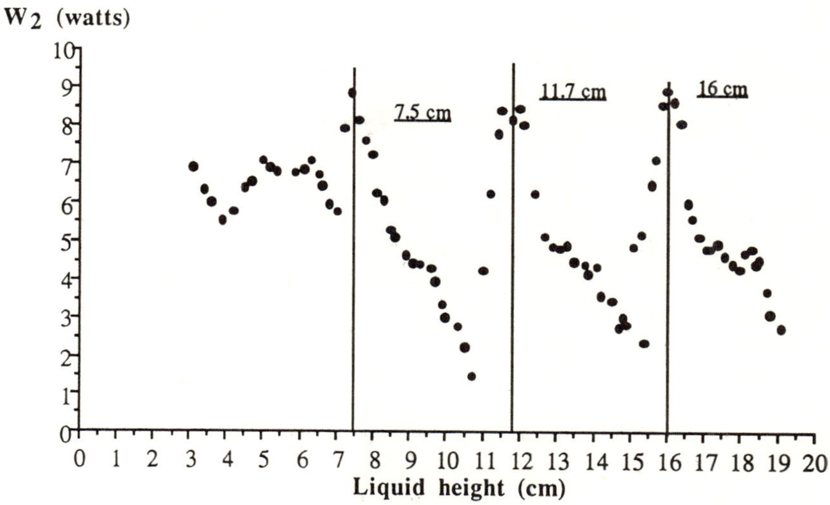

Figure 2. Input power to the transducer against liquid height (0.5 M aqueous NaOH) in a Sodeva cup-horn.

where e_{11} = piezoelectric constant; V_{eff} = input voltage; T = crystal thickness, and Z = acoustic impedance of the medium. Such equations are of limited interest since they do not take into account parameters such as aging of the transducer or the nature of the experimental assembly in which the transducer is mounted.

To determine W_t one could also measure W_2 and adopt a known power transformation ratio. This ratio of course depends on the acoustic load and has to be determined in each and every particular case. Again, such a method does not take into account any possible aging of the system, and calibration should be made from time to time. Furthermore, it will be of little use in chemistry as the acoustic impedance of the load will almost certainly change as the reaction proceeds.

Many other different methods have been devised to measure the transmitted power, and these have been reviewed from time to time [16–22]. All these methods rely on the measurement of a primary or secondary effect on the propagation of the

Table 1. Comparison of Literature and Experimentally Determined Values for the Half Wavelength of 20-kHz Ultrasound in Different Solvents

Liquid	*λ (cm) Literature*	*λ (cm) Experimental*
Water	7.48	7.4
Toluene	6.4	6.4
Ethylene glycol	8.29	8.2
Acetone	5.87	6.1

wave in the irradiated material. In gases or solids, this is a relatively simple problem, but the situation is much more complicated in liquids, or materials having large liquid content, including human tissue. In this case, propagation of the sonic wave results in highly complex nonlinear and interdependent phenomena, especially beyond the cavitation threshold, including acoustic streaming (micro and macromixing), shear forces, and jet streams together with stable and transient bubble oscillations. Depending on the measured effect and on the experimental device, these methods can be applied for global or local power measurements, in a free or restrained field, at high or low intensity or frequency.

These can be tentatively divided into four main groups, although some are interrelated thus a strong link exists between thermal effects, acoustic streaming, and cavitation.

Class 1–Thermal Methods. These utilize calorimeters [27,29], thermocouples or absorption probes [32,49], and the acoustic dilatometer [30,31].

Class 2–Measurements at the Transducer. These use electrical [2,19,144,53] or mechanical [22,54,56] methods of measurement.

Class 3–Methods Based on Direct Mechanical Effects. These include the use of acoustical probes [57–71], acoustic impedance measurements [72–75], acoustic fluxmeter [76], the measurement of radiation forces [17,21,77–112], the distortion of liquid surface [113–115], surface cleaning, dispersive effects, emulsification [116–118], erosion [19,22,119–125], mass transfer measurements (electrochemical probe) [26,129], absorption methods [93,132], particle velocity measurements [132], and optical methods [133–141].

Class 4–Methods Based on Secondary Effects of Sound Propagation and Cavitation. These include methods based on volume changes [142–144], acoustic output and noise measurements [145–150], conductance changes, electric and electrokinetic effects [151–156], sonoluminescence [157–171], and chemical probes [172–198].

Other classifications have been suggested in the past; thus Neppiras [16] proposed three classes:

1. Statistical methods based on measuring energy or radial velocity in the liquid in the cavitation field: thermal methods, acoustic output, and measurement of velocity associated with bubble oscillation.
2. Methods based on measuring the undissolved gas content of the liquid: direct measurement of the volume change; effect of undissolved gas on the electrical impedance of the source, transparency of the liquid to light (UV-visible), diagnostic ultrasound, X-rays; and effect on electrical or acoustic permittivity or conductivity.

3. Methods based on measuring primary effects of cavitation, i.e. effects occurring in the gas phase: sonoluminescence or effects occurring at the gas/liquid interface and\or entirely in the liquid phase, sonochemical effects; erosion, dispersion, accelerated dissolution; and biochemical effects.

Zieniuk [17] suggested that these three types of measurement could be distinguished as:

1. Methods giving absolute energy values: thermal measurements (calorimeter, thermal probes).
2. Methods based on acoustic pressure: capacitive or piezoelectric probes, optical methods, etc.
3. Methods based on nonlinear effects: e.g. radiation forces. He also introduced other subdivisions depending on whether the method could be used for total or local power measurements, under free or restrained field conditions.

3. THERMAL METHODS IN DOSIMETRY

Among the existing methods of measuring ultrasonic energy thermal methods are currently the most common. They can be divided into two classes: calorimetry and measurements which involve thermal probes. The former are used to get total power, mean intensity or energy density, while the latter are more specifically targeted at the measurement of local power at particular points in the acoustic field within the reactor. If the thermal probe measurements were integrated over the whole volume such values could also give the overall power.

3.1 Calorimeters

There are two different kinds of calorimeter: adiabatic (or quasi-adiabatic calorimeters) and non-isothermal, non-adiabatic calorimeters (often referred to as n-n calorimeters). The accuracy of measurements made using such methods will be high if:

- all the acoustic energy entering the system is transformed into heat; this requires good matching of the system and low reflections at any interfaces
- the temperature measurements are very accurate; temperature rises are greater at high intensities and this will also generally require that the time response factor of the thermometer must be rapid.

Calorimetric methods are quite general; they can be used under cavitating conditions and in either free or restricted ultrasonic fields. Essentially the technique consists of measuring the rate of temperature increase in the sonicated liquid and from this calculating the power input according to Eq. (7),

$$W = \frac{m\,c_{\mathrm{p}}\Delta T}{\Delta t} + w_{\mathrm{c}} \tag{7}$$

where m is the mass of heated liquid, c_{p} its heat capacity, ΔT the temperature increase during the sonication time Δt, and w_{c} a correcting term due to the heat absorption by the calorimeter and heat losses. Alternatively, one can plot the temperature versus time, determine the slope of the tangent to this curve at time zero $(\delta T/\delta t)_{t=0}$, either using a graphical method or by curve-fitting the data to a polynomial in t, and then calculate the acoustic power according to Eq. (8):

$$W = m\,c_{\mathrm{p}}\,(\delta T/\delta t)_{t=0} + w_{\mathrm{c}} \tag{8}$$

The accuracy of the method depends on the correct determination of the characteristics of the ultrasonic field, and accurate measurement of the temperature rise of an absorber of closely controlled energy equivalent. It is also essential that the absorbing medium has a high absorption coefficient and that the acoustic matching within the system is good. Castor oil is among the best liquids which have been used to calibrate transducers.

When the method used depends upon an estimation of the slope of the tangent at zero time, it is essential to provide good stirring of the liquid. In the absence of stirring the response of the thermal probe is far from simple.

Adiabatic or Quasi-Adiabatic Calorimeters

In an ideal system, the emitter should be acoustically coupled to and thermally isolated from the absorbing medium through a resonant entry made of a plate of a solid material having low acoustic loss and low thermal conductivity; in most cases a coupling liquid is added. Suitable solid coupling materials are fused quartz, some ceramics, or titanium alloys, and the coupling liquid is generally water or castor oil. The use of a resonant entry results in some drawbacks; for example it is only matched for a particular equipment set up and this will apply to only a single frequency. The resonant entry must be changed every time that the emitter frequency changes in order that the emitter can be made to function under optimized conditions. Alternatively, the emitter can be directly dipped into the calorimeter. In this case, however, the coupling with the load could well be less than ideal, and in this situation the power could be underestimated. The vessel must be lagged to prevent heat losses.

This methodology is not suitable for ultrasonic cleaners, but it can be used to calibrate horns or any emitter coupled to a device that can be fitted to the calorimeter. It does however suffer from several disadvantages including the need for corrections, which are difficult to determine, and the high thermal inertia of the calorimeter.

Sokollu [28] described a calorimeter made of a double-walled brass vessel, with internal stirrer and a conical metallic sound diffuser to scatter the beam in a random

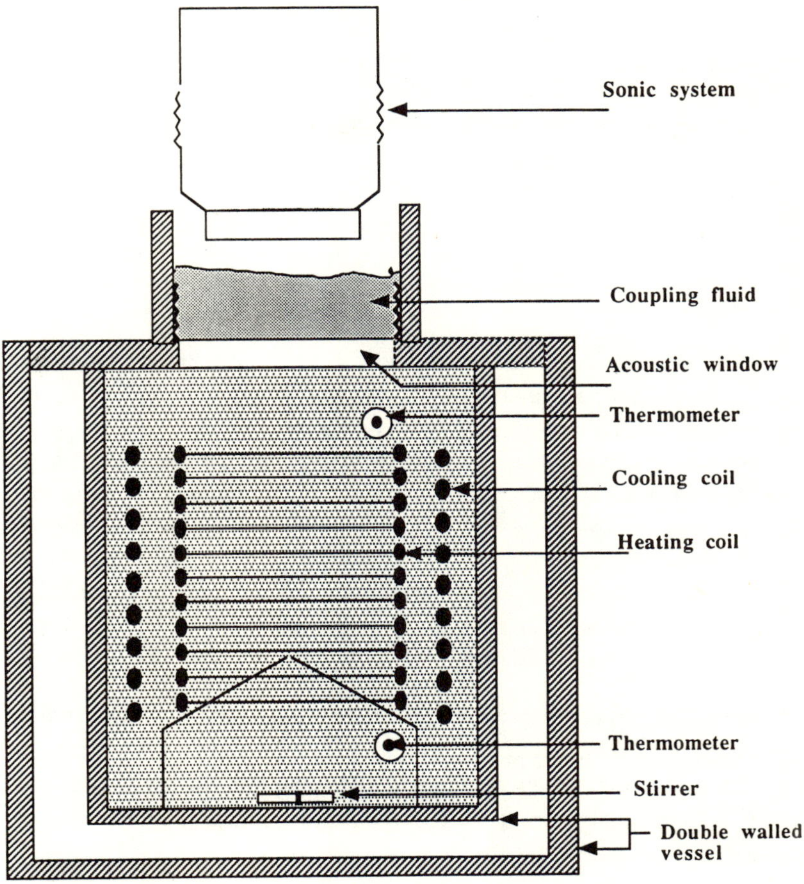

Figure 3. Design of an adiabatic calorimeter.

direction which prevents sound echoes returning to the transducer (Figure 3). The temperature was measured with two thermocouples to increase sensitivity and the system can be directly calibrated in joules with an electrical heating coil. A cooling spiral of polyethylene tubing allows for fast cooling of the system after each calorimetric experiment. Thus successive measurements can be made at short intervals of time with a reproducibility within a 4–5% range.

The Nonisothermal, Nonadiabatic (n-n) Calorimeter

A n-n calorimeter is schematically represented in Figure 4. In this system the thermal method of power estimation can be derived from Eqs. (9) and (10). Let T be the temperature of the jacket which will be kept constant; T_i the actual temperature inside the calorimeter, T_m the measured temperature by the thermometer, k the

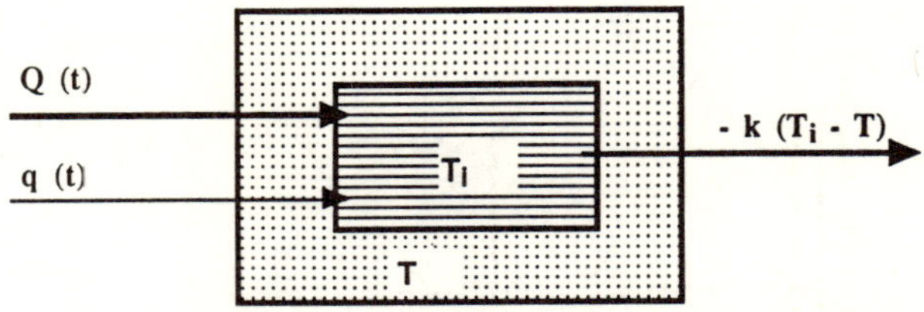

Figure 4. Schematic diagram of a nonisothermal nonadiabatic (n-n) calorimeter model.

cooling coefficient (the amount of energy lost by the calorimeter per second for one degree centigrade temperature difference with the jacket), C the energy equivalent of the calorimeter, and g the time constant of the thermometer.

The temperature inside the calorimeter depends on (a) the thermal energy input, that is ultrasonic power input $W(t)$ plus secondary heat sources (e.g. mechanical stirrer) $q(t)$, and (b) heat losses $-k(T_i - T)$. The following equations can be written:

$$W(t) + q(t) = C(\delta T/\delta t) + k(T_i - T) \quad (9)$$

$$(\delta T/\delta t) = -1/\gamma(T - T_i) \quad (10)$$

Solutions to these equations have been given by Zieniuck et al. who studied the influence of the different terms: Q, q, g, C [17]. They showed that the most suitable system for sonic power measurements should have the heat capacity C as small as possible, a high value of thermal losses k, and a small time constant of the thermometer γ.

A system with a small C has an additional advantage in that the system returns rapidly to its initial temperature so that many measurements can be made in a given time. Such calorimeter systems can be calibrated in two different ways: (1) by determining the thermal characteristics of the calorimeter which is a rather laborious method, or (2) more simply by calibrating the calorimeter with a secondary energy source, an electric current of known intensity passing through a heating coil of perfectly known resistivity.

A typical experimental device, described by Zieniuk [17] is illustrated in Figure 5. The calorimeter can be used not only for calibrating the transducer output, but also to measure the sonic energy dissipated in the irradiated medium.

Especially devised calorimeters can be used to calibrate transducers or check how a sonic system (generator and transducer) works. For a given power setting at the generator, the transmitted power in a sonicated medium depends on the acoustic load, i.e. on the nature of the sonicated medium and on the shape and size of the vessel.

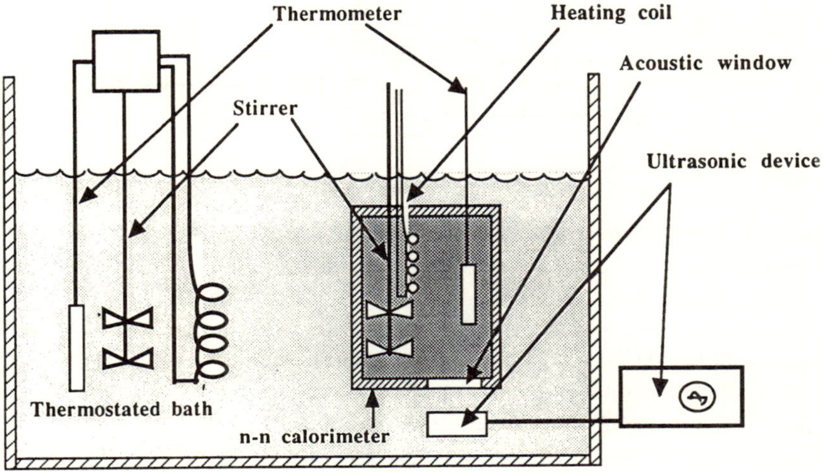

Figure 5. Experimental configuration for the use of an (n-n) calorimeter.

The data derived from calorimetric measurements reflect acoustic power delivery for fairly well matched loads. This is not always the case under normal working conditions. If the calorimeter is used as reaction vessel, and if a matching system is used, the difference in acoustical impedance between the medium inside the calorimeter and the coupling liquid must be known in order to introduce a correction factor.

If the transducer which has been tested with a calorimeter is used to sonicate another reactor, the calibration obtained with the calorimeter may give somewhat erroneous values of the delivered acoustic power. Separate calibrations should be made for each kind of condition thus the system should be considered as a n-n calorimeter and the methodology described above should be applied (i.e. the heat equivalent of the system should be determined together with a calculation of heat losses, or accurate independent calibration of the system with an electric heater). This approach can be very time consuming.

In practice however, a rough estimation of the acoustic power can be rapidly and easily obtained [18] using Eq. (8) and by measuring the temperature rise of a noncoated thermoprobe, under stirring. However, as will be seen below, the response of the thermoprobe depends on its nature, and a calibration is needed if more accurate determinations are needed. This method often gives an underestimate of the actual power, but is accurate enough in many cases and allows relative comparisons to be made. Its accuracy can be improved by calibrating the temperature rise with a heating coil as described previously.

Finally, as suggested by Byron [29], in the case of jacketed reactors power can be estimated by measuring the temperature difference of the cooling fluid in and out of the jacket using,

$$W = D\, c_p\, (T_{out} - T_{in}) \tag{11}$$

where D is the flow of cooling liquid, c_p its heat absorption coefficient, and $(T_{out} - T_{in})$ the temperature difference.

3.2 The Acoustic Dilatometer

Mikhailov and Shutilov [30,31] developed this device as illustrated in Figure 6. It is made of a horn shaped glass double walled Dewar vacuum cell (of about 0.1 liter capacity). The shape is designed to favor sound absorption and avoid reflected waves from the bottom of the tank. It is filled with degassed olive oil which has a high absorption coefficient. The system is oriented in a transverse direction to the direction of propagation of the wave which enters through an acoustic window made of thin nylon film and of known surface area S (in cm^2). A calibrated glass capillary is fitted at the top of the flask so that as the temperature increases the liquid expands into the capillary. The method consists in measuring the rate of thermal expansion Δh of the liquid in the capillary tube. It can be calibrated against the electric power W dissipated in the liquid by a heating coil. From the $W = f(\Delta h)$ plot, the sound intensity is given by,

$$I = W/S + I_c \text{ (W cm}^{-2}\text{)} \tag{12}$$

where I_c is a corrective term taking into account the reflection of the wave at the surface S.

Capillaries of different sizes are available. This allows power measurements to be made over a wide range, from 0.5 to 30 W cm^{-2} with an accuracy of 6–7%. One

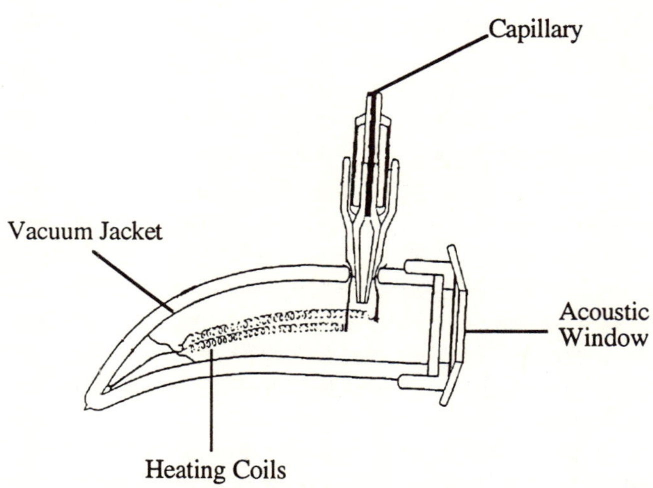

Figure 6. Acoustic dilatometer.

of the problems which arise is that the cell simulates a free field and this may not correspond to the actual conditions in the vessel under test. Under cavitating conditions it measures both thermal expansion and bubble volume. This system can be of interest to calibrate transducers or cleaning baths, but not for determining the transmitted power in a reaction vessel.

3.3 Thermal Probes

Thermal probes can be constructed quite easily and cheaply. Their response is non-directional, and very small devices with diameters between 0.5 and 1 mm can be fashioned. They are very simple to use with almost every kind of ultrasonic equipment and measurements can be made very rapidly. Several kinds of thermal probes have been described which are basically thermocouples or thermistors used bare or embedded in an absorbing medium. Bare probes are used to measure the actual temperature of the medium, just as in a calorimeter. Coated probes will generate internal heat under the influence of the sound wave and are used to determine local power dissipation in the absence of stirring. Coated probes are often used in conjunction with a bare probe, and the temperature difference between the two probes is then proportional to the acoustic power. Great care should be taken since the response of a coated probe strongly depends on its nature and geometry, and on the medium used.

The temperature rise of a coated probe follows the pattern illustrated in Figure 7 (curve a) [32,33]. When ultrasound is switched on there is an initial rapid rise (ΔT_i) caused by heat generation at the interface between the thermocouple and the treated medium due to viscous forces acting between the probe and the fluid medium. This phase of heating rapidly reaches an equilibrium and this is followed by a period (ΔT_2) when the temperature rises more slowly (ΔT_2) due to absorption of the wave within the coating. However the temperature does not start increasing until several tens of milliseconds (time t_1) after the sound was switched on. For castor oil this delay is 20 ms, for silicone rubber 20–30 ms, and for glue (UHU brand) 40–50 ms.

The typical thermal behavior shown in Figure 7 was observed by Martin and Law using UHU glue coating and castor oil [33]. These authors observed that the period Δt_1 is longer with probes of large diameter because of the reduction of heat losses. A coating thickness of 0.2 mm is recommended to give a linear response over a 150–180 ms period. The initial temperature rise due to shear viscosity is greater for thermistor probes than for thermocouples and extends over a longer period because thermistor elements are larger than thermocouple junctions.

The temperature increase inside an embedded thermocouple due to acoustic absorption has been studied in detail by Fry [32]. The initial temperature rise (ΔT_1) (Figure 7) is the same for an embedded thermistor as for a thermistor coated with a very thin layer but the latter rapidly reaches an equilibrium temperature due to heat losses (resulting from heat conduction and acoustic streaming) as can be

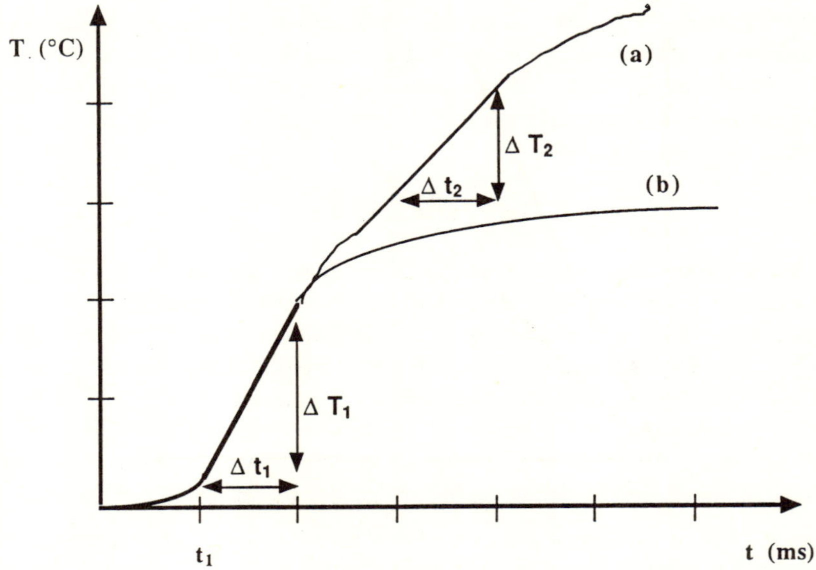

Figure 7. Typical response behavior for a coated thermocouple probe. (**a**) thermoprobe embedded in a disc of UHU glue; (**b**) coated with a thin layer of UHU glue.

seen in curve b. This behavior is similar in the case of noncoated thermocouples. The rise in temperature to an equilibrium value is much lower with noncoated thermoprobes and is critically dependent on the nature of the probe (size, shape, material). This is the main reason why bare thermal probes are mainly used to measure total sonic power under stirring, and coated probes are used to measure local power.

If the equilibrium temperature is to be used for the calculation of acoustic power a previous calibration of the system is required. Either of the temperature rises ΔT_1 or ΔT_2 can be used to measure sound intensity since there is a linear correlation between them as shown by Law (Figure 8) [33]. However, it can be seen that the slope of the line depends on the nature of the coating.

The typical patterns shown were generated using small probes. When large probes are used there may not be significant differences between temperature rises ΔT_1 and ΔT_2.

As already mentioned thermal probes of several different types have been used for acoustic power measurements. In 1954 Fry published a very detailed study on the determination of absolute sound levels and acoustic absorption coefficients [32]. The probe which was used is shown in Figure 9 and is made of a thermocouple junction (copper constantan or iron constantan; 0.0005 inch diameter) imbedded in a thin disk of absorbing liquid. The absorbing liquid is separated from the medium

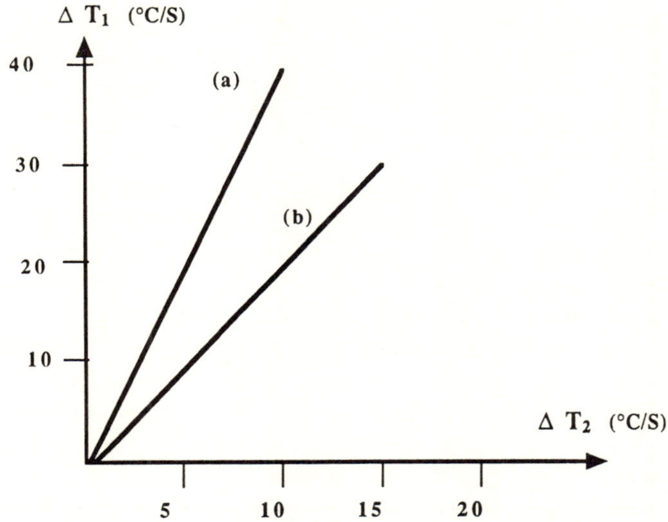

Figure 8. Plot of the rate of temperature rise during period t_1 against the rate of temperature rise during period t_2; (**a**) thermistor embedded in castor oil, at 1.5 or 2.5 MHz; (**b**) thermistor embedded with UHU glue at 1.5 or 2.5 MHz.

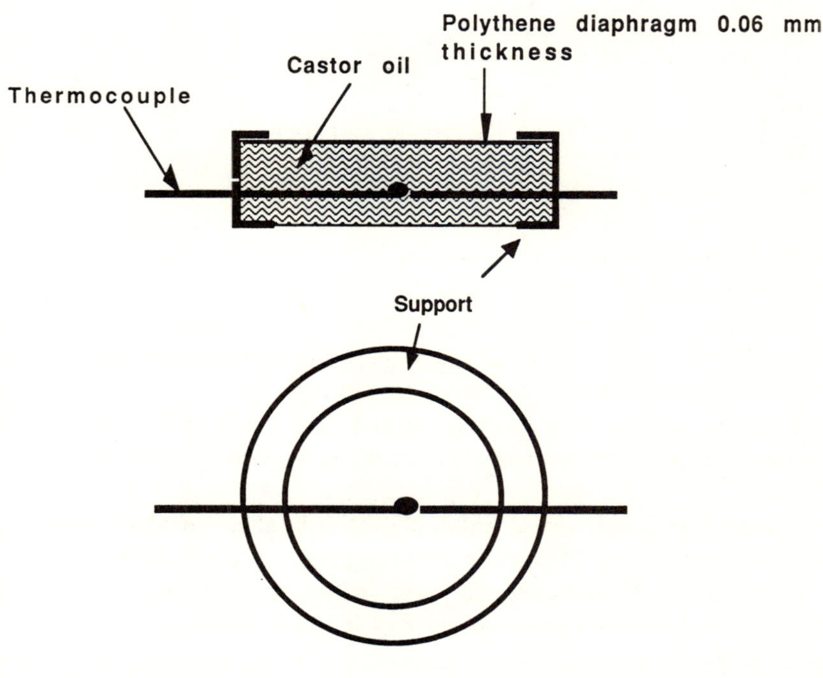

Figure 9. Construction of a disk type ultrasonic instrument.

in which the sound levels should be measured, by thin polyethylene sheets (0.003 inch thickness) mounted on a support. The source of ultrasound was a focusing irradiator (980 kHz) of the type used in medical applications. The beam width at half intensity was about 4 mm and pulses of 100 s were used. Sound intensity was in the range 20–30 W cm^{-2}.

The best imbedding liquid was found to be castor oil for measurements in water. Castor oil has an acoustic impedance close to that of water (so that the percentage of incident sound energy reflected at the interface is small) and a rather high absorbtion coefficient. The probe is placed in such way that the direction of propagation of the sound wave is perpendicular to the thermocouple wires. The response of the probe follows the pattern described above (see Figure 7). According to the authors, this type of probe is extremely useful because (i) it is small in size, (ii) it has a low input electrical impedance, and (iii) it is not sensitive to stray radio frequency fields.

The calculation of sound intensity requires a knowledge of the acoustic absorption coefficient of the imbedding material and its heat capacity per unit volume at the temperature at which measurements are made, according to,

$$\mu I = \rho C (\delta T/\delta t) \tag{13}$$

where I = acoustic intensity; μ = absorbtion coefficient of the medium; ρC = heat capacity of the embedding fluid per unit volume. If the sound intensity is known then this equation can be used to determine the absorption coefficient of the liquid.

This probe has been compared to a radiation pressure probe (barium titanate), and calibrated by comparison with a radiation pressure measurement. Its accuracy was estimated to better than 4%. The drawback of this device is its relative fragility and the need to employ complex electronics for accurate pulsing with square top

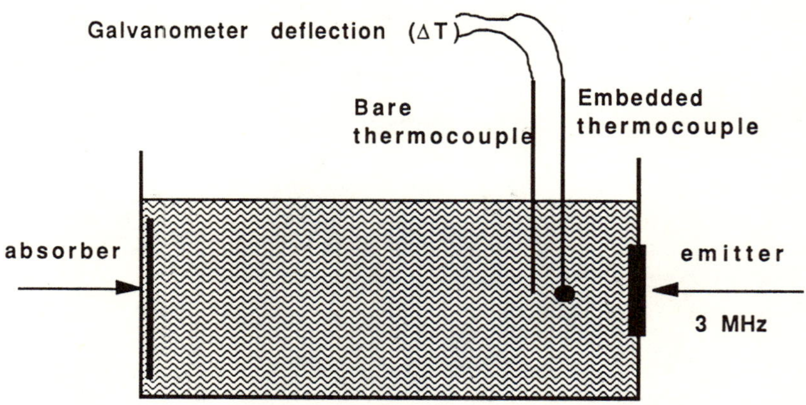

Figure 10. Experimental apparatus for power measurements employing both a coated and bare thermocouple.

pulses, but it makes an accurate standard to calibrate probes at frequencies above 200 kHz.

Palmer [34] carried out measurements using a system made of two sets of thermocouple wires (copper/constantan) in an experimental device illustrated in Figure 10. The frequency of the ultrasonic device was 2.5 MHz. It was mounted in one side of a box (Perspex; 19 × 17 × 38 cm) and the opposite side of the box was covered with paraffin wax in order to absorb the ultrasonic wave and simulate free field conditions. One set of thermocouples was left bare, and the other embedded in various materials, e.g. two kinds of plasticine, vacuum grease, paraffin wax, roughened glass, and mixtures of Durofix and glass wool. Among these, plasticine and vacuum grease were found to be the most successful. The thermocouples were connected to a galvanometer and the galvanometer deflection plotted against potential (V) and intensity (I) applied to the transducer. A linear relationship was observed, except with one kind of plasticine were there was a small departure from linearity. It is clear that the slope is greatly dependent on the nature of the coating (Figure 11).

Martin and Law [33] carried out accurate measurements using several probe designs made of thermistors elements 0.2 mm in diameter and 0.35 mm long with pair of leads 0.02 mm in diameter. The thermistors were either held by their connecting wires in the center of a ring (60 mm in diameter) to minimize disturbance of the sound field, or mounted at the end of a thin rod, or supported in a chamber filled with castor oil (a system similar to Fry's device), or embedded in

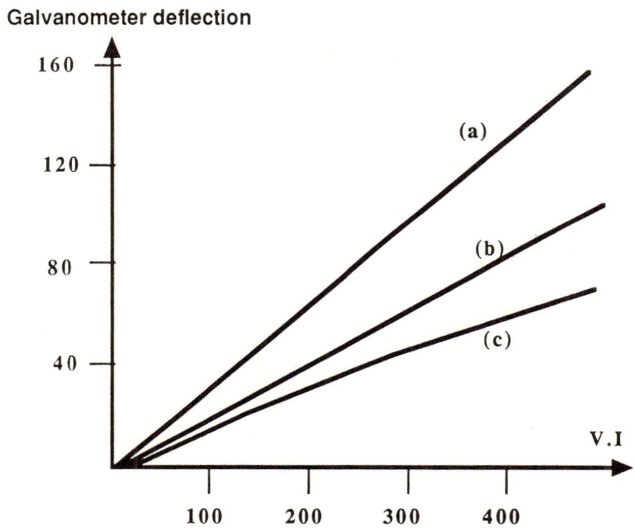

Figure 11. Probe response when coated with (**a**) paraffin wax; (**b**) plasticine; (*c*) old plasticine.

the center of a disc of UHU glue (30 mm in diameter, 0.5 mm thick). The responses of these systems followed similar patterns, but with different response times. The experimental device is similar to that used by Palmer and is illustrated in Figure 12. The ultrasonic emitter (0.78, 1.5, or 2.5 MHz, with diameters of 25, 25, and 10 mm, respectively) were mounted on one side of a perspex tank (45 × 30 × 135 cm) filled with water. The sound is reflected through an angle of 45° using a plate and is then absorbed in a cavity containing a brush with closely packed bristles which scatter and absorb the wave. The tank is also lined with carpet to further reduce sound reflection. The intensity of the sound wave was in the 0–10 W cm^{-2} range, with different irradiation times (from several ms up to 2 s).

Both of the periods of linear temperature rise ΔT_1 and ΔT_2 were used to calculate power (see Figure 8). The plots are normalized with respect to the maximum value using both periods and were shown to be very similar. Several different absorbing coatings were used: glues (polystyrene cement and UHU), epoxy resins, silicone rubber, polyurethane varnish, and cellulose lacquer. The diameter and length of the beads were varied within the ranges 0.5–0.8 mm and 0.68–1 mm.

Once again the response of the probe dramatically depends on the nature of the coating and, to a lesser extent, on the size of the bead. It also depends on the frequency. Selected results are shown below in Table 2.

It is interesting to observe that response differences between the coatings decrease when the frequency is increased. This is probably connected with the respective values of the wavelength and of the thickness of the layer in which absorption takes place which was estimated to be around 120 μm thick. The influence of the relative size of the probe compared to the wavelength was studied in standing waves fields [33]. As expected, maxima and minima were observed for the probe response. The ratio of the temperature rise at pressure maxima and minima

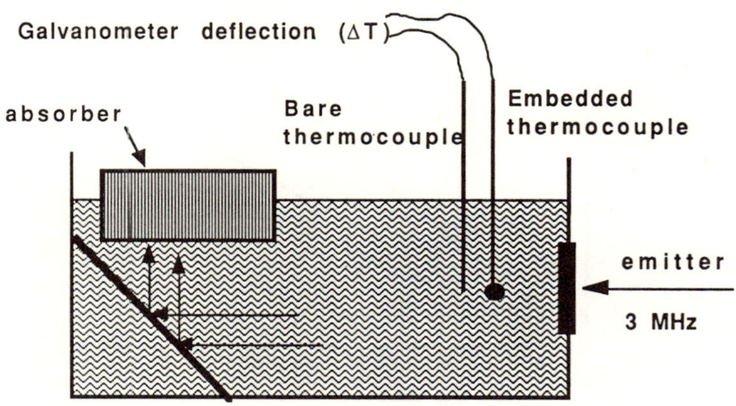

Figure 12. Alternative experimental apparatus for power measurement employing both a coated and bare thermocouple.

Table 2. Response of a Thermal Probe Coated in Different Materials

Absorbing Material	*Bead Diameter (mm)*	*Bead Length (mm)*	*Response at 0.78 Mhz (× 10^{-4}) °C/s/W m^2,*	*Response at 1.5 MHz (× 10^{-4}) °C/s/W m^2,*
cellulose lacquer	0.52	0.68	0.08	0.5
UHU glue	0.54	0.96	0.4	1.0
Silicone rubber	0.62	0.75	0.15	0.9

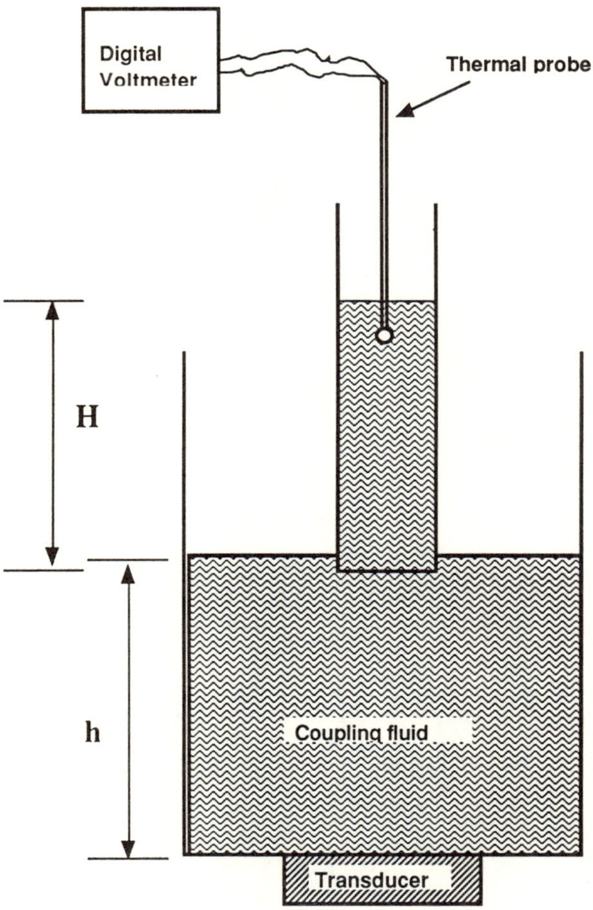

Figure 13. Experimental apparatus for power measurement at low ultrasonic frequency under restricted field conditions.

was determined at various frequencies. Using a probe consisting of a 0.81 mm diameter spherical bead of varnish it was observed that this ratio dramatically decreases when the thermistor radius equalled half of the wavelength.

The preceding studies were mainly carried out under free field conditions at high frequencies. Weber and Chon [35] carried out similar measurements at low frequency (21.5 kHz) under restricted field conditions. Their experimental device, similar to that used by Timmerhaus and Fogler [36], is illustrated in Figure 13. The emitter is a magnetostrictive transducer with base area 16 cm^2 attached by epoxy to the bottom of a 600 cm^3 beaker filled with water as coupling fluid. A tube

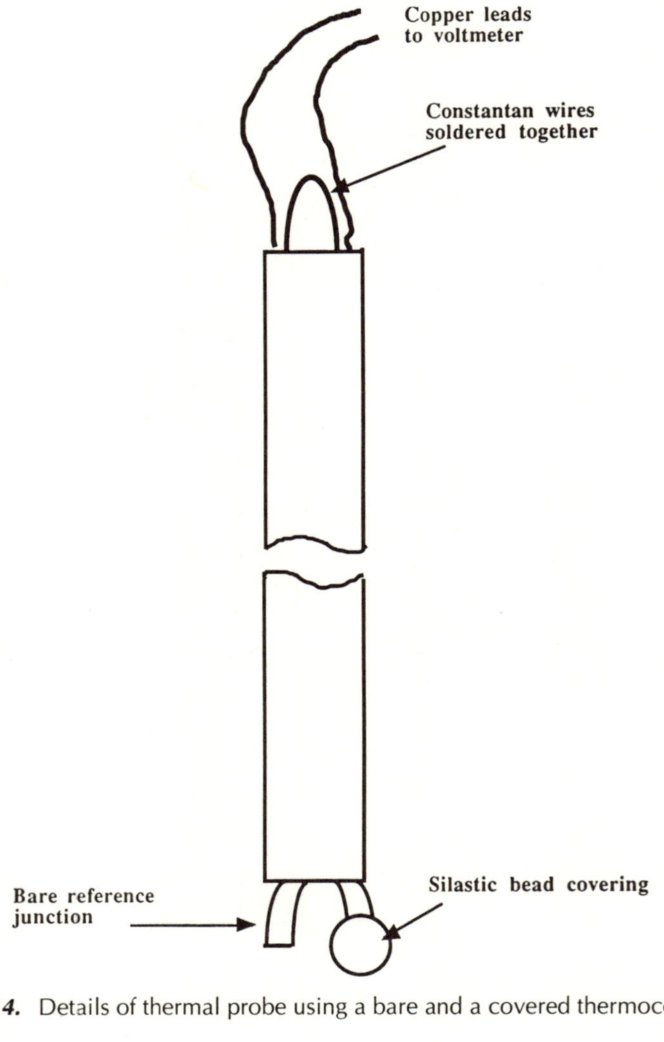

Figure 14. Details of thermal probe using a bare and a covered thermocouple.

containing the test liquid was dipped into the coupling fluid. Power was in the range 0–300 W. The height of liquid in the test tube (H) and in the beaker (h) were varied, and two kinds of test tube were used: one with a flat bottom (2.5 cm internal diameter), and one with a round bottom (3.6 cm internal diameter).

The probe (Figure 14) is made of two copper constantan thermocouples connected so that the difference in the emf generated between them can be measured, and is very similar to Palmer's system. One thermocouple was left bare, and the other embedded in silicone rubber (silastic). The probe output was read using a digital voltmeter.

The probe response in mV was plotted against the distance of the probe from the air–water interface, and typical results are shown in Figure 15.

Standing waves are observed using the flat bottom tube but not with the round bottom one, and the voltmeter signal at maximum intensity decreases when the height of coupling liquid increases. A theoretical treatment was carried out and the following equation was developed,

$$\frac{P}{P_o} = \sqrt{\frac{e^{-2\beta x} + e^{-2\beta(2H-x)} - e^{-2\beta H}\cos 2k(H-x)}{1 + e^{-4\beta x} + e^{-2\beta H}\cos(2kx)}}$$

where P and P_o = magnitude of sound pressure at x and $x = 0$, respectively; x = distance from the transducer face; β = sound absorption coefficient in cm^{-1}; k = wave number of sound (cm^{-1}) and H = height of the test liquid column.

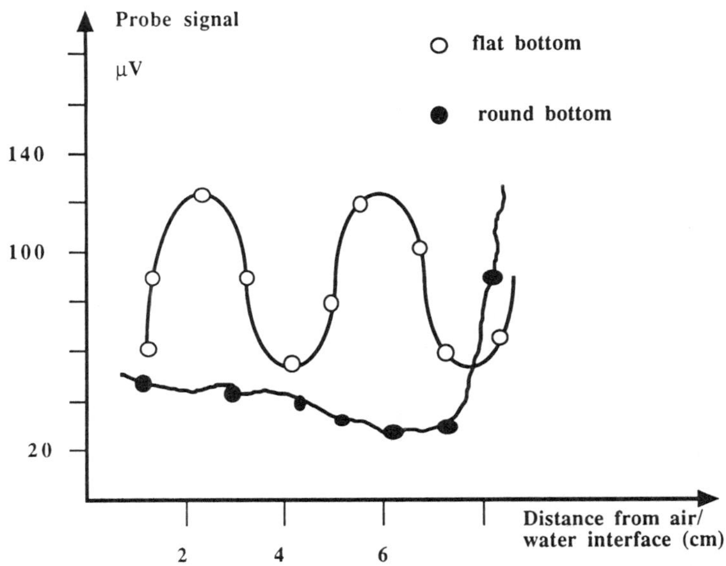

Figure 15. Thermal probe response vs. distance from surface for vessels of different shape.

According to this equation, the first maximum intensity should be at a distance one-fourth wavelength from the air–liquid interface, and the others at every subsequent half wavelength below. From the experimental study, the location of the first maximum is in good agreement with the predicted value, but the others exhibit peak to peak distances of ca. 2.2–2.8 cm, while the above equation predicts 3.6 (half wavelength) for reasonable values of β (<0.1 cm^{-1}). The reason for this discrepancy is not known.

In 1987, Pugin [37] published a similar study, but under conditions much closer to those normally used by chemists carrying out reactions. He mainly used flat or round bottom flasks immersed in ultrasonic cleaners. Two different cleaners were used: Laborette 17 from Fritsch GmbH (35 kHz) and Bransonic 220 (50 kHz). Power was varied using a variable transformer, and a hydrophone was used to confirm that the ultrasound frequency remained constant within a 40–220 V range. Pugin measured the temperature difference between the bath and the liquid inside the immersed flask (with a thermal probe located at different positions in the flask). The bath temperature was kept constant by circulating water through a thermostat. Here the thermal probe was a digital thermometer equipped with a thermocouple protected by a metallic sheath. The thermocouple was inserted into a hole drilled by a needle in a piece of silicon rubber for measurements in water, or of cork for organic solvents (since silicon swells the organic solvents). The ΔT difference was measured when the probe reached the temperature equilibrium of the system.

This gives only a rough estimate of acoustic power. Although the time necessary to reach temperature equilibrium of the probe was not given, it seems most likely that within this time some convective cooling of the probe due to streaming will

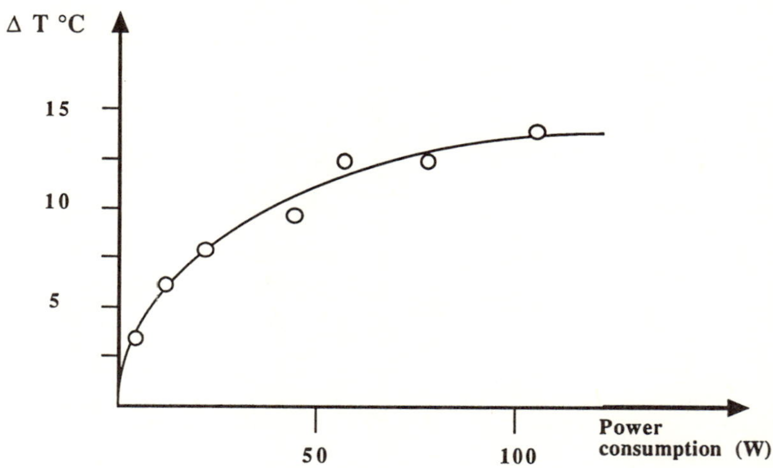

Figure 16. Temperature difference (ΔT) between sonicated liquid in a bath and a liquid inside vessel immersed in bath vs electrical power to the transducer (W).

have occurred. In fact a plot of ΔT against consumed power (measured with a wattmeter) shows that there is no linear correlation (Figure 16). Other phenomena may also contribute to this nonlinearity, e.g. nonlinear response of the sonic system, reflections and scattering of the wave by the walls, and the presence of cavitation bubbles.

This system is very useful however since it is robust and inexpensive and allows quite interesting studies of three-dimensional sound intensity distributions to be made. Pugin studied the influence of the water level in the bath, the ultrasound intensity, the shape of the flask, the liquid level in the immersed reaction vessel, and the addition of suspended solid particles. He also attempted to correlate his methodology with other techniques.

Pugin also carried out some measurements using a horn system as the source of ultrasound with a Branson B30 sonifier operating at 20 kHz. The horn was dipped in a 500–cm^3 beaker (18 cm diameter; water level 21.6 cm; distance from the horn to the bottom 18 cm). From Figure 17 it can be seen that at low power (curve a) the temperature rise is quite small. Two flat maxima are barely discernable (probably because the sensitivity of the probe is too low), but the distance between the two does not correspond to a half wavelength. At higher power levels however, clear maxima are observed, almost at the same position (curve b). When the power level is further increased these maxima are no longer observed (curve c) probably due to heat losses generated by acoustic streaming. Somewhat similar patterns have been observed when using an electrochemical probe (see later).

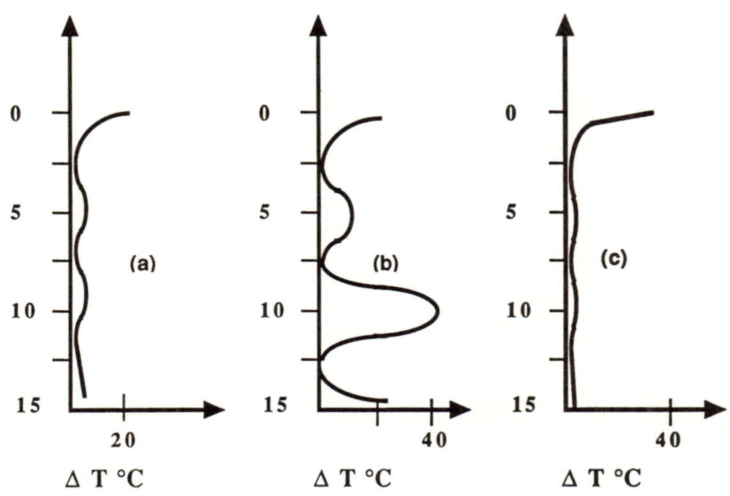

Figure 17. The effect of power on the temperature distribution from a horn (20 kHz) dropped into 500 cm^3 beaker containing water at (**a**) lower power (**b**) medium, and (**c**) high power.

Similar studies have been carried out by Romdhane [38]. He compared UHU glue and several kinds of silicone rubbers (Rhodorsil RTV 132 A and B, Rhodorsil 3B-2 and 3B-3). Silicone rubbers gave higher sensitivity, but could only be used for long periods of time in aqueous media because silicones absorb organic solvents. It is possible to use silicone rubbers in organic solvents provided that the contact time with the solvent does not exceed 5 to 10 mn. For all such measurements the probe was connected with a computer through programmable converter CVP 400 (response time: $70 \cdot 10^{-3}$ s) and an interface RS 232/ RS 422 IT 400. The temperature could be monitored every 0.1 s. From the plot of temperature against time the initial slope could readily be obtained. This slope was compared to the temperature difference between the equilibrium temperature reached by the embedded probe after some time, and the consumed power W_1 measured with a wattmeter. The coating of the embedded thermocouple was made with molds of different sizes, and the thermocouple (nickel-chromium type K; 0.5 mm in diameter; response time $25 \cdot 10^{-3}$ s) was located at different positions in the bead as indicated in Figure 18. Typical results are reported in Table 3.

It can be seen that maximum sensitivity is obtained when the thermocouple is located near the base of the bead. Heat transfer modeling was carried out. This system proved to be quite sensitive as shown by the fact that the system was able to detect power variations in an Undatim Sonoreactor during its automatic tracking sequence in searching for optimum resonance frequency. We have used this system for the investigation of the power distribution in various systems [25,38] including a cleaning bath, two horns (Undatim 20 and 40 kHz) dipped in a beaker, a Sodeva resonating tube and a hexagonal bath [24].

In our own laboratories we have been studying thermal responses from coated thermistors [39]. While it is straightforward to produce individual probes which give significant temperature rises when irradiated with ultrasonic energy in a fluid medium, it is extremely difficult to produce probes with predetermined characteristics. We have found that for any particular coating material the temperature rise measured depends not only on the coating thickness and size but also on exposure

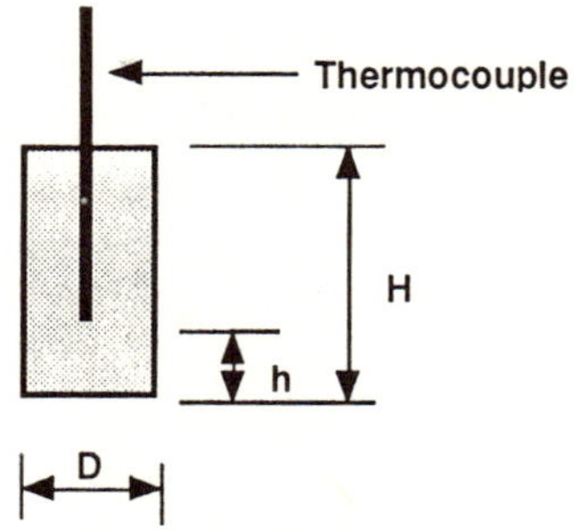

Figure 18. Dimensions involved in the construction of an embedded thermocouple.

Table 3. Effect of Changes in Construction Dimensions[a] on the Responses of an Embedded Thermocouple

H (mm)	*h (mm)*	*dT/dt* (°C s^{-1})	$T_e - T_0$ (°C)
5	2.7	2.1	4.2
	4.0	3.0	8.5
8	3.0	2.9	5.8
	5.0	4.1	7.9
	6.5	4.3	8.5
10.5	3.0	2.2	4.4
	6.0	3.7	7.2
	9.5	4.3	8.6

[a]See Figure 18.

time. In fact it would seem that the temperature rises measured using individual probes with a given irradiating power often vary with time and appear to be related to the state of "cure" or "setting" of the coating. Despite this we have achieved some promising results in terms of large temperature rises using silicone rubber coated

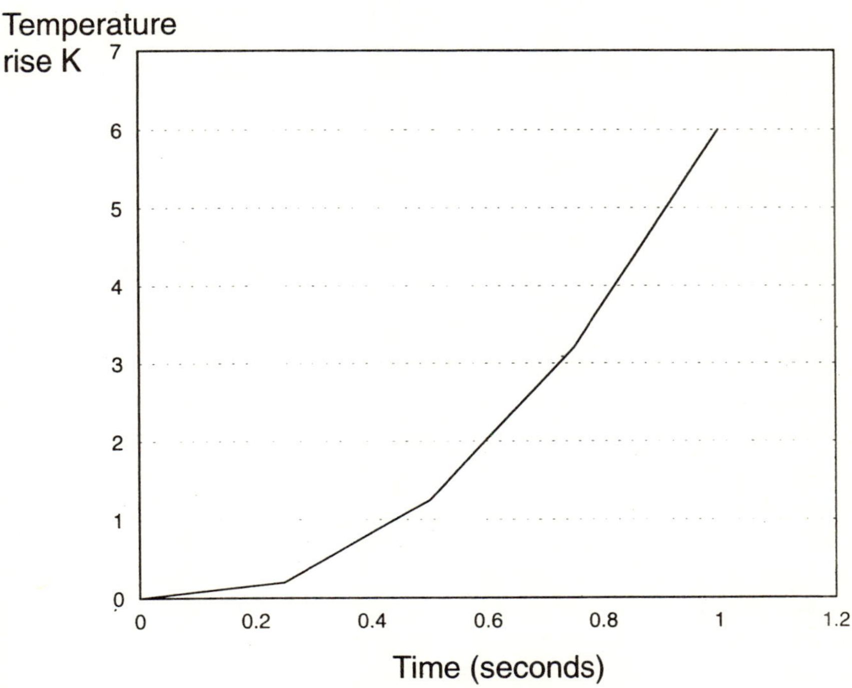

Figure 19. Temperature rise for silicon rubber coated thermistor. Coating 1.5 mm., ambient temperature, 6 W input power.

thermistors. Although not yet optimized, the potential for power measurement is clearly present with temperature rises registered of some 6 degrees in only one second.

Other probe designs have been described in the literature [40–49]. Almost all of these systems can detect standing waves and local field distributions.

3.4 Summary of Thermal Probes

On a relative scale thermal probe systems give reproducible and accurate measurements of ultrasonic power input provided that a few basic precautions are taken:

1. The preparation and nature of the coating should always be the same.
2. The size of the probe should be small compared to the half wavelength of sound in the media.
3. A reproducible size of the coating should be achieved, and when the size of the coating greatly exceeds the size of the thermocouple, the location of the tip of the probe inside the absorbing material should be carefully monitored.
4. The coating should be periodically inspected and renewed because its sound absorbing properties change with time. This is especially true after prolonged exposure to cavitation which results in a different sensitivity of the probe.

Thermal probe systems are inexpensive, easy to handle in almost all ultrasonic devices and particularly those used in sonochemistry. Field distributions and optimization of the geometry of the system can be rapidly obtained and the accuracy of the method is high enough to ensure reproducibility. Chemists who make use of ultrasonic equipment should, as a very minimum, consider this method to calibrate and optimize sonication conditions prior to carrying out sonochemical reactions.

Equation (8) can be easily applied to the determination of power (recognizing that heat transfer in the system should be minimized), but it should be appreciated that the measurement strongly depends on the response time of the probe (which should be as low as possible). However this kind of measurement strongly depends on good acoustic matching of the ultrasonic device to the reactor and, as a consequence of this, the nature of the sonicated medium is very significant. This was further demonstrated by Contamine [26] who determined the dissipated power W_t as a function of the consumed power W_1. W_t was measured with the system of acquisition used by Romdhame [38], but with a bare thermocouple, and by calculating the rise in temperature at $t = 0$ according to Eq. (8). The ultrasonic system used was a Sodeva cup-horn already described in Figure 1. The cup-horn was filled with various liquids and insulated against thermal losses; the liquid height was always equal to 9/4 wavelengths of the sound in the liquid under study. W_1 was measured with a wattmeter. A linear relationship between W_1 and W_t was observed, but the slope clearly depended on the nature of the liquid as can be seen in Table 4.

Table 4. Ratio of Dissipated Power (W_t) Consumed Power (W_1) for Various Liquids in a Cup-Horn Device

Liquid	W_t/W_1
acetone	0.28
chloroform	0.23
diethyl ether	0.15
toluene	0.34
ethanol	0.28
methanol	0.26
ethylene glycol	0.52
water	0.48

There are two possible explanations which could account for this observation. The first is that although the liquid height was always adjusted to the same multiple of the wavelength, it was not sure that matching of the system was optimized. It is also a fact that the liquids do not all have identical absorption coefficient, viscosity, or thermal conductivity [23]. It is thus quite clear that in the search for more accurate measurements and absolute values of sound intensity, further precautions should be taken. This may involve calibration using another method (e.g. heating coil, radiation forces). The method then becomes more lengthy, but is still useful.

4. ELECTRICAL AND MECHANICAL MEASUREMENTS AT THE TRANSDUCER

In principle, emitted power can be obtained from electrical or mechanical measurements at the transducer.

4.1 Electrical Impedance Measurements

The complex impedance is measured under resonant conditions and plotted using an Argand diagram. Power can be obtained by comparing measurements under loaded and unloaded conditions [50,51]. Unfortunately this method is not very reproducible when operating at very low intensities. At the other extreme using high or very high intensities "surface cavitation" tends to unload or decouple the vibrator from the liquid system and measurements using its fundamental frequency become unreliable. The equipment necessary for such measurements include an admittance or impedance bridge, a frequency counter, a wattmeter and, in principle, an anechoic chamber.

Within a 4–10 $W cm^{-2}$ range, in water at 20 kHz, reasonably reproducible results have been obtained, and they are in good agreement with calorimetric determinations of input power [19]. This method although quite accurate is time-consuming but it can be a good test for transducers through the measurement of the resonance–antiresonance curve [22].

4.2 Mechanical Measurements at the Transducer

The complex mechanical impedance can be obtained by measuring force, velocity, and their phase difference using probes or pickups attached to the transducer at suitable points [144]. This method is very convenient with solids [52] (sometimes the most accurate), but not with liquids, although in this case it can be a convenient method of checking the performance of a transducer. The output from an accelerometer or strain-sensitive pickup gives information on resonant frequency and vibrational amplitude. However subsequent correlation with the power transmitted to the sonicated medium is not straightforward.

4.3 Amplitude Displacement

This kind of measurement can easily be applied to horn systems. It involves the direct measurement of the amplitude at the vibrating tip of the horn, i.e. the actual mechanical motion being transmitted to the chemical reaction in which the probe is immersed. In all such measurements the shape of the horn must be taken into account since the "taper" or shape will influence the amplification of the vibratory motion of the transducer. The amplitude of vibration of the tip will generate a parameter that should be directly proportional to the acoustic power. The measurement of amplitude can be achieved simply by observing the actual ultrasonic vibration using a metallurgical microscope with a calibrated eyepiece. Since most transducers will generate amplitudes of at least 10 μm, quite accurate measurements can be made. A small spot of aluminum paint, when placed on the surface of the probe, will assist in this measurement since a single metallic fleck can then be focused in the graticule. On turning on the power the rapidly vibrating fleck appears as a "smear" the length of which is the amplitude of vibration. Naturally this method of measurement cannot be easily made when the probe is in use during a sonochemical experiment.

An electromechanical method is available however which does provide continuous monitoring with a display while the probe is in use. The alternating stress in a resonating horn is at a maximum in its center. If a strain gauge is bonded across this point then the output from the gauge will be proportional to the displacement or amplitude of vibration. The output signal can be displayed on a meter which can then be calibrated by the use of a microscope as described above. Vibrational amplitude is not an absolute measure but it does offer a very sensitive monitor for acoustic changes during a sonochemical reaction. Its major drawback is that a strain gauge is a somewhat fragile instrument, however amplitude does give an indication of the acoustic power output rather than the electrical power into the transducer. Using this method, it is only necessary to calibrate the meter once since any subsequent change in transducer amplitude due to loading will be accompanied by

Table 5. Motion Amplitudes for Acoustic Horn Systems

Frequency (kHz)	*Amplitude (μm)*	*Pressure Amplitude (atm)*	*Maximum Velocity*	*Maximum Acceleration (g)*
20	2.94	5.4	36.6	4,660
800	0.07	5.4	36.6	183,000

a proportional change in strain in the system. The meter will thus follow any induced changes of amplitude which occur either as a result of power input or changes in properties of the sonicated reaction.

Another method is to use an accelerometer fitted to the probe. It can give the displacement according to,

$$D_o = a_o / 4\pi^2 f^2 \tag{15}$$

where D_o = maximum displacement, a_o = maximum acceleration, and f = frequency.

As an example motion amplitudes and acceleration are given in Table 5 [23] for an intensity level of 10 W cm^{-2}.

It can be seen that for an increased frequency, but at the same intensity level, the maximum amplitude motion decreases, but this is accompanied by a dramatic increase in acceleration. This is an important parameter when considering the chemical effects of ultrasound. This type of measurement is also of great interest in welding and drilling systems. In principle, according to Eq. (2), this permits the calculation of the sound intensity. Unfortunately unless it is known how much of this motional amplitude is transmitted to the sonicated medium it is again difficult to correlate this value to the actual transmitted power. Furthermore, for a given amplitude, the effect of ultrasound depends on the surface area of the emitter immersed in the medium.

5. METHODS BASED ON DIRECT MECHANICAL EFFECTS

5.1 Acoustic Probes

There are many different kinds of acoustical probes including microphones [57–62], hydrophones, radiometers, and piezoelectric devices (most often small barium titanate transducers) [63–68], and the hot wire microphone (based on acousto-resistive effect) [63]. Their resonance frequency is generally very different from that of the ultrasonic field under study.

These probes can be used to measure the pressure amplitude in the system [19]. In principle the local acoustic power can be obtained by measuring the pressure amplitude P, the velocity v of an imaginary particle submitted to the field, and their phase difference $\acute{Y}$. The transmitted power can then be obtained by integrating $P \cdot v \cdot \cos \acute{Y}$ over the total volume.

This is a good method for local measurements, but rather tedious for overall power. Furthermore, to calculate the ultrasonic power one needs to measure the particle velocity, and this is not a trivial task. Indeed one might assume that the particle velocity is the same in the liquid and at the tip of the probe (see Section 4.2). This is almost never exactly true, but this assumption can lead to a reasonable estimate of the dissipated ultrasonic power.

High pressures generated by the oscillations of the bubbles and (or) their collapse cannot be directly measured, but an indication can be obtained by using very small microphones as probes [59–61].

Although these acoustical probes can be made very small they will always slightly disturb the ultrasonic field. Just as in the case of coated thermal probes, the response signal depends on the nature and size of the probe, thus it is important that the microphones are carefully calibrated. They are however widely used, especially to calibrate medical ultrasonic equipment. Recently, very small and sensitive devices using PVDF membranes [68,69] or fiber optics [70] have been described. PVDF has piezoelectric properties and miniature membrane hydrophones (about 0.5 mm in diameter) are available. Fiber optic probes can even be smaller and a spatial resolution of 0.1 mm has been claimed [70].

A further interest in the use of acoustical probes is that the Fourier transform of the $P = \mathrm{f}(t)$ signal gives a $P = \mathrm{f}(w)$ signal, that is the noise emitted in the sonicated medium. Noise measurements will be further described in Section 6 below.

5.2 Acoustic Impedance

A further use of acoustical probes is for the detection of changes in the acoustic impedance of the sonicated medium [71–74]. These devices have been mainly applied to the determination of cavitation threshold. The signal received by a probe microphone placed at some distance from the emitter changes abruptly when cavitation occurs because the acoustic wave is scattered by the cavitation bubbles generated near the surface of the emitter. This results in changes in acoustic impedance and the wave is strongly attenuated as it travels away from the emitter. This provides a sensitive determination of the cavitation threshold. If acoustic power is increased further, beyond the cavitation threshold, the changes in acoustic impedance are small and so this system cannot be used to determine the sound intensity with any great accuracy.

The change in acoustic impedance is a clear illustration of the screening effect of the bubbles near the surface of the emitter; this is the reason why power density is generally limited to a few W cm^{-2} when cavitation is generated in a large volume of liquid.

5.3 Acoustic Fluxmeter

This method was originally developed as an acoustic fluxmeter for power measurements in air [76]. It is based on having two microphones spaced apart at a small distance compared to the sound wavelength. It is possible that this instrument could be adapted for use in liquids, but both plane wave and anechoic conditions would be required.

5.4 Radiation Forces

At a boundary between two media having different acoustic impedance a radiation force *F*, or radiation pressure, applies when an acoustic wave reaches that interface. This should not to be confused with sound pressure. The magnitude of the force generated by radiation pressure depends on the intensity of the sonic or ultrasonic field, and on the shape of the field. When it impinges on a body (e.g. a target) immersed in a liquid it will also depend on the size and the shape of the material from which that body is formed. When the ultrasonic wave is completely absorbed the surface is submitted to a steady force [21],

$$F = aI/c \tag{16}$$

where I = ultrasonic intensity, a is a constant of proportionality, and c = sound velocity. Alternatively,

$$\vec{I} = \vec{F} \cdot \vec{v} \tag{17}$$

where I = sound intensity, F = radiation pressure, and v = velocity of an imaginary particle subjected to the sound field). The sound intensity can be computed by time averaging F and v.

The existence of this acoustic radiation force was established a long time ago by Dvorak in 1876, and Raleigh in 1878 gave the basis of a theoretical study. Since that time many papers have been published in this field [77,78], and it has been the subject of some controversy due to the complexity of the problem. For this reason only some of the elements relating to this topic will be discussed here.

Several theories have been suggested to explain radiation force. Under free field conditions, with a plane wave front, a very simple equation seems to apply,

$$F = I \cdot S \tag{18}$$

where F = radiation force applied to a target of surface area S perpendicular to the wave direction and I = energy density.

Under a restricted field other equations have been derived [79],

$$F = \tfrac{1}{2}(1 + \gamma)\, I \cdot S \quad (\text{and } \gamma = c_p/c_v) \tag{19}$$

or more recently [80,81],

$$F = \frac{1}{2}(1 + B/2A) I \cdot S \tag{20}$$

with $B/A = 2\,p \cdot c\,(\delta c/\delta P)_T + 2\,c \cdot \beta\,(\delta c/\delta T)_P$; $\beta = 1/V\,(\delta V/\delta T)_P$ and P = acoustic pressure and V = volume.

There still remains the controversial question as to what extent the radiation force depends upon the physical properties of the medium [82]. In fact it seems that the answer to this depends on the conditions, i.e. if there is a free or restricted acoustic field or if the wave front is planar or non planar.

Many different systems have been devised to measure the radiation force. They are based on the measurement of the displacement of a target which is suspended in a liquid. This target is the key point of the system. The actual target can be of a variety of designs: for example, a solid sphere [83–88]; a wedge [89]; a cone [90,91]; or a disc fashioned from thin metal foil with an air gap behind in order to provide a perfect reflector [92]. Kossov used a float as the target and measured the change in buoyancy when it was located at the interface between two immiscible liquids, the test liquid and a heavy liquid (e.g. CCl_4) [95].

It is most important that the shape of the target allows it to act as a perfect absorber. Almost always some additional systems are required in the chamber to absorb the sound wave and thereby avoid side influences from reflected or standing waves on the target. Typically paraffin wax or linen are used for this purpose.

There are two different techniques to measure the radiation force, compensated or uncompensated [17]. In the uncompensated methods the target, which is suspended by a thin wire, is allowed to move and its displacement is measured [89,93–95]. The displacement is then either the vertical displacement if the emitter is placed below the target (Figure 20a) or the deflexion if the emitter is placed perpendicular to the wire (Figure 20b). Methods for measuring the displacement of the target can be optical [92], photographic [101], or electrostatic [102] (e.g. a radiometer disk [112]). In the compensated methods [96–100] the target is maintained in its initial position with some kind of balance which can be mechanical or electromechanical.

Among the different possible shapes for the target, the sphere has been extensively used; in this case the radiation force is generally given by, [17]

$$F = I \cdot \pi \cdot r^2 \cdot Y_p \tag{21}$$

where r = radius of the sphere and Y_p = acoustic radiation force function of the sphere. Y_p depends on the material, and values can be found in the literature [94,101,103–107].

With a device similar to that illustrated in Figure 20b, Palmer [34] found a good linear relationship between the deflection of a steel ball and $V \cdot I$, the product of applied voltage and intensity at the transducer. He used Eq. (22),

$$F = mgd/l \tag{22}$$

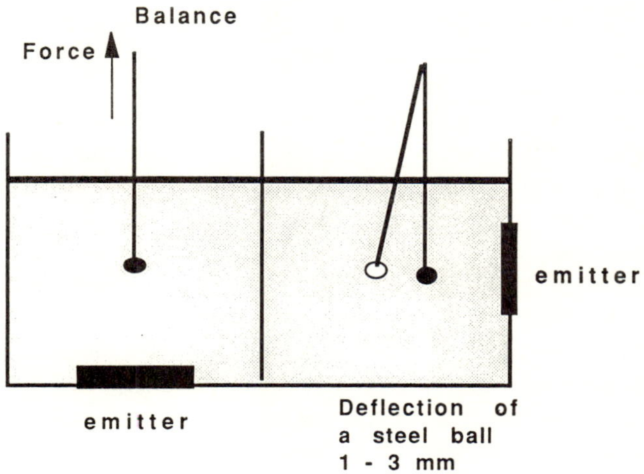

Figure 20. Different configurations for the measurements of radiation force.

where d = displacement of the ball; l = length of the nylon thread, $l >> d$; and m = mass of the ball, and Eq. (23):

$$F = 2I \cdot 10^7 \cdot 500 \cdot \lambda^2 / 4\pi^2 \cdot c \tag{23}$$

This equation was derived from that given by Fox [103],

$$F = 2\, I \cdot \Gamma / K^2 \tag{24}$$

where $K = 2\pi/\lambda$; Γ is a complicated function of $K \cdot r$; and r is the radius of the sphere.

A significant advantage of this method is that it can be used over a wide range of frequencies since the radiation pressure depends on the intensity of the wave and not on its frequency. Limitations to the method arise from the power involved. This technique should preferably be used under free field conditions, and below the cavitation threshold. Once beyond this threshold, the relation between power and radiation force becomes nonlinear, and the method no longer gives an accurate indication of power density. Its accuracy also depends on the shape of the field; in general this method gives a measurement of the total flow of energy unless free field and plane wave conditions are satisfied [108–110].

Another problem is that the radiation force is steady in nature, and can be masked by acoustic streaming. Sonication gives rise to a streaming flow of the liquid due to absorbtion, the force of which is proportional to the energy gradient. Here it has been assumed that the sum of the steady force (radiation pressure) and of that associated with the streaming flow is constant [110]. As the streaming flow increases with power this will result in an overestimation of the radiation pressure unless the influence of the flow onto the target is eliminated. This can be made by interposing a thin screen in the path of the beam which cuts off the flow and only allows the radiation force to pass [111,112].

Some experimental errors may be involved in using the suspension device, e.g. thermal effects and the reflecting properties of the target which can be less than ideal. Attempts to avoid the influence of the suspension device have been made by using falling drops and measuring their deflection [101]. There is a problem however in that the control of the falling drops not easy.

All of the methods which involve radiation force measurement have been extensively used to calibrate diagnostic ultrasound systems with acoustic intensities of only a few milliwatts per square centimeter for which radiation balances are readily applicable. An accuracy of 2% has been quoted and acoustic powers as low as 10 μw have been detected. Less commonly it has been used for therapeutic systems at much higher energy with outputs of the order of several watts, although in this case the accuracy is somewhat lower [17].

5.5 Distortion of Liquid Surface

It is clear from the foregoing that the propagation of a sonic or an ultrasonic wave in liquids results in acoustic streaming. When the wave reaches the liquid–air interface, total reflection occurs but due to its elasticity the surface is distorted. At high intensity a fountain, or bulge, is observed. Its height varies approximately linearly with the energy density and several workers have suggested that measuring the surface distortion (e.g. with a capacitance-plate pickup located close to the

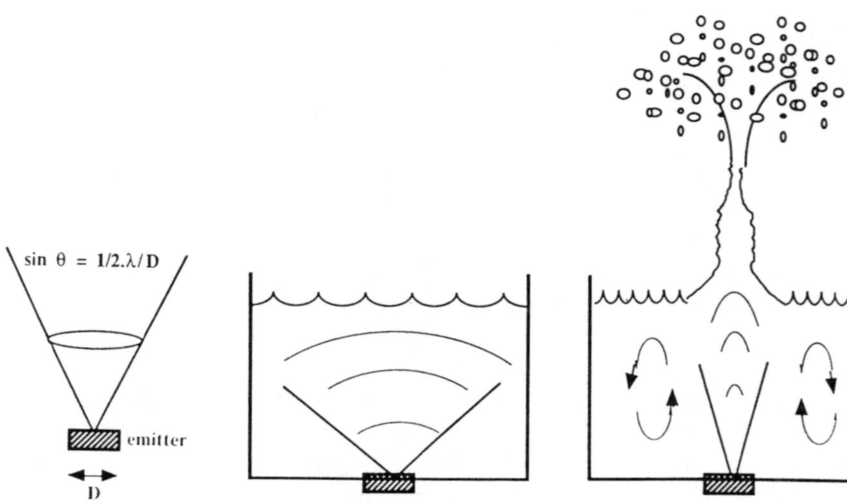

Figure 21. Movement of radiation force by disturbance of the liquid surface.

surface) could give an estimate of the transmitted power [93,113–115]. This method is mainly applicable to high-intensity systems. The height of the fountain also depends on the height of liquid, and on the frequency. For the same power density, the height of the fountain is increased when ultrasound frequency is raised. This is due to the fact that the spreading of the wave is proportional to *l/D*, where *l* is the wavelength and *D* the diameter of the radiating surface (Eq. (25)) [23] and is illustrated in Figure 21.

$$\sin\theta = \tfrac{1}{2}\cdot\lambda/D \tag{25}$$

5.6 Surface Cleaning, Dispersive Effects, Emulsification

Removal of coating or soil deposited at the surface of a solid, dispersion of solid particles, or emulsification are well known effects of sonication [1–12]. They can be used to determine the transmitted sonic power, as the rate, or the amount of removed soil directly depends on sound intensity [19] for a given sonication period. Cleaning and dispersion result from both cavitation and acoustic streaming. Neppiras reported that below the cavitation threshold but above a lower limit (the value of which depended upon the system) cleaning and dispersion rates increase linearly with acoustic intensity, i.e. with the square of acoustic pressure [16]. This indicates that acoustic streaming also participates in these effects. True cavitation is probably not essential for the stripping off of a layer, but the strong acoustic streaming occurring near oscillating bubbles, especially if they are resonant, probably strongly enhances the rate of cleaning. These effects decrease with frequency eventually falling to zero. However, for submicron particle size, it has been reported [116] that high frequencies (around 500 kHz) can be more efficient than low frequencies (20–50 kHz), indicating again a significant influence of acoustic streaming on these processes.

In order to obtain accurate and reproducible results, temperature and pressure should be carefully monitored. These mechanical effects tend to increase to a maximum and then decrease to zero with the upper limits defined by the boiling point of the solvent and a hydrostatic pressure at ambient equal to the peak acoustic pressure. This is also the normal pattern in sonochemistry [6,7,117,118].

Several kinds of coating have been used in these studies [16] including:

1. Radioactive tracers which provide a simple, accurate, reproducible, and rapid method.
2. Chemical coatings, e.g. zinc oxide which, when displaced from the surface, can be estimated by titration with dilute acid.
3. Chemical dyes which can be estimated by colorimetry and provide a safe, rapid, and inexpensive method have been widely used, although they have proved slightly less accurate than radioactive tracers.
4. Fluorescent or phosphorescent paints or dyes.
5. Photographic plates. Here using the property that sonication of photographic material in a sodium thiosulfate solution selectively dissolves silver bromide.

Some of the important parameters which affect surface cleaning are (a) the size and nature of the support material and the coating, (b) the surface conditions of the support, (c) the time of treatment, and (d) the method by which the coating was applied. This last point is of crucial importance since the consistency of the method of coating determines the reproducibility and the accuracy of the method. It should be carefully standardized to allow comparative studies. The size of the item to be cleaned is also important. If it is too small it will be difficult to measure the amount of removed soil, and if it is too large it will dramatically disturb the ultrasonic field. A method used by ultrasonic cleaner manufacturers to assess cleaning power is to remove Tipp-Ex fluid from the screw threads.

Another similar "cleaning" method for power measurement is the emulsion probe [19]. It consists in a fine mesh gauze fixed to a wire holder which is first immersed in oil (e.g. olive oil) and then dipped in water. The time needed to remove all the oil is measured. This method is not quantitative, but effects start at the cavitation threshold and this give a visual method for the determination of this threshold.

For each of these methods the actual sound intensity cannot be obtained directly, thus these methods need some form of calibration with another more quantitative methodology (e.g. radiation pressure or a thermal method).

5.7 Erosion Methods

The erosion of surfaces is a direct consequence of cavitation and therefore has the potential to be the basis of a dosimetry technique specific to cavitation. Weissler [119] assumed that the effect should be proportional to $\Sigma\, a_i n_i$ where i represents a class of cavitation events, and n the number of events per second and per unit volume. The most common example of the use of such a method is the erosion of aluminum foil, which is very commonly used to illustrate (at least qualitatively) the dramatic effects of ultrasound and is often employed to demonstrate the power of an ultrasonic cleaner. Quantitative measurements can be made [120] either by determining the weight loss after irradiation or by optical methods (by measuring the light intensity passing through the holes generated by sonication). These methods can be reasonably reproducible (to within 1% of area or weight and 2% of the time).

The erosion of pieces of other metals have been studied [120]. In some cases the amount of material removed is extremely small (e.g. 400 mg/h with lead). The better types of indicators are foils composed of aluminum or lead placed under slight tension. A further advantage is that the cavitation pattern is "printed" on the foil and in this way standing waves can be detected.

Erosion loss is a linear function of intensity under some conditions [121]. As a function of time, erosion rate is first low (especially with polished finishes) until sufficient pitting is established, and then increases to a constant value [121]. This is illustrated in Figure 22.

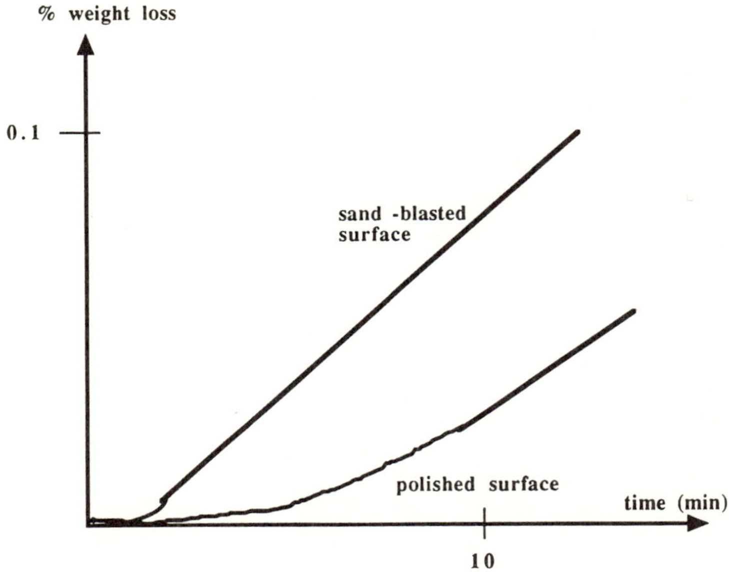

Figure 22. The relationship between the rate of surface erosion and acoustic intensity.

The slope of the linear part and the "incubation time" are very dependent on hardness, grain structure, purity, mechanical fatigue strength, surface finish, and conditions of the eroded material. All these parameters should be carefully monitored in order to obtain reproducible results which was also necessary in dosimetry methods described above which used cleaning effects. It is, of course, inevitable that the foil used as a solid test probe will interfere with the cavitation field to some extent. Erosion rates have, however, been linearly correlated to the total noise of cavitation [125].

Somewhat similar measurements could be based on solid disruption [18], polymer degradation [7], or accelerated dissolution. These well-known mechanical effects of ultrasound also derive from cavitation. Thus one might measure the rate of particle size reduction under sonication of some standard solid dispersed in a given fluid. Alternatively one could measure the rate of dissolution of a standard solid in a solvent, or the reduction in molecular weight of polymer chains. Here again the initial particle size and surface conditions, together with pressure and temperature, should be carefully monitored.

5.8 Mass Transfer Measurements: The Electrochemical Probe

The preceding methods are all in some way related to the mechanical effects of ultrasound. It is also possible to make direct measurements of mass transfer coefficients and this can be conveniently carried out at the surface of an electrode using an electrochemical probe. This technique was successfully used by H. Delmas [126] to measure the efficiency of conventional stirrers, and empirical correlations have been established between the calculated mass transfer coefficient and the electric power consumed by the stirrer itself [127].

A similar approach was recently initiated for the monitoring of mass transfer through sonication. The basic principle of the method is derived from Nernst equation. Under diffusion controlled conditions, the intensity of the limiting current is related to the mass transfer coefficient of the active species at the electrode by the relation,

$$I_\mathrm{l} = n \cdot F \cdot k_\mathrm{d} \cdot A_\mathrm{c} \cdot c_\mathrm{i}^{\infty} \tag{26}$$

where I_l is the diffusion limited current intensity, k_d the mass transfer coefficient, A_e the electrode surface area, and c_i^{∞} the bulk concentration of the electroactive species.

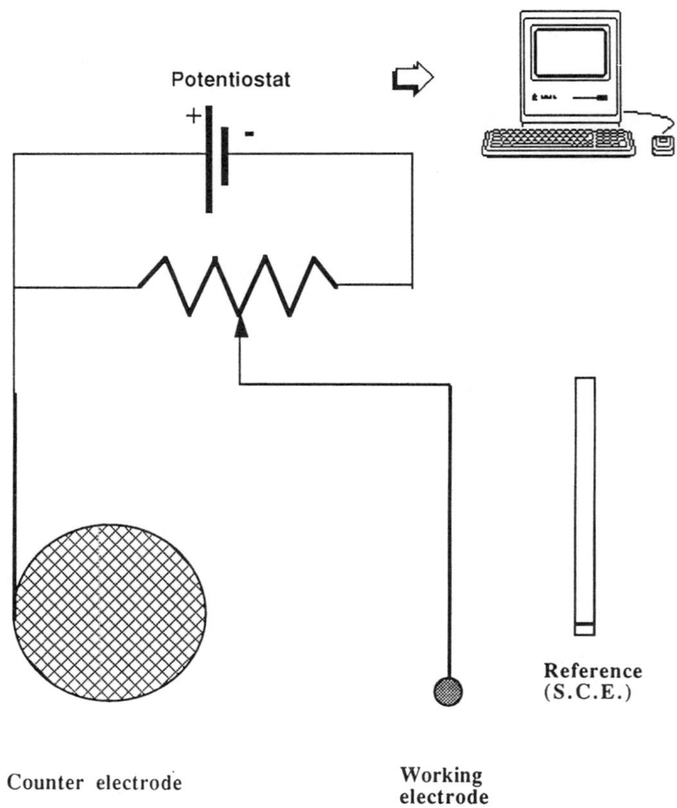

Figure 23. Schematic diagram of the electrochemical probe.

The experimental device is illustrated in Figure 23 and the following conditions were used. The redox system was an equimolar solution of potassium ferro- and ferricyanide (0.005 M each) in water with sodium hydroxide (0.1 M) as supporting electrolyte. This system was chosen due to its fast response time and its reversibility. The cathode (working electrode) is a gold sphere (1 mm in diameter) and the counter electrode is a disc (4 cm in diameter) of expanded titanium plated with platinum. The surface of the counter electrode is more than 100 times that of the working electrode in order to make sure that limiting diffusion occurs at the working electrode. The reference is a calomel electrode. Since the reduction of the ferricyanide occurs between –0.25 V and 0 V, the potential was monitored at –0.1 V/s.c.e. with a potentiostat-galvanostat (Tacusel P.J.T. 16.06) connected to a computer. The current intensity was measured every 10^{-2} s for 10 s and the average value was calculated from these.

Using this system, several ultrasonic devices were investigated. F. Contamine [26,128] studied the energy distribution in a Sodeva cup-horn as described in Figure 1. It was a jacketed cylindrical reactor (internal diameter 8 cm; total height 30 cm.) equipped at the base with a 20-kHz piezoelectric transducer (4 cm in diameter) connected to a Vibracell V1A Sonic & Materials generator. The working electrode was placed at different heights (x) from the emitter and moved along different axis parallel to the walls of the cylinder at a distance (y) from the center. The limiting current intensity was measured and the mass transfer coefficient calculated from this. The study was repeated for different total liquid heights (H), and at different power W_1. The power W_1 was measured at the generator with a wattmeter. Typical results are given in Figures 24 and 25.

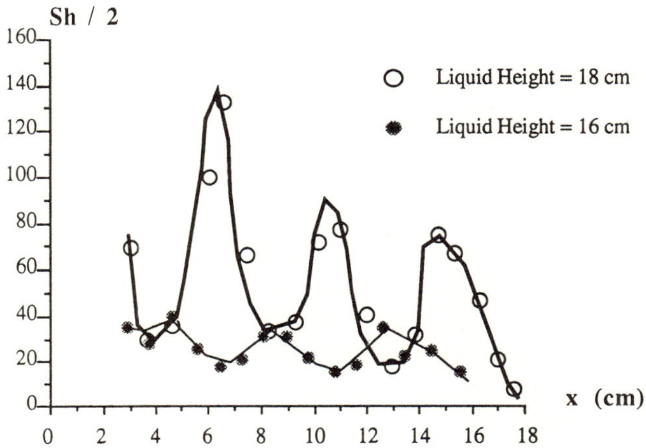

Figure 24. Relationship between Sherwood number (Sh) and position of electrochemical probe with respect to emitter surface (x).

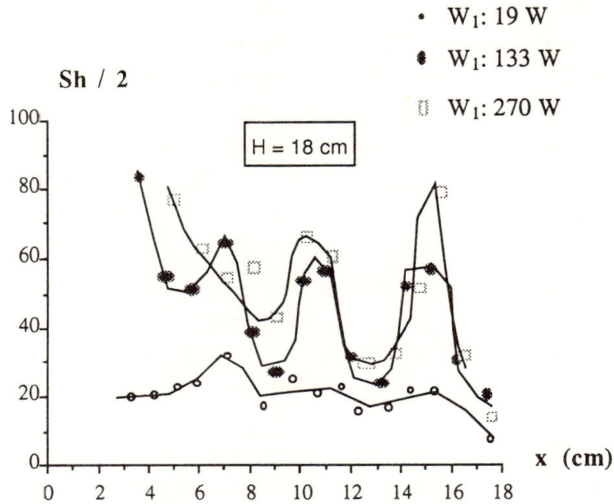

Figure 25. Relationship between Sherwood number (Sh) and position of the electrochemical probe with respect to emitter surface at different accoustic power.

In these figures the Sherwood adimensional number *Sh* is plotted against x at different total liquid heights (Figure 24) or at different power settings (Figure 25). Sh is related to the mass transfer coefficient by the relation $Sh = k_d \cdot r/D_{dif}$, where r = the radius of the electrode and D_{dif} = the diffusion coefficient of ferricyanide in the bulk solution. It can be seen that at low power settings a standing wave pattern is observed and the distance between two maxima corresponds to the half wavelength determined by the method described previously (Section 2.2). The magnitude of the mass transfer changes dramatically with the total liquid height as can be seen in Figure 24. The standing wave pattern also changes with the input power. At very low power the differences between maximum and minimum values of *Sh* are low, and at very high power (270 W) the standing waves pattern tends to disappear near the emitter, probably due to intense acoustic streaming.

A similar study was carried out by Faïd [129] with an immersed probe (diameter 12 mm). The reactor was a large beaker (25 cm in diameter). The probe was located at the center and the tip of the horn was immersed to a depth 2 cm below the surface of the liquid as shown in Figure 26.

As with the previous work, the influence of the liquid height and of the input power at the transducer on the spacial distribution of the mass transfer coefficient in the liquid were studied. Typical results are shown in Figures 27 and 28 where k_d is plotted against x (y = 0). It can be seen that at very low input power (< 20 W) a standing wave pattern is observed but this rapidly disappears as the power exceeds 20 W. This behavior is almost certainly due to acoustic streaming and the formation

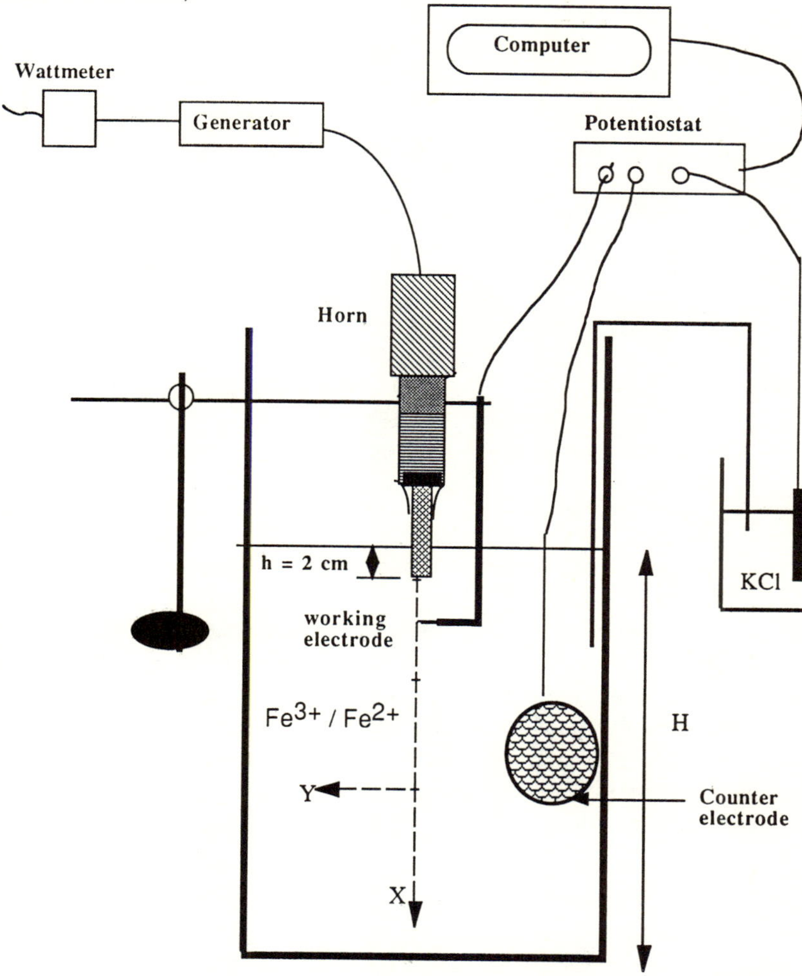

Figure 26. Experimental arrangement for the use of the electrochemical probe with an immersed horn system.

of a cloud of cavitation bubbles at the tip of the horn ("surface cavitation") which lowers transmission of the vibration into the bulk liquid.

Two further observations from this study are worthy of note. At low power, when the tip of the probe is at a nodal position, the mass transfer coefficient near the tip (x = 2 cm) dramatically increases as the power is increased. It is somewhat inexplicable that the distance between two maximum does not correspond to the half wavelength of sound in this medium as observed by Contamine using a cup-horn configuration.

The influence of the input power to the horn (measured with an oscilloscope) was then monitored with the probe located at different fixed places in the beaker.

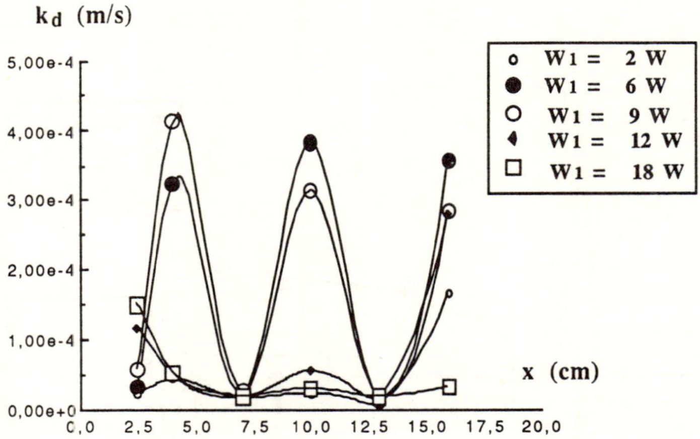

Figure 27. Mass transfer coefficient (k_d) variation with distance from emitting surface (x) at low power.

These results are shown in Figures 29 and 30 for particular points [x = 4, 10, 16 cm; y = 0] and for [x = 2.5, 7, 13 cm; y = 0] which correspond respectively to the maximum or minimum amplitude positions for k_d under standing wave conditions. A quite unusual observation has been made for this system at very low power, for a total liquid height of 19 cm when the probe is located at a maximum of the standing wave pattern. A very intense peak is observed around W_2 = 8 W, then k_d decreases sharply and starts increasing again more slowly and not linearly. This peak is not

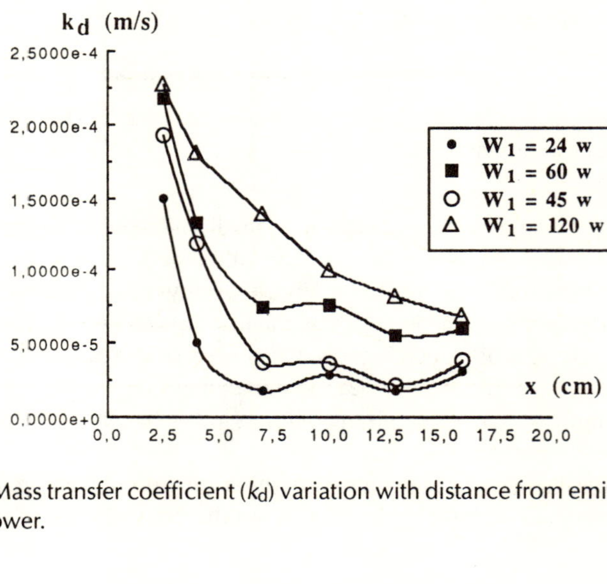

Figure 28. Mass transfer coefficient (k_d) variation with distance from emitting surface (x) at high power.

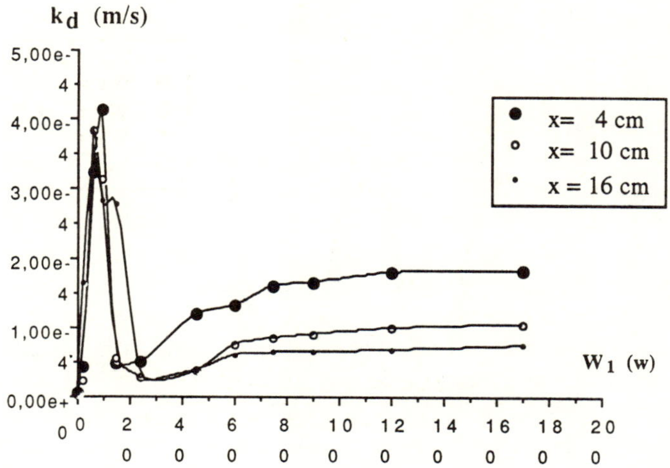

Figure 29. Influence of power on mass transfer at maximum amplitude positions of standing wave (x).

observed when the probe is at a minimum, and the mass transfer coefficient increases with W_2 probably due mainly to acoustic streaming.

Quite curiously the peak value of k_d at W_2 = 8 W is much higher than further values of k_d at much higher input power. A possible explanation is that at low power (< 10 W) acoustic streaming is low, and "permanent" cavitation at the surface of

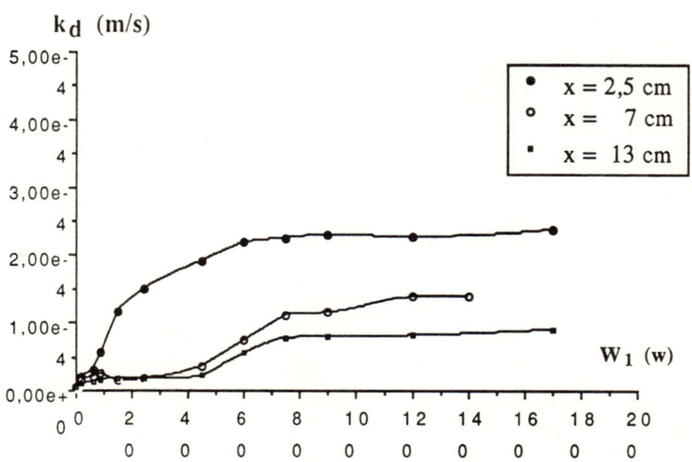

Figure 30. Influence of power on mass transfer at minimum amplitude positions of standing wave.

the probe electrode occurs under standing waves conditions. In other words, cavitation bubbles are able to remain attached to the electrode and their oscillations strongly increase the mass transfer at this surface. Whether it is gaseous or vaporous cavitation is not yet determined. When power is increased, standing waves are no longer observed for the reasons given above. Under these conditions it is likely that bubbles are swept away from the electrode during sonication and then the main contribution to the mass transfer increase under sonication is the result of acoustic streaming. As a result of this k_d dramatically decreases. At powers above 20 W—that is when the standing wave pattern is no longer observed—the magnitude of k_d is almost independent of x.

Utilization of ultrasound in the field of sonoelectrochemistry is well documented [11]. It is clear that both acoustic streaming and cavitation near or at the surface of an electrode accounts for increased mass transfer [130]. The relative contribution from each process cannot be easily estimated, but both are certainly related to the amount of dissipated power. Any dosimetry technique depending on the measurement of the mass transfer coefficient at an electrode surface should allow local and, by integration through space, overall power determination. Up to now attempts to establish quantitative correlations have failed.

It is difficult at this time to give a full interpretation of the results described and these problems are still under investigation. It is clear however that this technique provides interesting information on the energy distribution in a sonicated reactor, quantitatively similar and therefore complementary to those given by thermal probes.

5.9 Absorption Methods

As an acoustic wave propagates it is attenuated by absorption in the medium. This process results in a radiation pressure which is proportional to the absorbed energy and acoustic streaming (*vide supra*). The pressure can be measured from the height difference of liquids in two connected tubes, one being sonicated and the other not, at different values of z where z is the distance from the emitter surface. If we let I_z be the sound intensity at z and I_o the sound intensity at the surface of the emitter then,

$$I_z = I_o\,(1 - e^{-2\alpha z})/\rho \cdot g \cdot c \tag{27}$$

where ρ = liquid density, α = absorbtion coefficient, and c = the velocity of ultrasound in the medium. Thus it is possible to calculate (theoretically) values of α and I_o from I_z at different distances z. The accuracy of this method is limited because of bulk heating, which results in liquid expansion in the tubes, and because of volume change generated by cavitation. As we will see below most ultrasonic sources do not produce the single frequency specified by the manufacturer. This will effect the accuracy of the method since harmonics of the ultrasonic frequency used will have different absorption coefficients.

5.10 Methods Based on Particle Velocity

Direct measurement of the velocity or the amplitude of displacement of an imaginary particle submitted to an ultrasonic field is not easy. Filipczynski [132] suggested the use of a capacitance probe method in which the vibrations in the medium are picked up by a diaphragm. The displacement of the diaphragm is measured with an electrostatic microphone, and this is then related to the particle displacement. Sound intensity is given by the relation shown in Eq. (28) where r = particle displacement. The method can be used up to a frequency of 300 kHz.

$$I = \omega^2 \cdot r^2 / 2\,\rho \cdot c \tag{28}$$

Another method has been suggested by Neppiras [16] and is based on a secondary effect of cavitation bubbles oscillations. These oscillations induce an electric current in an electrically conducting liquid submitted to a strong magnetic field. These currents can be detected with a coil immersed in the liquid perpendicular to the field. Suitable liquids for the coil are mercury or aqueous salt. The response is a sinusoidal signal and is proportional to the strength of the magnetic field and the number of turns in the coil. In practice sufficient sensitivity can be obtained using small hand-held magnets. The coil current is detected just before the cavitation threshold (at the onset of cavitation) recorded for the subharmonic. This method suffers from the practical limitation that the probe itself interferes with the ultrasonic field.

A somewhat similar system was used by Mikhailov to measure absolute acoustic intensity in solids [31].

5.11 Optical Methods

Optical and light scattering methods have been extensively used for the detection of ultrasonic fields [133]. They can be used for material testing or to get an estimate of the intensity of the sound. The ultrasound beam generates periodic changes in the refractive index of the medium and this effectively produces a grating which diffracts a light beam through an angle $\acute{U}$ (Eq. (29)) [20,134–137].

$$\sin\theta = k \cdot n \cdot \lambda_i / \lambda \tag{29}$$

where k is a constant, λ_i = the light wavelength, λ = the sound wavelength, and n is the nth diffraction order.

The presence of undissolved gas and of cavitation bubbles affects the transparency and refractive index of a liquid. Thus when a sonicated liquid is irradiated with light, X-rays, γ-rays, or even high-frequency ultrasound, the attenuation and (or) refraction of the wave can be used to detect both the cavitation threshold and bubble density, and their variation with time. This is possible even within a very short period of the order of one ultrasonic cycle [138,139].

These methods have several advantages over the methods previously described including: (a) the absence of distortions of the ultrasonic field which might be engendered by an invasive probe system; (b) they can be used in a wide range of frequency and ultrasonic power, below or above the cavitation threshold; and (c) they can even be used with solid materials by studying the reflected beam at the surface of the material [140].

The main drawback to systems which use optical methodologies is that they require somewhat sophisticated equipment and a complex mathematical treatment of the data [141]. Naturally they also require that a transparent pathway is available through the walls of the sonicated cell and that the medium itself should be transparent to the diffracted wave.

6. METHODS BASED ON THE SECONDARY EFFECTS OF SOUND PROPAGATION AND CAVITATION

The preceding methods are mainly based on the primary mechanical effects of ultrasound during which cavitation is most often present. In contrast to this, the following methods are connected with either the secondary effects of cavitation and/or with the tremendous local accelerations reported in Table 5.

6.1 Volume Changes

At the point at which the cavitation threshold has been passed, the generation of cavitation bubbles results in a sudden increase of the volume of the sonicated liquid. This volume change can be detected and measured with a system similar to Mikhailov's acoustic dilatometer, by measuring the liquid rise in a capillary tube connected to the sonicated vessel [119,142–144]. This method can be very sensitive and is a good and simple way to determine the cavitation threshold. In principle, it could also be used for quantitative measurements but it suffers from two drawbacks. First, it does not distinguish between gaseous cavitation, vaporous cavitation, and "dead" bubbles which may become occluded at surfaces. Second, part of the liquid rise in the capillary might also be the result of expansion caused by the temperature increase which accompanies sonication of the liquid.

6.2 Acoustic Output and Noise Measurements

The response of hydrophones or microphones can be calibrated in terms of either sound level (dB) or acoustic pressure either of which are related to ultrasound intensity. The Fourier transform of these give a frequency-dependent signal, which is also related to sound intensity. It is possible therefore to obtain a plot of sound level (and thus theoretically ultrasound intensity) against frequency. However, as already mentioned, ultrasound power measurements using acoustic probes are not straightforward and require preliminary calibration of the probes with another

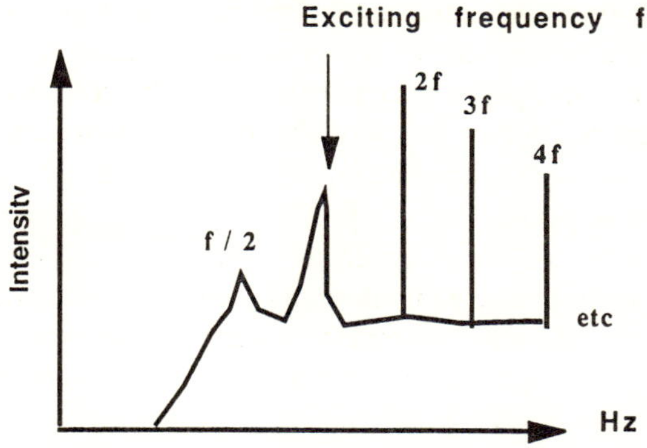

Figure 31. Typical acoustic spectrum in which only the main frequencies are shown.

technique. Noise measurements do however provide very interesting information on cavitation.

A typical acoustic spectrum is given in Figure 31 [145–147]. It includes subharmonics and harmonics of the exiting frequency and white noise. The main subharmonic, at half the fundamental frequency, is caused by forced nonlinear pulsation of bubbles at twice their resonant frequency [16].

In his pioneering work, Esche [145] suggested that the appearance of continuous components in the spectrum occurs at the onset of cavitation. Akulichev and Il'Ichev [148] separated the noise intensity of the fundamental frequency P_f from the total amplitude of noise P_n. They observed that the ratio P_n / P_f increases sharply with the input power near the cavitation threshold, and is a better criterion for the detection of cavitation than the increase of P_n alone. This was confirmed by Neppiras [16] who also studied variations in both fundamental and subharmonic signal intensities with the input power the results of which are shown in Figure 32.

From the response of a microphone located far away from the source, Neppiras observed that the fundamental signal increases and then decreases as the cavitation threshold is approached and then passed due to the screening effect of the bubbles at the surface of the emitter. At the same time, the subharmonic increases sharply at the threshold, almost at the same time as white noise, and passes through a maximum and then decreases due to cavitation between the source and the microphone. It is possible that this maximum could provide a test to detect the optimum volume coverage of cavitation.

Quantitative correlations between the intensities of fundamental, subharmonic, or harmonic intensities and the overall ultrasound intensity are not clear since the signal received depends on the nature and shape of the probe used for measurement. Preliminary calibration is required and this is certainly possible. For instance, it has been shown that the erosion of a metal foil increases linearly with the total noise [149], with the white noise output and with the square of the acoustic pressure

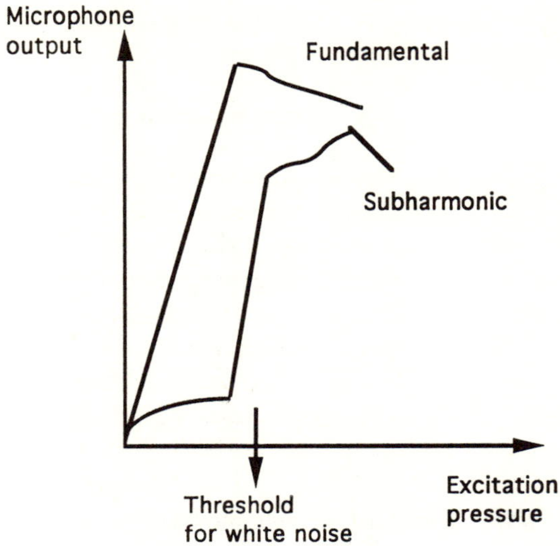

Figure 32. Microphone response to fundamental and subharmonic frequencies.

amplitude [150]. Niemczewski [118a] published a very useful study on the comparison of ultrasonic cavitation intensity in liquids using a cavitation intensity meter (Model 200 manufactured by Branson Instruments Inc., U.S.). The meter operates by the measurement of the cavitation white noise [118b]. These measurements can be made relatively easily and rapidly, and are quite useful for the detection of the cavitation threshold.

As has already mentioned, acoustic probes can be made very small and very sensitive; however their use requires somewhat sophisticated equipment together with accurate calibration which is not easy to achieve.

6.3 Conductance Changes, Electric and Electrokinetic Effects

The propagation of an ultrasonic wave and cavitation can produce several electrical effects which include:

1. Changes in conductance or permittivity of liquids [19,72,151].
2. Electric [152] and electrokinetic effects in solution and at liquid–liquid or solid–liquid interfaces (Debye effect [153], U-effect [154]).
3. Piezoelectric effects in liquid crystals [155,156].

Although these effects are related to ultrasound intensity, to date there have been no attempts to derive quantitative correlations. These effects could be useful to get at least a relative estimate of the electrical effects of ultrasound the importance of which are still a matter of debate (see below).

6.4 Sonoluminescence

When a liquid is submitted to sonication, under certain conditions light is emitted in the UV and visible bands. This phenomenon, which is called sonoluminescence, is very complex, and has been extensively studied [6,157–165]. It can be enhanced by the addition of certain organic compounds such as luminol. The origins of sonoluminescence are still a matter of controversy. The most popular explanation is related to the "hot-spot" theory, i.e. the generation of local very high temperatures and pressures during the collapse of cavitation bubbles [6,159,163]. Alternative views suggest that sonoluminescence arises through the discharge of local intense electric fields during bubble collapse [157,166,167] or the formation of local plasmas [168]. This extensive subject will not be treated in detail within this chapter except from the point of view that sonoluminescence measurements can be used as a dosimeter.

The intensity of sonoluminescence can easily be measured with photocells [159,163] or fiber optics [169] connected to a photomultiplier in darkened surroundings. This measurement is not invasive, and has been suggested as a standard [19]. In principle the sonoluminescence intensity could be correlated to the ultrasonic power, but at this time no direct theoretical correlation has been established. It has been used to determine the areas of maximum cavitational activity in a reactor. Any empirical correlation with power would necessitate preliminary calibration with another method, e.g. with thermal probes. Some care should be exercised when using a sonoluminescence probe for the following reasons:

1. Sonoluminescence strongly depends on experimental conditions [117,170].
2. The gas content of the liquid (nature, concentration), thermal conductivity, viscosity, temperature, hydrostatic pressure, frequency of the sound wave, and shape of the reactor. As a consequence the utmost care should be taken in monitoring experimental conditions to obtain good repeatability.
3. Sonoluminescence mainly occurs inside the bubbles (or in their immediate surroundings) and thus cannot be representative of what happens in the liquid phase where most of the events used in ultrasound application occur. In particular the temperature and frequency dependence of sonoluminescence is quite different from that of other effects, thus luminescence decreases with temperature, while liquid-phase effects usually increase to a maximum, and then decrease [19].

These problems are illustrated by the following example from the work of Petrier et al. [169]. Sonoluminescence and thermal probes measurements were compared at 500 kHz in a cylindrical cup-horn cell. The influence of liquid height and of

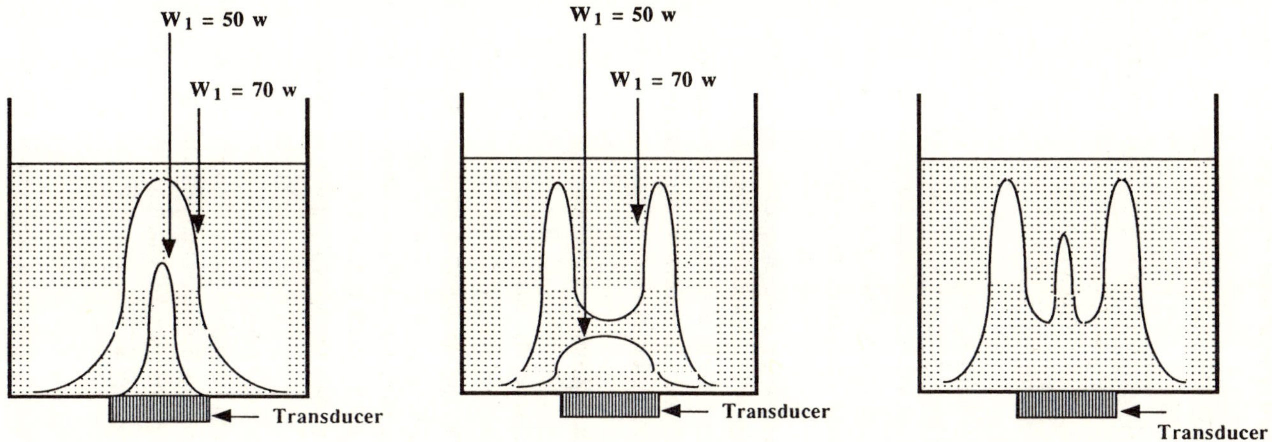

(a) Temperature (b) Sonoluminescence (c) Mass transfer

Figure 33.

ultrasonic power were studied in terms of their effect on temperature (measured with a coated thermocouple) and sonoluminescence (measured with a fiber optic) distributions (Figure 33).

As can be seen in Figure 33a, that with the thermal probe a maximum effect (maximum temperature rise of the probe) is detected at the center of the reactor, and near the liquid–air interface. This is also the case for sonoluminescence intensity at low power ($W_1 < 50$ W), but at higher power two symmetrical maxima are observed one on each side the center of the reactor (Figure 33b). Again, maximum intensity is observed near the liquid–air interface for liquid heights in a range 2.5–7.5 cm.

A similar study was made using the electrochemical probe described previously, and a typical result [171] is shown in Figure 33c. In this case, three maxima are observed, with the greatest effect again occurring near the liquid–air interface. The reason why maximum effects are observed near the liquid–air interface and not near the surface of the transducer is not clear. A tentative suggestion is that it results from the combination of the acoustic wave and the vibration of the liquid surface, together with possible inclusion of gas bubbles due to acoustic streaming, and the generation of intense gas cavitation and electrical effects.

This study highlights the fact that the dosimeter is probably responding to some selected effects of ultrasound, and not always of all the transmitted power. This point will be explored in more detail later.

6.5 Chemical Probes

The stimulation of chemical reactions has been known for a many years [1–15] and it has been suggested that some of them might be used as standards for the measurement of the efficiency of ultrasonic systems. Unfortunately, as is the case in the use of sonoluminescence as a probe, there seems to be no theoretical correlation between chemical effects and ultrasonic power. Nevertheless it is an undeniable fact that when sonochemistry is reported in the literature it would be extremely useful if the response of the system used to a standard sonochemical reaction could be included.

For absolute measurements, however, a preliminary calibration of the chemical probe would be required. Indeed since chemical effects strongly depend on applied and transmitted power, but a sonochemical yield *SY* (analogous to the quantum yield in photochemistry) has to be defined [24],

$$SY = Y/W \tag{30}$$

where Y can be the conversion or yield of the reaction under study in mole per second, W is the ultrasonic power (which can be W_1 or W_2 or W_t). *SY* depends on experimental conditions. Just as in sonoluminescence, the chemical effects of ultrasound depend on many parameters [8–14].

1. The characteristics of the medium, homogeneous or heterogeneous (e.g. viscosity, vapor pressure, the nature and size of suspended particles, the nature and concentration of any dissolved gas, thermal conductivity etc.).
2. The frequency, shape of the wave, nature of the ultrasonic source used.
3. The size and shape of the reaction vessel.
4. The temperature and hydrostatic pressure.

All of these parameters have to be carefully monitored in order to obtain reproducible results, and it is quite clear that calibration of a chemical effect can only be sustained for a fully described ultrasonic system and reactor. Any change in the nature of the device will most likely result in a change in the *SY*. Furthermore, the relation between ultrasonic power and chemical yield or reaction rate will not be linear within the whole range of ultrasonic power. An optimum in reaction yield is quite often observed. Numerous examples have been given during the past few years where optimum yields are obtained with other variable parameters such as bulk temperature, external pressure, and gas content.

Chemical probes suffer from similar limitations to sonoluminescence measurements. They are however quite attractive to sonochemists since, unlike sonoluminescence measurement, they do not require sophisticated and expensive equipment, and they offer the possibility of a direct comparison with other chemical effects.

In his classification of sonochemical reactions, Luche refers to "false" and "true" sonochemistry [172]. The former occurs when the cavitation effects produce purely mechanical effects, and this is most often the case in heterogeneous systems where rate enhancements derive from surface cleaning, in heterogeneous catalysis, the enhancement of mass transfer, and in liquid–liquid or solid–liquid phase transfer reactions. True sonochemistry, on the other hand, is the result of effects derived directly from the "hot spots" of cavitational collapse energy. This is observed in homogeneous systems, but also in solid–liquid reactions. It is thought that true sonochemistry is associated with the ultrasonic stimulation of single-electron transfer reactions. In some cases "sonochemical switching" is observed (i.e. the products obtained under conventional conditions are different from those obtained under the influence of ultrasound). Accordingly, the choice of a model reaction for use as a chemical probe will be of crucial importance since it can be representative of certain selected effects of ultrasound. Currently there is no a complete understanding of all of the complex phenomena associated with sonochemistry. It is quite clear, however, that great care should be taken in the interpretation of the results when using a chemical dosimeter as a semiquantitative determination of ultrasonic power.

Although in principle almost any chemical reaction could be used as a standard for sonochemical dosimetry, we will focus attention on those few examples of classical reactions which have been studied as potential dosimeters. They are listed in Table 6.

Table 6. Various Types of Chemical Dosimeter

	Reaction	*References*
Homogeneous solutions	$H_2O \rightarrow H_2O_2$	173,174
(aqueous or organic	$I^- \rightarrow I_2$	175,176,177
solvent mixtures)		178,179,186
	$CCl_4 \rightarrow Cl_2$ etc.	178,180,181
	$CH_3COOCH_3 + H_2O \rightarrow CH_3COOH + CH_3OH$	182,183,184
	DPPH æ $DPPH_2$	185,186
	Fluorescence of generated hydroxy terephthalate	187,188,189
	$Fe^{2+} \rightarrow Fe^{3+}$	190,191
Organic solutions	$Fe(CO)_5 \rightarrow Fe_3(CO)_{12}$	6
Solid–liquid reactions	$RBr + Li \rightarrow RLi + LiBr$	37
	Phase transfer	192
	Michael reaction	

Reactions in Aqueous Medium

The hydrolysis of methyl acetate is only weakly stimulated by ultrasound [182–184] and so this reaction would seem to be a poor contender in the pursuit of a chemical dosimeter. Despite this, Fogler and Barnes [183] have used this hydrolysis to investigate sonochemical reaction conditions. With a cup-horn type reactor they observed a temperature dependent optimum power input for this system (56 W at 40 °C, 61 W at 35 °C, and 67 W at 30 °C). The reaction itself has not been used more generally as a dosimeter.

The most commonly cited reactions derive from sonolytic decomposition of water which generates H• and HO• radicals and leads to the formation of H_2O_2 (Scheme 2). In the presence of oxygen this reaction proceeds further (Scheme 3).

Highly oxidative species such as HO• (and in the presence of oxygen HO_2•) are formed, either inside the bubble, according to the "hot spot theory" [6], or at the gas–liquid interface of the bubble, according to the electrical [166,167] or "plasma theory" [168]. Their "activity" will depend on their lifetime compared to that of the bubbles themselves [23], thus if the lifetime of the bubble is short enough, the radicals generated during collapse will be rapidly released in the bulk liquid and will induce chemical reactions. If the lifetime of the bubble is longer than that of

$$H_2O \rightarrow H\bullet + HO\bullet$$

$$2H\bullet \rightarrow H_2$$

$$2HO\bullet \rightarrow H_2O_2$$

Scheme 2

$$H\bullet + O_2 \rightarrow HO_2\bullet$$

$$O_2 \rightarrow 2O\bullet$$

$$2HO_2\bullet \rightarrow H_2O_2 + O_2$$

$$O_2 + O\bullet \rightarrow O_3$$

Scheme 3

the radicals, they will mainly recombine, and there will be less chance of chemical effects taking place in the bulk liquid. It has also been proposed that any oxygen present could act as a radical carrier for H• [185].

A dosimetry method based on the detection of the H• radical involves its reaction with DPPH to generate $DPPH_2$ which can easily be monitored by UV spectrometry. With this method, Verrall et al. reported a linear correlation between percent degradation of DPPH and free energy of cavitation [185].

Several methods exist for the identification and quantification of the HO• and HO_2• radicals generated by the sonolysis of water. These species can oxidize ionic moieties e.g. Fe^{2+} into Fe^{3+} (the Fricke dosimeter) and I^- into iodine. In addition, either can dimerize to form hydrogen peroxide (Schemes 2 and 3), which can then be titrated using conventional techniques. The HO• will also react with terephthalate anion in aqueous solution to produce hydroxyterephthalate anion, a fluorescent material which can then be estimated using fluorimetry.

The decomposition of KI in aqueous solution has been used quite frequently as a chemical dosimeter. The iodine generated can be titrated conventionally with sodium thiosulfate or, more generally, estimated by UV spectrometry (by measuring the absorbance of the colored I_3^- species at around 350 nm). If the aqueous KI solution is saturated with CCl_4 (Weisslers solution) the generation of iodine is intensified due to decomposition of the halocarbon to Cl• and then chlorine (Scheme 4). The discovery of this reaction dates back to 1938 when Liu and Wu found that adding small amount of CCl_4 to a KI aqueous solution increases the rate of formation of iodine [193]. Subsequently Weissler reported that if a KI solution is kept saturated with CCl_4, the rate of iodine released during sonication is 15 times higher than with pure KI [194].

$$CCl_4 \rightarrow Cl\bullet + CCl_3\bullet \rightarrow \text{etc.}$$

$$2Cl\bullet \rightarrow Cl_2$$

$$Cl_2 + 2I^- \rightarrow I_2 + 2Cl^-$$

Scheme 4

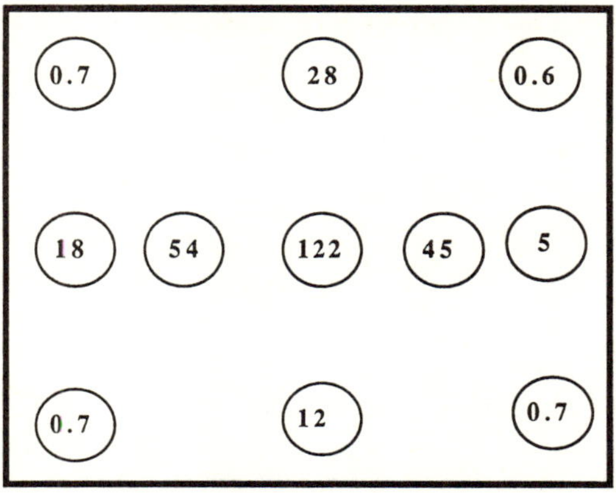

Figure 34. Relative amount of chlorine (arbitrary units) generated at various positions in an ultrasonic bath in a test tube containing water saturated with CCl_4.

The formation of iodine during sonication can be visualized by adding soluble starch to the medium and observing the blue color which appears. This technique has been used to identify the positions in a reactor where cavitation is most intense [19]. In this way standing waves can be detected since the blue color appears in the zones of maximum amplitude (maximum sonochemical activity).

In the case of halocarbon, saturated aqueous solutions an alternative estimation of cavitation can be obtained through the titration of liberated chlorine itself by adding orthotoluidine [178]. This method has been used by Weissler to study the relative amounts of cavitation in a 28-kHz cleaning bath [178]. A glass test tube containing a saturated solution of carbon tetrachloride in water was dipped in the bath filled with water and sonicated for 10 s. The test tube was placed at different locations in the bath, keeping constant the liquid heights in the tube and in the bath, and the distance between the bottom of the bath and that of the tube. The results are illustrated in Figure 34 where the reported numbers represent the relative amounts of chlorine generated (in arbitrary units).

Weissler also studied the influence of water height in the tank, and showed a periodic dependence of the amount of chlorine which can be attributed to the presence of standing waves. He also studied the effect of prolonged use of the ultrasonic tank on the liberation of chlorine. The test tube containing the solution was sonicated for 10 s, withdrawn, and the amount of chlorine determined. The ultrasonic bath was left operating continuously, and every few minutes the test tube was sonicated again but with fresh solution. The amount of chlorine was plotted against time (Figure 35).

It can be seen that the amount of chlorine decreases with time, indicating some change in sonication conditions, i.e. the amount of energy delivered by the bath to the reaction vessel. If the ultrasonic bath was filled with fresh water (saturated with

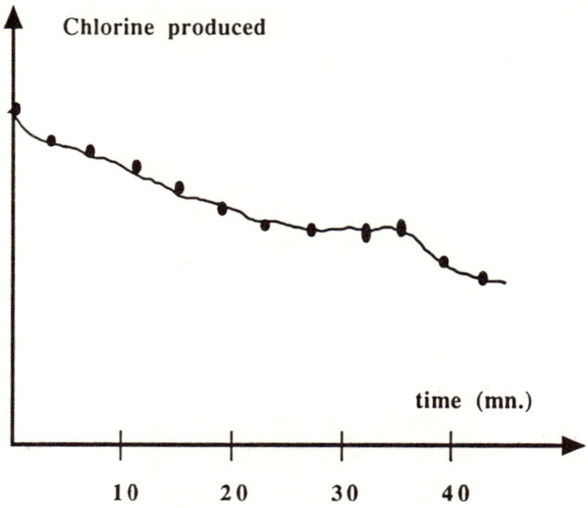

Figure 35. Showing the diminution in chlorine yield from water saturated in CCl_4 after long exposure to ultrasonic irradiation.

air), and the experiment repeated again, the amount of chlorine generated, which had fallen to a low level, was increased over the first 10 min and then decreased as before. Such a change can be ascribed to the degassing of the water in the tank. This study shows that the reaction conditions for sonochemistry need careful monitoring. More particularly, the transmission of sound through a liquid and thus the extent of the sonochemical effects observed strongly depend on the concentration and physical characteristics of dissolved gas.

Weissler's reaction is a quite attractive chemical probe, due to the ease of handling and the fact that it works well over a wide range of frequency, from 20 kHz up to several MHz. It should be noted, however, that its rate is very frequency-dependent being quite small at low frequency and sharply increasing as the frequency is increased.

Despite the fact that a 10% reproducibility has been reported [178], often the results are somewhat erratic due to its great and not always totally understood dependence on experimental conditions. If a series of consecutive experiments are carried out on the same day, fairly consistent results are usually obtained. However it often happens that the same experiments performed on different days give results exhibiting variations by a factor of 2 or more. This is the reason why some authors prefer the terephthalate probe [187–189] (referred to below as the TA probe), although it will be seen that it also suffers from similar drawbacks.

The terephthalate chemical dosimeter can be prepared by dissolving terephthalic acid (TA) $1.5.10^{-3}$ mol^{-1} and NaOH 5.10^{-3} mol^{-1} in a phosphate buffer at pH = 7.4. The fluorescence generated by the product (hydroxyterephthalate) is measured at 425 nm with an excitation wavelength of 315 nm. The TA dosimeter solution can

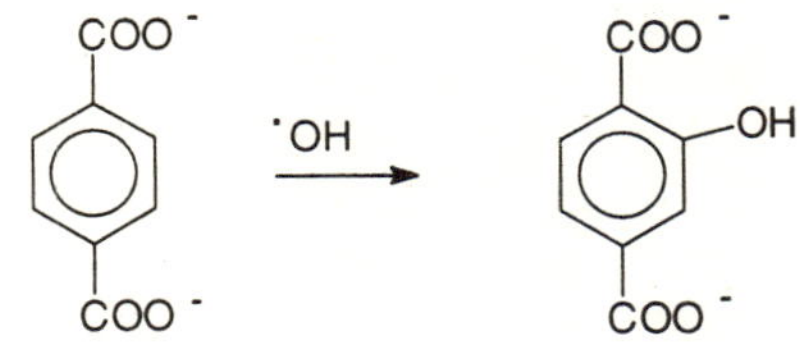

Scheme 5. Reaction of terephthalate ion with hydroxy radical.

be calibrated by exposing it to a known strength radioactive source (e.g. Co-60[187]) which produces a known concentration of HO• in aqueous solution. In water, hydroxyl radicals, generated by ultrasound or any other means, react with terephthalate anion to give hydroxyterephthalate according to the following reaction (Scheme 5).

McLean and Mortimer [187] have studied the variations in HO• free radical production during the sonication of aqueous solutions at different powers at 970 kHz. A typical curve is given in Figure 36. From this it is clear that a threshold exists for radical production, after which there is a linear correlation with acoustic power up to a limiting value which probably corresponds with "surface cavitation". Acoustic power was calibrated with a radiation balance and a PVDF hydrophone. Repeatability on experiments performed on the same day was less than 15%, but day-to-day variations could be as much as 50%, probably mainly due to small uncontrolled changes in the alignment of the reaction chamber (a test tube dipped in a water tank) with the ultrasonic source which was an acoustic horn.

Price and Lenz [188] compared the TA probe with the Fricke dosimeter ($Fe^{2+} \rightarrow Fe^{3+}$), using a Sonics and Materials VC 600 horn operating at 20 kHz.

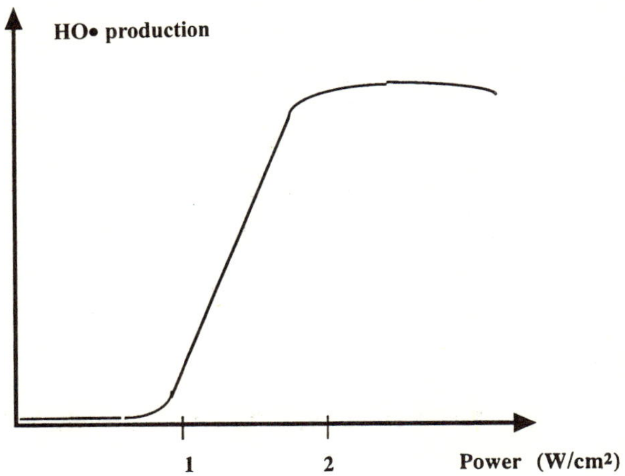

Figure 36. Effect of power HO• radical production at 970 kHz.

Output intensities were measured calorimetrically by comparison with an electrical heating coil. It was concluded that for the Fricke dosimeter at a given ultrasonic power, changes of absorbance at 350 nm vary linearly with time and with ultrasonic power in the 10–30 W cm^{-2} range. An observation from previous workers that deoxygenation reduced the rate of the oxidation process was confirmed. In the case of the TA probe subjected to ultrasonic irradiation at a single frequency, the fluorescence intensity increased with increased TA concentration up to 1.5 10^{-3} mol^{-1} after which fluorescence intensity remained constant or decreased slightly, as illustrated in Figure 37.

This suggests that below 1.5 10^{-3} mol^{-1}, there is not enough TA present to react with all of the HO• produced, but beyond this concentration there is no advantage in increasing TA concentration, and there could be some drawbacks as TA may act as a fluorescence inhibitor. Price and Lenz confirmed that the power dependence of fluorescence intensity was similar to that reported by McLean and concluded that the TA dosimeter is much more sensitive than the Fricke system.

Mason et al. reported for the first time the response of the TA dosimeter with different ultrasonic sources and frequencies. They employed an ultrasonic cleaning bath (Kerry Pulsatron 55 operating at 38 kHz) with different immersed reactors (flat bottom Erlenmeyer and round bottom flask) and the Undatim Sonoreactor with 20-, 40-, or 60-kHz horns. Ultrasonic power measurements were monitored using the calorimetric method described previously.

It was reported that under constant sonication conditions the measured fluorescence is directly proportional to exposure time. Within the power ranges studied, the yield of HO• radicals was proportional to the power input, and that at constant

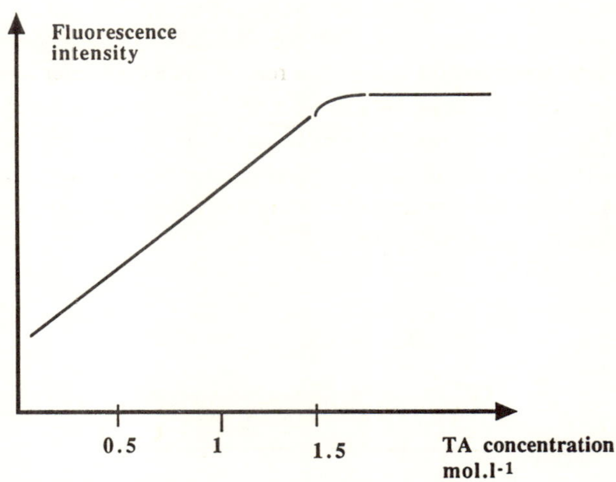

Figure 37. The effect of increased terephthalate ion (TA) concentration on fluorescence intensity at 20 kHz.

power fluorescence yield increases from 20 to 60 kHz. The fluorescence yield was found to increase with TA concentration in the range $0.5–2\cdot10^{-3}$ mol^{-1}, a slightly different result from that of Price and Lenz who reported $1.5\ 10^{-3}$ mol^{-1} as the upper limit.

A more surprising result was obtained when comparing an Erlenmeyer with a round bottom flask dipped in the cleaning bath. Although the energy input, measured calorimetrically, was roughly the same (respectively 17.4 and 14.4 W), a great difference was found in fluorescence yields (respectively 4.5 and 1.93). At this time this discrepancy is not clearly understood, but illustrates once again that the determination of ultrasonic power is not an easy task.

In conclusion it is important to note that the above chemical dosimeters do not measure the same effects. The TA probe is a specific dosimeter for HO• radicals, while the others are more general—thus both I^- and Fe^{2+} can also be oxidized by HO_2•, H_2O_2, or indeed other species and such processes do not occur at the same rate (e.g. the rate of production of I_2 from I^- oxidation and the formation of H_2O_2 in water can be monitored independently and are not the same [174]). Chemical dosimeters are strongly frequency-dependent, thus the production of iodine in air saturated KI solutions is 6 times faster at 514 kHz than at 20 kHz [174]. They are also strongly dependent on experimental conditions, especially with respect to the gas content.

Reactions In an Organic Medium

Cavitation has mainly been studied in water, although some studies have been carried out in organic solvents [2,6]. One possible dosimeter for nonaqueous conditions could be the decomposition of iron pentacarbonyl in hydrocarbon solvents [186]. In different alkanes it was possible to demonstrate the inverse relationship between sonochemical effect (i.e. the energy of cavitational collapse) and solvent vapor pressure. The use of this reaction as a dosimeter was developed by Suslick who studied the sonolysis of $Fe(CO)_5$ which produced $Fe_3(CO)_{12}$ together with finely divided iron (the proportion of each depending on the solvent vapor pressure) (Scheme 6). In fact this decomposition provided a significant result in sonochemistry since it differed from both thermolysis (which gave finely divided iron) and UV photolysis [which gave $Fe_3(CO)_9$].

$$Fe(CO)_5 \rightarrow Fe_3(CO)_{5-n} + nCO$$

$$Fe(CO)_5 + Fe(CO)_3 \rightarrow Fe_2(CO)_8$$

$$2Fe(CO)_4 \rightarrow Fe_2(CO)_8$$

$$Fe(CO)_5 + Fe_2(CO)_8 \rightarrow Fe_3(CO)_{12} + CO$$

Scheme 6

Reactions In a Heterogeneous Medium

All of the preceding dosimeters for sonochemistry (both chemical and physical) are applicable in, and have been studied under, homogeneous conditions. On the other hand, most of the potential industrial applications of ultrasound concern solid–liquid mixtures. The use of such dosimeters under heterogeneous conditions could lead to some discrepancies, however, since the presence of a suspended solid may result in scattering and dampening of the wave. For this reason the search for accurate dosimeters working under heterogeneous conditions is of considerable interest.

Pugin [37] monitored the formation of an organolithium compound (butyl lithium in THF) and compared it to thermal and erosion measurements. He found a linear correlation between the rate of reaction for this process and the temperature rise of a coated thermocouple. This is an interesting result, but the slope of the line will almost certainly depend on the surface condition of the lithium pieces, on their size, and on the location of the solid relative to the ultrasonic source. These parameters should be carefully controlled in order to get reproducible results using this dosimeter.

Another possible probe could be the Michael addition of diethyl malonate to chalcone (Scheme 7) [192]. This dosimeter operates in a toluene solution of chalcone and diethyl malonate with suspended powdered potassium hydroxide. Under sonication and in the presence of a phase transfer agent this reaction is too fast, but its rate can be conveniently controlled by adjusting the reagent concentrations in the absence of the phase transfer agent. Interestingly, the size of the KOH particles is not an important parameter since this solid is rapidly disrupted down to an average size of ca. 60 µm.

The reaction can easily be monitored by HPLC, NMR, or by simple weighing of the addition product [197]. The rate increase with ultrasound not only depends on the mechanical effects (mass transfer improvement) but also on some "electronic effects" as it has recently been shown that the reaction mechanism involves a single electron transfer step which can be stimulated by ultrasound [198]. Hence the development of this "chemical probe" could provide a very good dosimetry system since it involves both the mechanical and sonochemical effect of ultrasound.

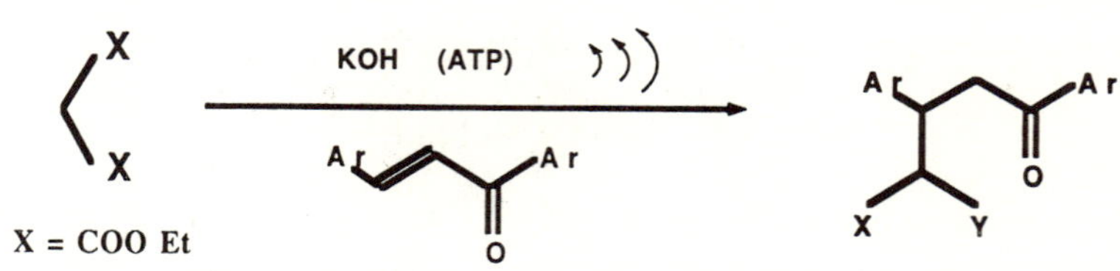

Scheme 7. Michael addition reaction used as a probe for power measurement.

7. SUMMARY

There are two reasons why a sonochemist should be interested in dosimetry. The first is that it might be possible to calculate the power required to perform an operation in order that the economics of scale-up can be assessed. The second reason relates to the need to "normalize" results obtained in different laboratories or with different equipment. If a suitable dosimeter is chosen then it should be possible to perform sonochemical reactions anywhere under precisely comparable conditions.

In any discussion on dosimetry it is very important to recognize that the particular method of power measurement adopted should be suitable for the given application. A few of the methods described are more general, thus input power measurements at the generator or at the transducer are very easy to perform. Such a measurement should therefore be made from time to time with a standard load since it permits a check to be made on how the ultrasonic system is performing through the W_2/W_1 ratio. Furthermore the input power to the transducer W_2 can be controlled and monitored and this can be applied in order to carry out sonications under reproducible conditions. Commercial ultrasound generators as used by sonochemists have been designed to deliver different powers by changing a power setting as represented on a dial; this provides a basis for reproducibility within a given reaction study. If the type of acoustic horn is changed or a different reaction mixture is sonicated then switching on the generator at the same power setting does not necessarily mean that exactly the same power is delivered to the system.

In order to measure the actual transmitted power W_t into the sonicated medium, many different methods have been devised. Since these methods are based on measuring changes in different parameters of the system, both physical and chemical, such methods may not be directly comparable. The decision as to which of these methods is the most accurate really depends upon which is the best adapted to the sonication system under study in terms of both the ultrasonic device employed and the application.

An attempt to compare the sensitivity, reproducibility, and accuracy of some methods has been made by Zieniuk and Chivers [17]. In our own survey above, the dosimetry techniques quoted will have advantages and limitations a summary of which appears below. Only a few methods allow the direct and absolute measurement of transmitted power and these include thermal methods, radiation pressure measurements, and electrical or mechanical measurements at the transducer.

7.1 Thermal Methods

From a practical point of view, the most generally applicable and the easiest dosimeters to use are those based on thermal methods, especially those using thermal probes. These probes have almost no limitations since they can be used (a) in any reaction vessel below or beyond the cavitation threshold, (b) in free or restrained acoustic fields, and (c) they allow local and global power measurements. Thermal methods are reasonably sensitive (0.2 mW) and can be used over a wide frequency and power range. Accuracy (generally better than 7%) can easily be improved further by calibration with a heating coil.

Some care should be exercised when using coated thermocouples since the response of the probe strongly depends on the coating. Although a calorimeter is important for the basic calibration of transducers, the calorimeter itself of little interest for sonochemical studies, unless it is used as the reactor.

7.2 Radiation Force Measurements

Methods which have been developed for the monitoring of radiation forces can be very sensitive (up to 1 mW). They are better suited in free-field and noncavitating conditions, but involve errors due to acoustic streaming. The uncertainty for such methods ranges from 2.2% at 1 MHz to 12% at 30 MHz, and accuracy depends on the shape of the ultrasonic wave. Radiation force measurement provides a good method for the calibration of transducers in specifically devised chambers, but its use in chemical reactors of defined geometry could prove to be difficult.

7.3 Electrical and Mechanical Measurements at the Transducer

These measurements can be very accurate, but require specialized equipment. The main drawback is that to calculate the transmitted power it is necessary to know the coupling coefficient with the load; that is to know how much of the energy generated at the transducer is transmitted to the medium. This may prove to be difficult in liquid processing and requires a calibration with another method. However, knowing the amplitude of the vibration may be of interest since for a given amplitude the sonochemical effects will depend on the surface area of the emitter.

7.4 Other Physical Methods

Other methods for the measurement of transmitted power normally require some form of preliminary calibration. Sound intensity and pressure measurements have the advantage that they respond precisely to the net flow of acoustic power, although they are directional. They can be used in any reaction vessel, but require somewhat sophisticated equipment and/or time-consuming measurements and calculations to derive power readings. They are better suited to local measurements but give interesting information about the cavitation threshold. In our laboratories we have used a hydrophone to "map" the energy distribution in a vessel subject to probe sonication with promising results [39]. The experimental setup consists of a 5-L beaker with a centrally positioned acoustic horn (20 kHz) placed 45 mm below the surface at a power of 2 W. For a water depth of 165 mm (3.5 L) the acoustic pressure

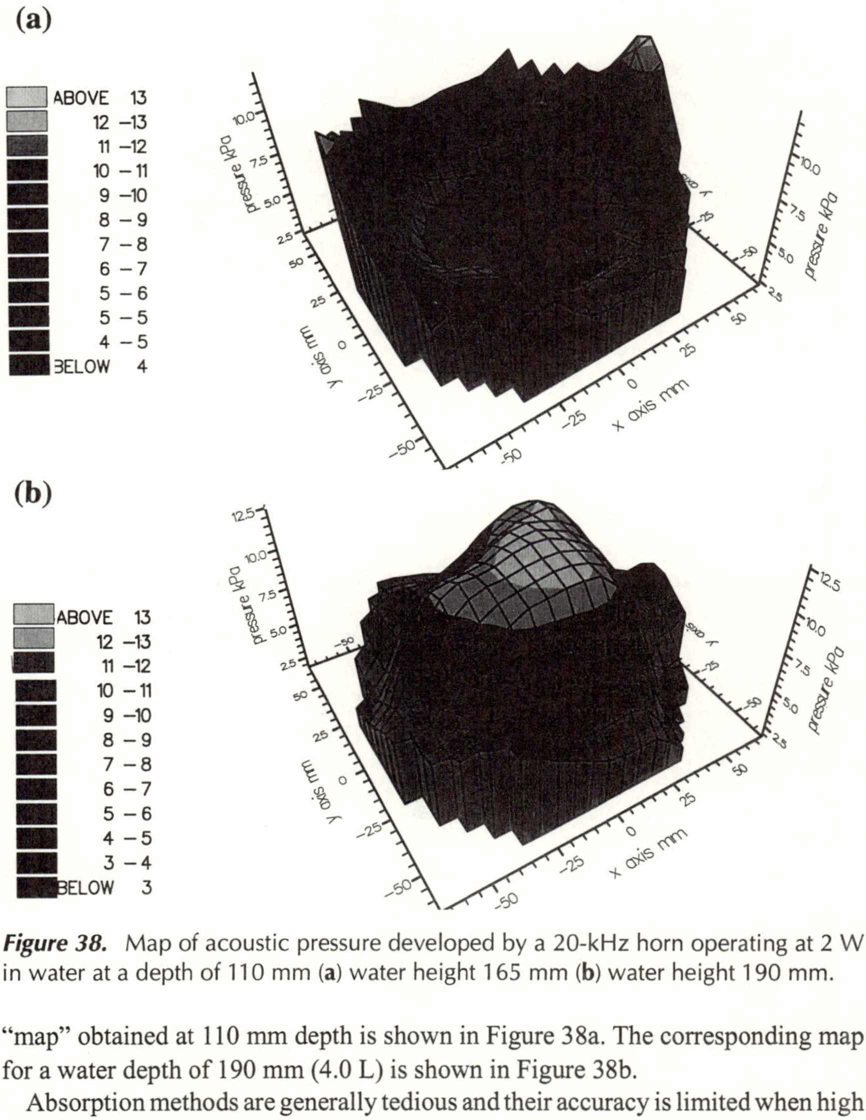

Figure 38. Map of acoustic pressure developed by a 20-kHz horn operating at 2 W in water at a depth of 110 mm (**a**) water height 165 mm (**b**) water height 190 mm.

"map" obtained at 110 mm depth is shown in Figure 38a. The corresponding map for a water depth of 190 mm (4.0 L) is shown in Figure 38b.

Absorption methods are generally tedious and their accuracy is limited when high frequency ultrasound sources are used.

Methods based on volume changes, erosion and surface cleaning, and dispersion are rapid but not very accurate and reproducibility is difficult to achieve. Data scattering may be up to 40%. However dosimetry based on erosion and surface cleaning are of particular interest in the study of systems involving of solid disruption.

Optical methods have the distinct advantage that they are noninvasive in contrast with the other methods, and therefore do not disturb the ultrasonic field. The drawback is that they require some sophisticated equipment and mathematical interpolation. Furthermore interpretation of the results may be difficult. Optical methods agree well with those employing radiation forces below the cavitation threshold. The accuracy of this type of dosimeter at high acoustic power is considerably reduced due to the scattering of the incident light wave by cavitation bubbles. As with radiation force methods, optical dosimetry through optical measurements cannot easily be operated in standard sonochemical reactors.

Mass transfer measurements using an electrochemical probe are still under investigation. They only allow local measurements and the system accuracy is not yet known. These methods do however allow interesting observations on acoustic streaming and standing waves conditions.

7.5 Chemical Dosimeters

Since this chapter appears in a volume devoted to sonochemistry, chemical probes would appear to be the most attractive since they could allow direct comparisons with other chemical reactions. Chemical dosimeters are generally used to test the effect of an ultrasonic device on the total volume of the reactor. Local measurements can however be made with very small cells containing the dosimeter which could be moved inside the reaction vessel as with a coated thermocouple. Most of these chemical probes are derived from reactions carried out in an homogeneous medium, e.g. Weissler's solution, the Fricke dosimeter, or the oxidation of terephthalate anions. Among these the latter shows promise in that despite the fact that to date it has been much less used than Weissler's reaction it seems to have higher sensitivity and better reproducibility.

Ideally when a chemical dosimeter is used to test or assess an ultrasonic device, care should be taken to match the system under study with the dosimeter type. The optimum conditions determined for a reactor using a chemical probe may well not be the same optimum as that required for the chemical system under investigation. Similar observations apply to the use of sonoluminescence.

Fortunately accurate and absolute measurements of the transmitted power are not always required to assess the effectiveness of a given ultrasonic treatment. In many instances, the important thing is to produce a dosimetry method which is reproducible and easy to handle and will achieve relative measurements. Furthermore the convenient method of choice can be calibrated with a more accurate, but probably more tedious one, and from that point on it can be used with more confidence.

The two main drawbacks to chemical dosimetry are that they have low sensitivity at low power and are often strongly frequency-dependent. For accurate measurements they should also be calibrated with another method, e.g. with a thermal probe.

7.6 Comparative Studies

Several comparative studies of various types of chemical probes have been reported [32,33,199,200] and good agreements obtained in relative terms. For

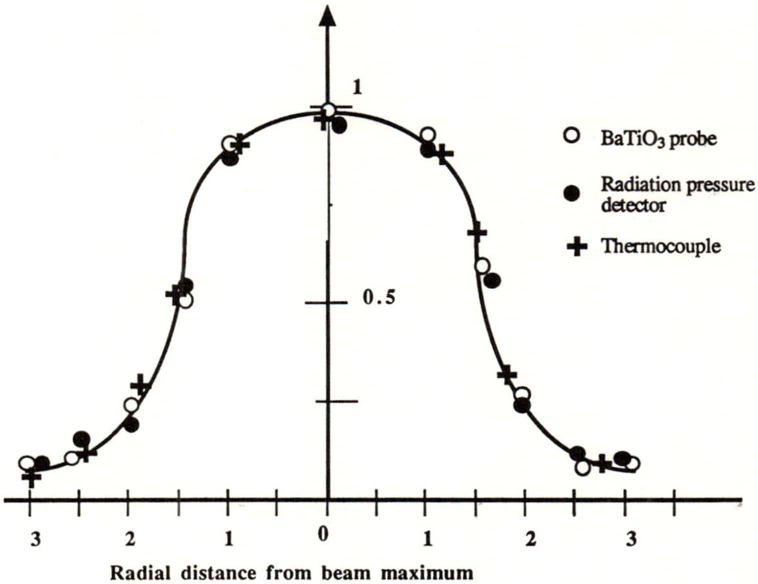

Figure 39. Normalized beam patterns obtained from a single ultrasonic source (980 kHz) using three different probes.

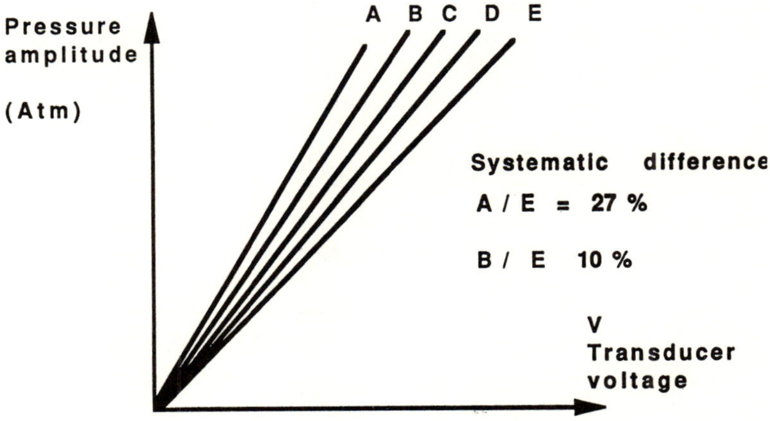

Figure 40. Calibration curves for a 990-kHz ultrasound source obtained by five different methods. (**A**) Calculated from transducer characteristics; (**B**) thermocouple; (**C**) optical image broadening; (**D**) light diffraction; (**E**) decrease of transmitted light intensity.

example Figure 39 shows the beam patterns taken with a thermocouple probe, a piezoelectric probe ($BaTiO_3$), and a radiation pressure detector [22]. These plots are normalized at 1 for the maximum response of the probe. The source was a 980-kHz focusing irradiator. A similar observation has been reported by Martin and Law [33] using a 0.8-mm thermistor probe coated with UHU glue and a 1-mm diameter hydrophone using a 780-kHz source. Both studies report very good agreements in a relative scale between the different methods.

Breazeale and Dunn [200] reported the calibration of a 990-kHz transducer with five different methods which included the calculated output power from the characteristics of the quartz transducer. Quite good agreement was observed as illustrated in Figure 40. The major discrepancy occurs with the calculated power, but it should be remembered that because of the assumptions made in the calculation the value obtained is the maximum value possible under ideal conditions and optimum matching.

8. CONCLUSION

In conclusion, since no single technique is perfect, a combination of several methods would give better confidence in monitoring sonochemical reactions and ultrasonic studies. Measurements of input power, combined with a thermal method or a previously calibrated chemical reaction, should always been carried out to allow an accurate description of experimental work and a reasonable reproducibility from one ultrasonic device to another. Even though they may not represent the best dosimetry method, thermal techniques are becoming widely used in sonochemical studies [18]. This is quite encouraging for sonochemistry in general since up to this stage of development power measurements have been neglected and indeed are often absent from published papers. If sonochemistry is to become an exact science, practitioners must adopt at least this simple dosimetry technique and, hopefully, move on to a standardization of methodology in the future.

9. REFERENCES

[1] Brown, B. and Goodman, J.E. *High Intensity Ultrasonics*, ILiffe Books Ltd, London, 1965.

[2] El'Piner, I.E. *Ultrasound, Physical, Chemical and Biological Effects*. Consultants Bureau, New York, 1964.

[3] Rozenberg, L.D. *High Intensity Ultrasonic Fields*. Plenum Press, New York, 1971.

[4] Rozenberg, L.D. *Physical Principles of Ultrasonic Technology*. Plenum Press, New York, 1973, Vol. 1.

[5] Rozenberg, L.D. *Physical Principles of Ultrasonic Technology*. Plenum Press, New York, 1973, Vol. 2.

[6] Suslick, K.S. *Ultrasound, Its Chemical, Physical, and Biological Effects*. VCH, 1988.

[7] Mason, T.J. and Lorimer, J.P. *Sonochemistry, Theory, Applications and Uses of Ultrasound in Chemistry*, Ellis Horwood, U.K., 1988.

[8] Ley, S.V. and Low, C.M.R. *Ultrasound in Synthesis*. Springer-Verlag, 1989.

[9] Mason, T.J. *Advances in Sonochemistry*. JAI Press, London, 1990, Vol. 1.

[10] Mason, T.J. *Sonochemistry, The Uses of Ultrasound in Chemistry*. Royal Society of Chemistry, 1990.
[11] Mason, T.J. *Chemistry with Ultrasound*. Published for the SCI by Elsevier Applied Science, 1990.
[12] Mason, T.J. *Practical Sonochemistry, User's Guide to Applications in Chemistry and Chemical Engineering*. Ellis Horwood, 1991.
[13] Mason, T.J. *Advances in Sonochemistry*. JAI Press, London, 1991, Vol. 2.
[14] Price, G.J. *Current Trends in Sonochemistry*. The Royal Society of Chemistry, 1992.
[15] Mason, T.J. *Advances in Sonochemistry*. JAI Press, London, 1993, Vol. 3.
[16] Neppiras, E.A. *IEEE Trans. Sonics Ultrasonics,* SU-15 (2) (1968) 81.
[17] Zieniuk, J. and Chivers, R.C. *Ultrasonics,* 14 (1976) 161.
[18] Mason, T.J., Lorimer, J.P., and Bates, D.M. *Ultrasonics*, 30 (1992) 40.
[19] Neppiras, E.A. *Ultrasonics* 3 (1965) 9.
[20] Saskena, T.K. *J. Acoust. Soc. India* VIII (1980) 13.
[21] Goswami, S., Ghosh, P.N., and Basumallick, S. *IE(I) Journal-ET* 69 (1988) 12.
[22] Welcowitz, W. *IRE National Convention Record,* 6 (1958) 199.
[23] Boucher, R.M.G. *British Chem. Eng.*, 15 (1970) 363.
[24] Berlan, J. and Mason, T.J. *Ultrasonics*, 30 (1992) 203.
[25] Berlan, J. and Delmas, H., unpublished results.
[26] Contamines, F. *Thèse de l'Institut National Polytechnique de Toulouse*, 6 (1993).
[27] Breazeale, M.A. and Dunn, F. *J. Acoust. Soc. Amer.*, 55 (1974) 671.
[28] Sokollu, A. *Bulletin Electroacoustique du Laboratoire d'Electroacoustique de l'Université de Liège*, 9 (1966) 23.
[29] Byron, J.J. *I.E.E.E. Transaction on Sonics & Ultrasonics*, SU 16, 2 (1978) 76.
[30] Mikhailov, I.G. and Shutilov, V.A. *Sov. Phys. Acoust.*, 3 (1957) 410; *ibid* 5 (1959) 385.
[31] Mikhailov, I.G. *Ultrasonics,* 2 (1964) 129.
[32] Fry, W.J. and Fry, R.B. *J. Acoust. Soc. Amer.*, 26 (1954) 294; *ibid* 26 (1954) 311.
[33] Martin, C.J. and Law, A.N.R. *Ultrasonics*, 21 (1983) 85; *ibid* 18 (1980) 127.
[34] Palmer, R.B.J. *J. Sci. Instruments*, 30 (1953) 177.
[35] Weber, M.E. and Chon, W.Y. *Can. J. Chem. Eng.*, 45 (1967) 238.
[36] Fogler, H.S. and Timmerhaus, K.D. *A. I. Ch. E. J.,* 12 (1966) 96.
[37] Pugin, B. *Ultrasonics*, 25 (1987) 49.
[38] Romdhane, M. *Thèse de l'Institut National Polytechnique de Toulouse*, 26 Novembre (1973).
[39] Barr, A., Martin, D.C., and Mason, T.J., unpublished results part of the UK/France ALLIANCE programme 1994–1995.
[40] Weiderheilm, C. *Rev. Sci. Instruments.*, 27 (1956) 540.
[41] Labartkava, E.K. *Soviet Physics-Acoustics*, 6 (1971) 468.
[42] Morita, S. *J. Phys. Soc. Japan*, 7 (1952) 214.
[43] Zieniuk, J.K. *Ultrasonics*, 4 (1966) 136.
[44] Dunn, F. and Fry, W.J. *I.R.E. Trans. Ultrasonic Eng.*, 5 (1957) 59.
[45] Szilard, J. *Proc. 8th Int. Conf. Acoust.* London, 1974, p. 352.
[46] Hawley, S.A., Breyer, J.E., and Dunn, F. *Rev. Sci. Instruments*, 33 (1962) 1118.
[47] Moran, G., Menton, M., and Lejeune, G. *Comptes Rendus,* 257 (1963) 1018.
[48] Wolley, P.F., Barnet, R.J., and Pond, J.B. *Ultrasonics*, 13 (1975) 68.
[49] Degrois, M. *Ultrasonics*, 4 (1966) 38.
[50] Hunt, F.V. *Electroacoustics*, Harvard Monographs in Applied Science, Wiley, New York, 1954.
[51] Hufter, T.F. and Bolt, R.H. *Sonics*, Wiley, New York, 1955.
[52] Rozenberg, L.D. and Sirotyuk, M.G. *Soviet Physics-Acoustics*, 8 (1962) 4.
[53] Kikuchi, Y. and Shimizu, H. *J. Acoust. Soc. Amer.*, 31 (1959) 1385.
[54] Teumin, I.I. *Soviet Physics-Acoustics,* 8 (1963) 291.
[55] Mikhailov, I.G. and Shutilov, V.A. *Soviet Physics-Acoustics*, 10 (1964) 77.
[56] Hill, C. *J. Acoust. Soc. Amer.*, 52 (1972) 667.

[57] Lanhan, T.E. *IBM Technical Disclosure Bulletin Vol. 25*, No. 11A (1983) 5497.
[58] Hufter, T.F. and Bolt, R.H. *Sonics* Wiley, New York, 1955, p. 151.
[59] Koppelmann, J. *Acustica*, 2 (1952) 92.
[60] Romanenkov, E.V. *Soviet Physics-Acoustics*, 3 (1957) 364.
[61] Mellen, R.H. *J. Acoust. Soc. Amer.*, 28 (1956) 447.
[62] Yeager, E., Dietrick, H., and Hovorka, F. *J. Acoust. Soc. Amer.*, 25 (1953) 456.
[63] Richardson, E.G. *Proc. Roy. Soc. London*, A146 (1936) 56.
[64] Palmer, B. *J. Sci. Instruments*, 30 (1953) 177.
[65] Fry, W.J. and Fry, R.B. *J. Acoust. Soc. Amer.*, 26 (1954) 311.
[66] Filipczynski, L. *Acoustica*, 3 (1969) 137.
[67] Lewin, P.A. *Ultrasonics*, 19 (1981) 213.
[68] Shotton, K.C., Bacon, D.R., and Quilliam, R.M. *Ultrasonics*, 18 (1980) 123.
[69] De Reggi, A.S., Roth, S.G., Kenney, J.M., Edelman, S., and Harris, G.R. *J. Acoust. Soc. Amer.*, 69 (1981) 853.
[70] Staudenraus, J. and Eisenmenger, W. *Ultrasonics*, 31 (1993) 267.
[71] Mikhailov, I.G. and Shutilov, V.A. *Soviet Physics-Acoustics*, 5 (1959) 385.
[72] Briggs, H.B., Johnson, V.B., and Mason, W.P. *J. Acoust. Soc. Amer.*, 19 (1947) 664.
[73] Strasberg, M. *J. Acoust. Soc. Amer.*, 31 (1959) 163.
[74] Brown, B. *British Commun. Electr.*, 12 (1962) 918.
[75] Awaya, K. and Kariya, S. *Convention Record of the Acoust. Soc. Japan,* No. 2-2 -(1957) 12.
[76] Nicolas, J. and Lemire, G. *J. Acoust. Soc. Amer.*, 78 (1985) 414.
[77] Gol'dberg, Z.A. In Rozenberg, L.D. (ed.), *High Intensity Ultrasonic Fields*, Plenum Press, New York, 1971, pp. 75–133.
[78] Post, E.J. *J. Acoust. Soc. Amer.*, 25 (1953) 55.
[79] Lord, R. *Phil. Mag.*, 3 (1902) 338.
[80] Beyer, R.T. and Letcher, S.V. *Physical Ultrasonics*. Academic Press, New York, 1969, pp. 65–67.
[81] Rooney, J.A. and Nyborg, W.L. *Am. J. Phys.*, 40 (1972) 1825.
[82] Rooney, J.A. *J. Acoust. Soc. Amer.*, 54 (1973) 429.
[83] Fox, F.E. and Griffing, V. *J. Acoust. Soc. Amer.*, 20 (1948) 352.
[84] King, L.V. *Proc. Roy. Soc. A.*, 147 (1934) 212.
[85] Yoioka, Y. and Kawasima, Y. *Acustica*, 5 (1955) 167.
[86] Hasegawa, T. and Yosioka, K. *J. Acoust. Soc. Amer.*, 46 (1955) 1139.
[87] Embleton, T.F.W. *Can. J. Phys.*, 34 (1956) 276.
[88] Embleton, T.F.W. *J. Acoust. Soc. Amer.*, 26 (1954) 40.
[89] Newell, J.A. *Phys. Med. Biol.*, 8 (1963) 215.
[90] Tarnocy, T. *Magyar Fizikai Folyroirat*, 2 (1954) 159.
[91] Cseko, A.K. and Veress, E., *Proc. 7th Int. Conf. Budapest*, 1971, p. 561.
[92] Wells, P.N.T., Bullen, M.A., and Freundlich, H.F. *Ultrasonics*, 2 (1964) 124.
[93] Fox, F. and Griffing, V. *J. Acoust. Soc. Amer.*, 21 (1949) 352.
[94] Hasegawa, T. and Yosioka, K. *J. Acoust. Soc. Amer.*, 58 (1975) 581.
[95] Kossoff, G. *Acustica*, 12 (1962) 84.
[96] Hill, C.R. *Phys. Med. Biol.*, 15 (1970) 241.
[97] Wells, P.N.T., Bullen, M.A., Follett, M.A., Freundlich, H.F., and Angell, J.J. *Ultrasonics*, 1 (1963) 106.
[98] Wemlen, A. *Med. Biol. Eng.*, 6 (1968) 159.
[99] Whittingham, T.A. *Ultrasound in Med. and Biol.*, 1 (1975) 475A.
[100] Bindal, V.N. and Kumar, A. *Acustica*, 46 (1980) 224.
[101] Hasegawa, T., Yosioka, K., and Omura, A. *Acustica*, 22 (1970) 145.
[102] Reynier, G., Garnier, J.L., and Gazanhes, C. *Centre de Recherches Physiques*, note n° (1964) 783.
[103] Fox, F.E. *J. Acoust. Soc. Amer.*, 12 (1940) 147.
[104] Maidanik, G. *J. Acoust. Soc. Amer.*, 29 (1957) 738.

[105] Maidanik, G. and Westervelt, P.J. *J. Acoust. Soc. Amer.*, 29 (1957) 936.
[106] Feilder, G. U.S. Patent 2531,844 (1950).
[107] Laufer, A.R. and Thomas, G.L. *J. Acoust. Soc. Amer.*, 28 (1956) 951.
[108] Richardson, E.G. *Ultrasonic Physics*, Second Edition. Elsevier, Amsterdam, 1942.
[109] Beissner, K. *Acustica*, 57 (1985) 1; *ibid.* 58 (1985) 17.
[110] Sirotyuk, M.G. *Soviet Physics-Acoustics*, 10 (1964) 398.
[111] Rosenberg, L.D. *Soviet Physics-Acoustics*, 14 (1965) 100.
[112] Borgnis, F.E. *J. Acoust. Soc. Amer.*, 25 (1953) 546.
[113] Seidl, F. *Acustica*, 2 (1952) 45.
[114] Oyama, H. *J. Inst. Elec. Eng. Japan*, 55 (1958) 560.
[115] Bogorodskii, V.V. and Romanov, V.N. *Soviet Physics-Acoustics*, 8 (1963) 326.
[116] McQueen, D.H. *Ultrasonics*, 24 (1986) 273.
[117] Chenke, P.K. and Fogler, H.S. *Chem. Eng. J.*, 8 (1974) 165.
[118] (a) Niemczewski, B. *Ultrasonics*, 18 (1980) 107. (b) U.S. Patent 3443797.
[119] Weissler, A. Paper presented at Institute of Radio Engineers' Symposium on Sonics and Ultrasonics, 1962.
[120] Crawford, A.E. *Ultrasonics*, 2 (1964) 120.
[121] Bebchuck, A.S. *Soviet Physics-Acoustics*, 3 (1957) 90; *ibid.* 3 (1957) 95; *ibid.* 3 (1957) 395; *ibid.* 4 (1958) 372.
[122] Antony, A.O. *Ultrasonics*, 1 (1963) 194.
[123] Chon, W.Y. and Wong, S.W. Paper presented at ACHEMA 67, Frankfurt, June 1967.
[124] Hickling, R. and Plesset, M.S. *Physics of Fluids*, 7 (1964) 7.
[125] Rosenberg, L.D. *Ultrasonic News*, 4 (1960) 4.
[126] Delmas, H. Thèse de docteur ingénieur INP Toulouse, 1983.
[127] Hinze, J.O. *Turbulence*. McGraw Hill, New York, 1959.
[128] Delmas, H., Berlan, J., Wilhelm, A.M., and Contamine, F., to be published.
[129] Faïd, F. Thèse de doctorat l'Institut National Polytechnique de Toulouse, 1994.
[130] Klima, J., Bernard, C., and Degrand, C. *J. Electroanalytical Chem.*, 367 (1994) 297.
[131] (a) Kossov, G. *J. Acoust. Soc. Amer.*, 38 (1965) 880. (b) Stuehr *J. Tech. Chem.*, 6 Pt. 2 (1974) 237. (c) Parthasarathy, S. and Pancholy, M. *An. der Physik*, 17 (1956) 417.
[132] Filipczynski, L. *Acustica*, 3 (1969) 137.
[133] Scruby, C.B., Dewhurst, R.J., Hutchins, D.A., and Palmer, S.B. In R.S. Sharpe (ed.), *Research Techniques in Non-Destructive Testing*. Academic Press, New York, 1982, Vol. 5, Chapter 8, pp. 281–327.
[134] Nagai, S. *Ultrasonics*, 23 (1985) 77.
[135] Riley, W.A. *J. Acoust. Soc. Amer.*, 67 (1980) 1386.
[136] Klein, W. and Cook, B. *IEEE Trans. Sonics and Ultrasonics,* SU 14 (1967) 123.
[137] Breazeale, M.A. and Hiedemann, E.A. *J. Acoust. Soc. Amer.*, 31 (1959) 24.
[138] Briggs, H.B., Johnson, V.B., and Mason, W.P. *J. Acoust. Soc. Amer.*, 19 (1947) 664.
[139] Strasberg, M. *J. Acoust. Soc. Amer.*, 31 (1959) 163.
[140] Noui, O.L. and Dewhurst, R.J. *Ultrasonics*, 31 (1993) 425.
[141] Reibold, R. and Kwiek, P. *Ultrasonics*, 31 (1993) 308.
[142] Akulichev, V.A. and Rosenberg, L.D. *Soviet Physics-Acoustics*, 11 (1965) 246.
[143] Mikhailov, I.G. and Shutilov, V.A. *Soviet Physics-Acoustics*, 5 (1960) 383.
[144] Neppiras, E.A. *Soviet Physics-Acoustics*, 8 (1992) 4.
[145] Eche, R. *Acustica Beihefte*, 2 (1952) 4AB 208.
[146] Bohn, L. *Acustica*, 7 (1957) 201.
[147] Mellen, R.H. *J. Acoust. Soc. Amer.*, 26 (1954) 356.
[148] Akulichev, V.A. and Il'Ichev, V.L. *Soviet Physics-Acoustics*, 4 (1963) 372.
[149] Rozenberg, L.D. *Ultrasonic News*, 4 (1960) 4.
[150] Berchuk, A.S. *Soviet Physics-Acoustics*, 4 (1958) 372.

[151] Gerdes, E. *Wissenschlaftliche Zeitschrift der Universtät Rostock*, 10 (1961) 17.
[152] Booth, F. and Enderby, J.A. *Proc. Phys. Soc.*, 65 (1951) 321.
[153] Yeager, E., Bugosh, J., Hovorka, F., and McCarthy, J. *J. Chem. Phys.*, 17 (1949) 411.
[154] Yeager, E. and Hovorka, F. *J. Acoust. Soc. Amer.*, 25 (1953) 443.
[155] Meyer, R.B. *Phys. Rev. Letters*, 22 (1969) 918.
[156] Ioffe, I.V. *Sov. Phys. JETP*, 53 (1981) 534.
[157] Frenkel, J. *Acta Phys. Chim., (USSR)* 12 (1940) 317.
[158] *Sonoluminescence and the Chemical Effects of Ultrasound*. Mullard Research Laboratories England MRL Rep., 1951, p. 136.
[159] Neppiras, E.A. and Noltingk, B.E. *Proc. Phys. Soc.*, B63 (1950) 674.
[160] Kutruff, H. and Plass, K. *Acustica*, 11 (1961) 224.
[161] Negeshi, K. *J. Phys. Soc. Japan*, 16 (1961) 1450.
[162] Flynn, H.G. In Mason, W.P. (ed.), *Physical Acoustics 1* B. Academic Press, New York, 1964.
[163] Suslick, K.S., Dokytcz, S.J., and Flint, E.B. *Ultrasonics*, 28 (1990) 203; *ibid.* 28 (1990) 280.
[164] Didenko, Y.T. and Pugach, S.P. *Ultrasonics Sonochemistry*, 1 (1994) S9.
[165] Roy, A. *Ultrasonics Sonochemistry*, 1 (1994) S5.
[166] Margulis, M.A., in ref. 9, p. 39.
[167] Margulis, M.A. *Soviet Physics-Acoustics*, 22 (1976) 310.
[168] Lepoint, T. and Mullie, F. *Ultrasonics Sonochemistry*, 1 (1994) S13.
[169] Renaudin, V., Gondrexon, N., Boldo, P., Pétrier, C., Bernis, A., and Gonthier, Y. *Ultrasonics*, 1995, in press.
[170] Didenko, Y.T., Nastich, D.N., Pugach, S.P., Polovinka, Y.A., and Kvochka, V.I. *Ultrasonics*, 32 (1994) 71.
[171] Berlan, J., Trabelsi, F., and Delmas, H., unpublished results.
[172] Luche, J.L. *Ultrasonics*, 30 (1992) 156.
[173] Lindstrom, O. *J. Acoust. Soc. Amer.*, 27 (1955) 654.
[174] Pétrier, C., Jeunet, A., Luche, J.L., and Reverdy, G. *J. Amer. Chem. Soc.*, 114 (1992) 3148.
[175] Renaud, P. *Bull. Soc. Chim. Fr.*, (1950) 1044.
[176] Bennett, G. *J. Acoust. Soc. Amer.*, 24 (1952) 470.
[177] Kosoff, G. *Acustica*, 12 (1960) 84.
[178] Weissler, A. and Hine, E.J. *J. Acoust. Soc. Amer.*, 34 (1962) 130.
[179] Heinglein, A. *Ultrasonics*, 25 (1987) 6.
[180] Liu, S.P. *J. Acoust. Soc. Amer.*, 38 (1965) 817.
[181] Aerstin, F.G.P., Timmerhaus, K.D., and Fogler, H.S. *AIChE Journal*, 13 (1976) 453.
[182] Chen, J.W. and Kalback, W.M. *I&EC Fundamentals*, 6 (1967) 175.
[183] Fogler, H.D. and Barnes, D. *I&EC Fundamentals*, 7 (1968) 222.
[184] Couppis, E.C. and Klinzing, G.E. *AIChE Journal*, 20 (1974) 485.
[185] Sehgal, C., Yu, T.J., Sutherland, R.G., and Verrall, R.E., *J. Phys. Chem.*, 86 (1982) 2982.
[186] Suslick, K.S., Schubert, P.F., and Goodale, J.W. *Chemical Dosimetry of Ultrasonic Cavitation*. Ultrasonics Symposium, 1981, pp. 612–616; Suslick, K.S., Hammerton, D.A., and Cline, R.F. Jr. *J. Amer. Chem. Soc.*, 108 (1986) 5641.
[187] McLean, R.J. and Mortimer, A.J. *Ultrasound in Medicine and Biology*, 14 (1988) 59.
[188] Price, G.J. and Lenz, E.J. *Ultrasonics*, 31 (1993) 451.
[189] Mason, T.J., Lorimer, J.P., Bates, D.M., and Zhao, Y. *Ultrasonics Sonochemistry*, 1 (1994) S91.
[190] Fricke, H., Hart, E.J., and Smith, P. *J. Chem. Phys.*, 6 (1938) 229.
[191] (a) Miller, N. *J. Chem. Phys.*, 18 (1950) 79. (b) *Trans. Faraday Soc.*, 46 (1950) 456.
[192] Ratoarinoro, N., Wilhelm, A.M., Berlan, J., and Delmas, H. *Chem. Eng. J.*, 50 (1992) 27.
[193] Liu, S.C. and Wu, H. *J. Amer. Chem. Soc.*, 60 (1938) 1497.
[194] Weissler, A. *J. Acoust. Soc. Amer.*, 25 (1953) 651.
[195] Prudhomme, R.O. and Grabar, P. *J. Chim. Phys.*, 46 (1949) 323.
[196] Prudhomme, R.O. and Busso, R.H. *C.R. Acad. Sci. Paris*, 235 (1952) 1486.

[197] Bates, D. The Effect of Ultrasound and Other Physical Parameters on the Reactivity of Powders and Catalysts, Coventry University, Ph.D. Thesis (1992).
[198] Berlan, J., Ratoarinoro, N., Wilhelm, A.M., Contamine, F., and Delmas, H. *Chem. Eng. J.*, (1995) in press.
[199] Tschiegg, C.E., Greenspan, M., and Eitzen, D.G. *Journal of Research N.B.S.*, 88 (1983) 91.
[200] Breazeale, M.A. and Dunn, F. *J. Acoust. Soc. Amer.*, 55 (1974) 671.

NUCLEAR MAGNETIC RESONANCE SPECTROSCOPY COMBINED WITH ULTRASOUND

John Homer, Larysa Paniwnyk, and
Stuart A. Palfreyman

OUTLINE

Advances in Sonochemistry
Volume 4, pages 75–99

ISBN: 1-55938-793-9

ABSTRACT

Following an overview of the relevant aspects of nuclear magnetic resonance spectroscopy, the natural dependence of the technique on ultrasound is discussed. The fundamental role of ultrasound in nuclear spin–lattice relaxation is explained, and the underlying principles are used as a base for the discussion of classical and new experimental methods that rely on NMR studies in the presence of ultrasound. Following a treatment of the well-known phenomenon of acoustic nuclear magnetic resonance in solids, some experimental evidence is advanced that indicates that the technique may also be possible for liquids. Subsequent coverage includes the acoustic modification of the nuclear spin–lattice relaxation times in both solids and liquids, the NMR detection of acoustically induced changes to molecular conformations and the ESR detection of radicals produced by *in situ* ultrasound. Particular emphasis is placed on a new technique, SINNMR, that enables the sonically induced narrowing of the NMR spectra of solids.

1. INTRODUCTION

The uses and applications of ultrasound are many and diverse, with some being of longstanding. Ultrasound can now be used in areas which range from its use as "observer" in the diagnostic testing of materials or in ultrasonic imaging techniques to that of "manipulator" in the relatively young area of sonochemistry. These uses have inspired the development of new techniques and new syntheses. Whereas for those interested in, but unfamiliar with, the field of sonochemical reactions an excellent review [1] is available, very little is available regarding the new subject that embraces the combined use of nuclear magnetic resonance spectroscopy (NMR) and ultrasound. The present contribution focuses on "ultrasound the manipulator" and its application to liquid and solid state NMR spectroscopy.

2. THE RELATIONSHIP BETWEEN NMR AND ULTRASOUND

2.1 NMR: An Overview

In order to develop certain ideas essential to the following discourse it is convenient to start with a brief résumé of the subject of NMR spectroscopy: for further detail the reader is referred to Akitt's [2] lucid text. For the purpose of NMR, nuclei may be categorized using their spin quantum numbers (I) as (i) those that are not active ($I = 0$), (ii) the dipolar active nuclei ($I = 1/2$), and (iii) the quadrupolar active nuclei ($I > 1/2$). By way of introduction, the present discussion is limited mainly to the classical description of NMR when observing dipolar nuclei (those nuclei which have a nuclear spin quantum number of 1/2) in solution.

For the simplest of NMR active nuclei, with $I = 1/2$, their spin angular momentum $[(I(I+1)^{1/2})\hbar$, where $\hbar$ is the reduced Planks constant $(h/2\pi)]$ and spherical charge distribution results in the nucleus having a magnetic moment (μ). The latter can interact with an externally applied magnetic field to result in just two accessible energy states: for the general case of a nucleus of spin I there are $2I + 1$ energy states. For nuclei with $I = 1/2$, the energy difference between the two allowed states is given by,

$$\Delta E = \hbar\gamma B_o \tag{1}$$

where γ is the magnetogyric ratio, which has a different fixed value for each nuclide, and B_o is the applied magnetic field. Due to the energy difference between the quantized states, an assembly of nuclei assumes a Boltzmann distribution between the states that is governed by the equation,

$$\frac{N_\beta}{N_\alpha} = e^{-\frac{\Delta E}{kT}} \tag{2}$$

where the number of nuclei in the low energy state is N_α and the number of nuclei in the high energy state is N_β. Each of the nuclear moments precess at the characteristic Larmor frequency about the applied magnetic field (B_o) direction. It is, therefore, necessary to use a weak orthogonal rotating magnetic field (B_1), derived from a sinusoidally varying signal at the Larmor frequency, to apply an appropriate torque to cause transitions between the allowed energy states. The probabilities of both the absorption and emission transitions caused in this way are identical. Consequently, the net effect of the NMR experiment is to cause a reduction in the excess number of nuclei in the lower energy state with an overall absorption of energy and a net transfer of some nuclei into the higher energy state. The required frequency (ν) of electromagnetic radiation to cause this transition is given by:

$$\nu = \frac{\gamma B_o}{2\pi} = \frac{\mu B_o}{Ih} \tag{3}$$

The detection of energy at this transition frequency is the basis of NMR spectroscopy. The actual detection of NMR signals, however, is made possible through the bulk magnetization (M) of the nuclear system that arises from the resultant of the individual nuclear magnetic moments that are distributed between the various energy levels. The rotating components (x and y) of μ transverse to the direction (z) of B_o at nonresonant equilibrium have no phase coherence and $M_x = M_y = 0$, whereas under the same conditions M_z is finite (M_0) and reflects the population distribution of nuclei between the allowed energy states. At resonance, M_0 (in general M_z) adopts some angle relative to the B_0 direction and the decay (see later) of the now finite components M_x and M_y permit the detection of NMR signals through emf's induced in detector coils placed along the x or y axes.

The utility of NMR spectroscopy stems from the fact that all nuclei of a particular isotope do not have the same fundamental resonant frequency as required by Eq. 3. In fact a spectrum of resonance absorptions may be observed for a particular nuclide in a given compound. These may be related directly to the chemical structure of the compound under investigation.

The implication of Eq. (3) is that the magnetic field at the nucleus is the same as that generated by the magnet used. In fact this is not true. For various reasons a nucleus is shielded from the external magnetic field so that chemically different nuclei of the same isotope in a particular compound experience slightly different local magnetic fields. As a result of this, the magnetically different nuclei resonate at slightly different frequencies. Equation (3) has, therefore, to be rewritten as,

$$\nu = \frac{\gamma B_0(1-\sigma)}{2\pi} \tag{4}$$

where σ, the 'screening constant', is determined directly by the chemical structure. The difference between the resonant conditions for two nuclei, i and j, are characterized by the chemical shift, δ, which is given by

$$\delta_{ij} = \sigma_i - \sigma_j \tag{5}$$

Detailed examination of chemically shifted resonances can reveal fine structure splitting that is governed by the scalar spin–spin coupling constant (J) that reflects the energetic interaction between magnetically different nuclei. Again J provides information about molecular structures.

In the absence of other mechanisms, a situation would quickly arise during the NMR experiment where there would be an equal number of nuclei in the upper and lower energy states. This would result in the system being saturated and no longer capable of absorbing energy. Naturally, this would correspond to a thermodynamically nonequilibrium situation that is not sustainable in nature. Of necessity, therefore, there must be a mechanism by which the nuclei which have been excited into the higher energy state (a nonequilibrium state) can transfer back down into the lower energy state. By the same token there must be a way by which the sample reaches the Boltzmann equilibrium position after it is first placed in a magnetic field. The mechanism is the same in both cases and is called spin–lattice relaxation. However, unlike in the NMR experiment the absorption and emission transition probabilities are not equal, and, to be consistent with the laws of thermodynamics, the latter probability must be greater than the former. These naturally occurring transitions are said to arise from spin–lattice relaxation processes. There is also another relaxation process that is known as the spin–spin relaxation mechanism that facilitates the interchange of energy between like spins. Of these two processes only the former influences the population distribution of nuclei between available energy states, and it is on this that attention is focused in the present context.

Spin–lattice relaxation occurs in a characteristic spin–lattice relaxation time, T_1, such that equilibrium and nonequilibrium macroscopic magnetizations are related by,

$$M_o - M_z = (M_o - M_i)e^{-\frac{t}{T_1}} \tag{6}$$

where M_z is the component of magnetization of the sample at time t following recovery from some perturbed initial magnetization M_i, and M_o is the value at equilibrium.

Inspection of Eq. (6) reveals that it takes ca. $5.3 \times T_1$ (the value of t when $M_z = 0.995\, M_o$ and $M_i = 0$) for the system to recover to within 0.5% of the Boltzmann equilibrium condition after the system has become saturated. It is for this reason that the duration between pulses in Fourier transform NMR spectrometers (most modern spectrometers) should be at least $5 \times T_1$ for the production of quantitatively meaningful NMR spectra.

2.2 Spin–Lattice Relaxation

Lattice Phonons in Solids

The thermal motion naturally found within a solid matrix results in the propagation of acoustic "waves" from lattice nodes. Comparable with the photon description of electromagnetic radiation these acoustic waves can be described as phonons and are similarly quantized. The density of phonons p_ω of a particular frequency ω ($\nu/2\pi$), in a solid continuum that contains N atoms in a sample of volume V has been deduced by Abragam [3] to be,

$$p^\omega = \frac{3V\omega^2}{2\pi^2 v^2} \tag{7}$$

where v is the velocity of propagation.

At each lattice frequency there is a phonon density that is characteristic of the material considered. The phonon spectral density vanishes above a characteristic "cut off" frequency that is typically 10^{13} Hz for many solids. The cut off point of the phonon density is some orders of magnitude greater than the Larmor frequencies of nuclei currently found in NMR spectroscopy, and so there is a significant phonon density at these NMR frequencies to influence nuclear transitions.

Relaxation Transition in Solids

Direct and indirect interactions of lattice phonons with a nucleus are possible and indeed provide the route by which a nonequilibrium nuclear system can 'relax' back to its original equilibrium state.

Direct. Of the vast number of phonons propagating through the solid, a small number have a frequency identical to the Larmor frequency of a nucleus in the sample. It is, therefore, possible for these phonons to act directly on such a nucleus causing, if the nucleus is in an energized state, the emission of energy and the return of the nucleus to its ground state: they can, of course also cause absorption transitions.

Due to the extremely low density of phonons at the Larmor frequency the probability of causing a directly stimulated transition is almost negligible.

Indirect. As there are only a very small number of pairs of phonons whose energies combine to match the energy which corresponds to the Larmor frequency, ω_0, of a nucleus, those indirect relaxation processes sequentially involving two phonons are highly improbable.

The main mechanism for indirect stimulation of relaxation is via Raman processes, and these process are responsible for the vast majority of all relaxation transitions in solids. In a Raman process the change in the energetic state of the nucleus is accompanied by the simultaneous absorption and emission of phonons. If the absorbed phonon has a frequency of ν, the energy of the emitted phonon may be either $h(\nu + \nu_0)$, where the nucleus contributes its excess energy to the phonon allowing the nucleus to relax, or $h(\nu - \nu_0)$ where the energy of the phonon is absorbed and the nucleus is promoted to a higher energy state. Obviously, absorption can only occur when the frequency of the phonon is above that of the Larmor frequency. However, emission can occur for any frequency of phonon. It is for this reason that emission relaxation transitions are more probable than absorption transitions.

The above simplistic description of relaxation in the solid state reveals an implicit dependence of NMR on natural sound. Naturally, therefore, it raises the possibility that if sound is introduced artificially into a solid lattice the normal NMR characteristics of the material may be modified.

Relaxation Transitions in Liquids

Relative to solids, it is the increased molecular motion, particularly translational and rotational, that provides the dominant relaxational pathways in liquids. Whereas the detailed molecular structures of solids are relatively easy to characterize, it is far more difficult to do this for liquids. Consequently, in the latter case it is convenient to leave the phonon and quantum mechanical approach behind and revert to a classical description of the system. For such a description of a liquid system the constituent species can be considered to generate positionally dependent and randomly fluctuating electric and magnetic field within the sample. It can then be seen that it is possible for the magnetic field to have a component, at a particular nucleus, which varies with the same frequency and sense as the Larmor frequency of the nucleus. Thus if the nucleus is in an excited state, its coupling to the rest of the sample (the lattice) via the random variations in the magnetic field allows it to transfer energy to the lattice and the nucleus to relax.

As for solids, the (exponential) rate at which the relaxation occurs is characterized by the spin–lattice relaxation time, T_1, which is a constant for a particular nucleus in a particular magnetic environment. T_1 can be considered to have components originating in the translational and rotational motion of the sample such that:

$$\frac{1}{T_1} = \frac{1}{T_1^{rot}} + \frac{1}{T_1^{trans}} \tag{8}$$

3. THE APPLICATION OF ULTRASOUND DURING THE NMR STUDY OF LIQUIDS

3.1 Acoustic Nuclear Magnetic Resonance (ANMR)

As indicated above, for solids it is possible to transfer energy between nuclear spins and the lattice via phonons generated within the sample. Kastler [4] and Altershulter [5–7] proposed that it should also be possible to utilize this path in reverse, with applied phonons (ultrasound) causing a nuclear spin system to experience a net absorption of energy, and result in detectable acoustic nuclear magnetic resonance (ANMR) spectra.

The phonons generated by an ultrasonic source usually are quite intense over a small frequency range, with less contribution at other frequencies except at appropriate harmonic frequencies. Provided these phonons have a frequency which corresponds appropriately to the Larmor precessional frequency of a nucleus, the latter and phonons can couple energetically. Unlike the natural case where the nucleus is coupled to phonons of low intensity, the nucleus can be coupled experimentally to high-density phonons. This causes the 'relaxation' mechanism to be driven in reverse causing the spins to absorb energy and transfer from a low to a high energy state. This is the basis of ANMR.

It is possible to detect ANMR without the use of rf (photon) irradiation by measuring directly the loss of acoustic energy to the spin system [8,9]. However, this is by no means easy due to low acoustic adsorption coefficients.

It is more common to measure the ANMR effect by observing acoustic spin saturation while performing standard NMR experiments. Usually the ANMR effect is detected by monitoring the intensity of a normal NMR resonance from a sample while it is irradiated with the appropriate frequency of ultrasound. In such cases the spin system will absorb energy and so drive the system closer towards its saturation point. This has the affect of showing a reduction in the intensity in the photon-stimulated NMR signal from the sample when compared to that of the corresponding signal obtained when using no ultrasonic irradiation. In fact, the first demonstration of ANMR by Proctor et al. [10,11] relied on this method, as did further investigations by Homer and Patel [12,13].

ANMR in solids is by now well documented, but its demonstration in liquids has not until recently been shown to be possible.

Homer and Patel [12,13] saturated the ^{14}N signal (Larmor frequency of 6.42 MHz) of a solution of *N,N*-dimethylformamide with varying intensities of ultrasound at different frequencies that were generated by piezoelectric transducers suspended in the NMR sample. No effect was observed at any intensity when 1.115-MHz ultrasound was applied to the sample. At 6 MHz there was some evidence of signal suppression, the effect increasing with ultrasonic intensity. At 6.42 MHz the saturation of the ^{14}N resonance was almost complete even with low acoustic intensity. When the frequency was raised to 10 MHz the signal suppression stopped and the ^{14}N resonance was once again observed. Similar results were also obtained with *N,N*-dimethylacetamide.

3.2 Promoting Spin–lattice Relaxation Using Ultrasound

During ANMR studies on a colloidal aqueous As_2S_3 system, Bowen [14,15] noted a reduction in T_1. If reductions of this type could be achieved routinely it would have a major impact on magnetic resonance studies of the normally very slowly relaxing nuclei that occur in some solid materials. The implications of the preceding discussion are that such reductions might be optimized through the *indirect* Raman process using acoustic frequencies less than the Larmor frequency.

There is no reason why ultrasound should not be used to reduce T_1's in liquids if sound can be used to induce additional relative molecular motions and hence alter the spectral density of the fluctuating magnetic field in the sample.

Extensive investigations on the effects of ultrasound at various frequencies on the T_1 of ^1H, ^{13}C, and ^{14}N in a variety of liquids and liquid mixtures have been conducted by Homer and Patel [12,16]: only the main conclusions of this work will be outlined. While changes to T_1 were observed when ultrasound in the MHz region was used, no effect was observed using low frequency ultrasound at 20 kHz. The changes in T_1 were observed only for liquid mixtures. This suggests that ultrasound causes relative motion of different molecular species, and that it modifies the translational contribution to the relaxation process.

Some selected results are presented for a cyclohexane/trimethylbenzene/chloroform-d mixture in Tables 1–3: when the ultrasound frequency was increased to 6 MHz little change in the data was observed. Table 1 shows the change in T_1 for the ^1H nuclei with increasing ultrasonic intensity; Table 2 shows the change in T_1 for the ^{13}C nuclei; and Table 3 shows the change in T_1 with temperature.

These data illustrate several points. First, ultrasound can be used to modify (beneficially) T_1 relaxation rates in liquids, with reductions in excess of 60% relative to the normal values being detected. Second, as the intensity of the applied ultrasound is increased progressively the T_1's first decrease from their normal

Table 1. Effects of Ultrasound at 1.115 MHz on the ^1H T_1 Values for a Sample Containing an Air-Saturated 1:1:2 Molar Mixture of 1,3,5-Trimethylbenzene, Cyclohexane, and Chloroform-d[a]

Ultrasonic Intensity/ W cm^{-2}	T_1/s		
	aryl-H	CH_3	C_6H_{12}
0	3.6/3.53	2.7/2.8	3.37/3.52
2	3.55/3.48	2.61/2.69	3.32/3.47
4	3.16/3.05	2.18/2.21	2.66/2.71
8	2.79/2.64<24%>	1.66/1.68<39%>	2.02/2.04<41%>
19	2.97/2.85	1.81/1.82	2.53/2.56
38	3.48/3.35	2.42/2.54	2.69/2.92

[a]Results separated by a slash are from two separate measurements; values between < > are average decreases relative to the normal values.

values [in the opposite sense to temperature-induced changes (Table 3)] to a minimum, and subsequently they increase. Third, the maximum decreases in T_1 are always larger for ^1H than for ^{13}C: this is consistent with the well-known fact that translational motion influences the latter less than the former.

The fact that the T_1 values pass through a minimum is not addressed here and the authors direct the reader to refs. 12 and 13 for a speculative explanation of this effect.

It must be noted that the presence of paramagnetic species, such as molecular oxygen, reduces T_1. Interestingly, therefore, it was observed that the effects of MHz ultrasound on T_1 are greater for deoxygenated than oxygenated samples: due to the deoxygentating effects of ultrasound the implication is that the ultrasonically

Table 2. Effects of Ultrasound at 1.115 MHz on the ^{13}C T_1 Values for a Sample Containing an Air-Saturated 1:1:2 Molar Mixture of 1,3,5-Trimethylbenzene, Cyclohexane, and Chloroform-d[a]

Ultrasonic Intensity / W cm^{-2}	T_1/s			
	=C<	=C<$_H$	$-CH_3$	C_6H_{12}
0	7.9	4.2	4.6	8.8
2	7.5	3.9	4.4	7.8
4	6.9	3.6	4.2	6.5
8	6.4<19%>	3.4<19%>	3.5<24%>	6.1<31%>
19	7.6	3.6	3.8	6.5
38	7.8	3.6	3.9	6.8

[a]Average values from two experiments: the average spread of each pair of measured values is ±0.23 s.

Table 3. Effect of Temperature on T_1 Values for ^1H in an Air-Saturated Equimolar Mixture of 1,3,5-Trimethylbenzene and Cyclohexane

	T_1/s	
T/°C	*aryl-H*	*cyclohexane*
34.5	3.45	3.03
39	3.97	3.45
47	4.18	3.63
53	4.79	4.75
61	5.27	5.94
72	5.82	6.98

induced changes to T_1 that were observed largely for oxygenated liquids are in fact less than the absolute effect of ultrasound on the T_1's of the major molecular components of the mixtures studied.

Returning to the possible origin of the acoustic reductions in the T_1's of dipolar nuclei, it is constructive to draw comparisons with the corresponding behavior for quadrupolar nuclei. For the latter, the contribution to T_1 is largely intramolecular due to the interaction between nuclear quadrupoles and the electric field gradients in their vicinity. If the speculation is correct that the ^{13}C and ^1H T_1 data indicate that ultrasound induced modifications to T_1 are largely due to modulation of the intermolecular contribution to T_1, ultrasound should have little effect on the T_1 of quadrupolar nuclei. This appears to be the case from studies of ^{14}N ($I = 1$). The interesting point here is that ultrasound was found to have no effect on the ^{14}N T_1 in the conformationally rigid CH_3CN. On the other hand, small changes were observed for *N,N*-dimethylacetamide and *N,N* dimethylformamide due, undoubtedly, to the transfer of ultrasonic energy to the conformational potential energy associated with the internal rotation about the N–C bond: induced conformational changes will modify the internal electric field gradient and hence T_1.

3.3 Conformational Changes

As implied immediately above, the introduction of ultrasound to a liquid sample should be capable of causing conformational changes to appropriate constituent molecules of the sample. One example of this has been observed by Homer and Patel [12] for *N,N*-dimethyl-acetamide.

In compounds of the type mentioned, rotation about the C–N bond is restricted at room temperature. The *N,N*-methyl groups are thus in two completely different magnetic environments and indeed have different ^1H chemical shifts. When irradiated with 20-kHz ultrasound the two resonances were found to merge. Homer and Patel suggested that the introduction of 20-kHz ultrasound to the sample induces

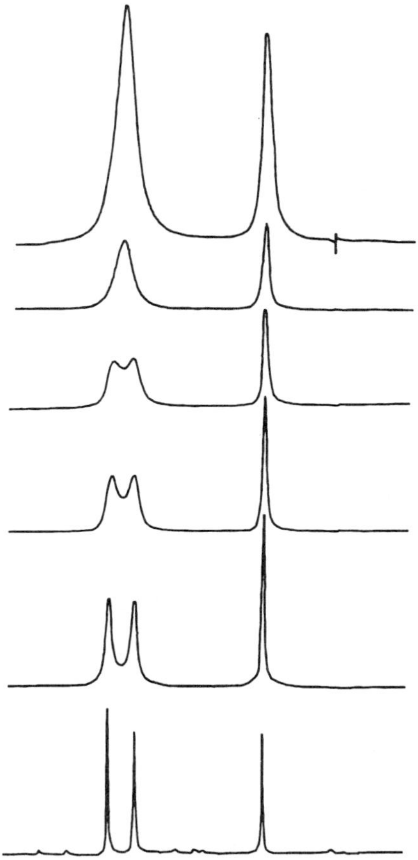

Figure 1. The effects of ultrasound at a frequency of 20 kHz with increasing ultrasonic intensity on ^{1}H NMR signals observed for *N,N*-dimethylacetamide with an ultrasound horn tip amplitude (**a**) of 10 μm, (**b**) of 6 μm, (**c**) of 4 μm, and (**d**) of 2 μm. (**e**) Shows the spectrum from the same sample when electrically heated to 44 °C with the ultrasonic probe in the sample but switched off, and (**f**) at room temperature with the ultrasonic probe out of the sample.

freer rotation about the carbonyl C–N bond causing the averaging of the chemical shifts to one value.

Figure 1 shows the effect of progressively increased conformational motion as the applied ultrasonic intensity is increased. It also shows the NMR spectrum of the sample when heated to 44 °C. The sample had to be heated electrically to over 100 °C before an equivalent ^{1}H CH_3 resonance coalescence occurred. After the ultrasound experiment the final temperature was ca. 30 °C.

4. THE APPLICATION OF ULTRASOUND DURING THE NMR STUDY OF SOLIDS

4.1 NMR Theory of Solids

The simplified overview of NMR in the previous sections revealed that the NMR of liquids can yield a wealth of chemical information. In solids the resonances are much broader, and often cannot be resolved, so that less chemical information can be derived directly from such spectra.

In the liquid state a number of interactions are averaged to zero by the random motion of the molecules that is not usually found in solids. It is, therefore, necessary to introduce and explain other features of NMR that are particularly significant in the study of solids; viz. 'chemical shift anisotropy,' 'dipolar interactions,' and 'quadrupolar interactions.'

Chemical Shift Anisotropy

If a carbonyl group and, in particular, its electron density distribution is considered, it is clear that the group can be traversed by passing through the carbon atom and regions of high electron density (i.e. along the bonds), or by passing through a region of lower electron density (i.e. perpendicular to the axis of the bonds).

These spatial differences in electron density cause the screening constant σ to become directionality dependent. It is for this reason that the chemical shift of the carbon nucleus is dependent on the orientation of the carbonyl group with respect to an external magnetic field. In liquid state NMR spectroscopy the immediate surroundings of the nucleus change so quickly that the screening constant is averaged to give a time and directionally independent value, and this results in the observation of the true isotropic chemical shift of the nucleus. In solid state NMR spectroscopy the lack of random motion causes the chemical shift to be dependent on the orientation of the sample with respect to B_o, and it is unlikely that the shift observed at the intensity maximum of the resonance will correlate with the isotropic chemical shift that is observed for a solution of the solid. Consequently, resonances from solids can be broadened due to the effects of chemical shift anisotropy (CSA).

Dipolar Interactions

NMR active nuclei generate magnetic fields which can interact with other nuclei in their vicinity. This, of course, is the basis of the dipolar interaction for $I = 1/2$ nuclei. In solution the rapid motion of the molecule results in the complete averaging of the dipolar interaction to zero. This is not the case in solids where a particular nucleus will interact with other NMR active nuclei, typically up to a distance of three bond lengths away.

The magnetic field (B_β) at a nucleus 'b', resulting from the interaction with a nucleus 'a' at a distance r from 'b' is directionally dependent and is given by,

$$B_{\beta} \propto \frac{3\cos^2\theta - 1}{r^3} \tag{9}$$

where θ is the angle between the direction of the external polarizing magnetic field and the radial vector connecting nucleus 'a' to nucleus 'b'. In solids, where for a given value of r a range of values for θ may exist, dipolar interactions are another source of the resonance of a particular nucleus becoming broadened and dependent on the orientation of solids in a magnetic field. It should be noted, however, that at the so-called magic angle of 55°44′ ($3\cos^2\theta$-1) becomes zero and so the dipolar interaction also becomes zero (as found in solution).

Most solid samples are studied as amorphous or polycrystalline powders. Each crystal in a powder generates its own orientationally dependent NMR spectrum, the resonances being affected by the chemical shift anisotropy and the dipolar interactions. These spectra superimpose on one another causing a static solid spectrum to appear as a single broad hump which often covers more than the whole of the normal isotropic chemical shift range, i.e. the signal may have a width (full width at half maximum height (FWHM)) of say 10,000 Hz rather than ca. 1 Hz as found for liquids. Presented with such an NMR spectrum it is extremely difficult to determine directly, and in detail, any information about the sample: the isotropic chemical shift of a single resonance may not even correspond with the value at the position of maximum intensity in the spectrum (due to the asymmetric nature of the signal produced).

Quadrupolar Interactions

The quadrupolar moments of nuclei with $I > 1/2$ may interact with electric field gradients in their environments. Depending on the symmetry of the nuclear environment, these interactions can lead to considerable line broadening due to both first- and second-order terms (for a detailed discussion see ref. 17). While both first- and second-order terms each contain angular (relative to the direction of B_o) terms only the former contains $3\cos^2\theta - 1$, like the dipolar interaction term.

Solid State NMR

Although the tacit implications of the preceding sections are that it is unprofitable to study solids using NMR, because of the inherent lack of spectral resolution, the opposite is true. In fact, those interactions (chemical shift anisotropy, dipolar coupling etc.) that are averaged to zero in the liquid sample are present in the solid sample resulting in the spectrum being overwhelmed with information. Nevertheless, while theoretical tools are available to abstract structural information from the broad resonances from solids, the usual route is to employ experimental techniques to narrow the resonances, so that in some cases the resulting spectra resemble those of liquids.

Various methods are employed which attempt to narrow line widths and increase the resolution of solid state spectra. The most used are those based on spinning samples of solids about "magic angles" that are chosen to remove the appropriate orientationally dependent line-broadening effects. For example, in the case of the now classical MAS NMR technique due to Andrew et al. [18] the sample is spun very rapidly about an axis at 54°44′ to the B_0 direction to remove dipolar broadening.

The rotation of samples at high speeds was first employed by Andrew et al. [18] and Lowe [19] as early as the mid 1950s. The rapid rotation of the sample causes the vector connecting the two interacting dipolar nuclei to be effectively averaged along the axis of rotation. If the axis of rotation makes an angle of 54°44′ to the external magnetic field direction this causes the dipolar interactions to be removed because the term $(3 \cos^2 \theta - 1)$ in Eq. (9) becomes zero. The spectra taken under such conditions are, however, still usually broad when compared to liquid spectra. Nevertheless, resolution can be improved enough to observe some chemical shift fine structures and allow reasonable interpretation of the spectra. Even so, spectral interpretation is often hindered further by the presence of spinning sidebands in the spectra.

For the success of the MAS techniques, one obvious difficulty to overcome is the ability to spin the sample at a rate in excess of the inverse of the FWHM of the static solid (often in excess of 10 kHz, particularly for nuclei with large dipole moments such as ^1H and ^{19}F) and precisely at the magic angle. Small imbalances in the packing of the MAS rotor may prevent the sample spinning fast enough to cause the required degree of averaging. This is overcome by using highly ground material to increase the uniformity of the sample. This is obviously a problem for tough samples such as metals and some inorganics, and for those which undergo a chemical change when a grinding pressure is applied to them.

By spinning samples simultaneously about the magic angles of 54°44′ and 30.6°, appropriate to quadrupolar interactions, both these and the dipolar interactions can be removed through the elegant DOR technique due to Pines et al. [20]. This technique, however, remains prone to some difficulty in spectral interpretation due to the proliferation of unwanted spinning sidebands.

Obviously, the magic angle spinning techniques described briefly above rely on coherent averaging out of resonance line-broadening interactions. An alternative approach is to cause solids to assume the incoherent motion of molecules in the liquid phase. Essentially, two such methods exist. The first is the so-called ultra-fine particle NMR (UFPNMR) method which was proposed originally by Yesinowski [21] and developed further by Kimura [22,23]. The second is the sonically induced narrowing of the NMR spectra of solids (SINNMR) that was demonstrated recently by Homer et al. [24,25].

4.2 Ultrafine Particle Nuclear Magnetic Resonance (UFPNMR) [21–23]

This technique relies on Brownian motion to cause very small particles suspended in a liquid to undergo the rapid incoherent rotational motion necessary to average out line-broadening effects. The preparation of appropriate suspensions is difficult and usually involves a combination of mechanical grinding and sonication with power ultrasound. The technique is, therefore, restricted to a few materials. Interestingly, the averaging process is thought to be only effective on small colloidal particles with dimensions of a few nanometers. This raises two points. The first is the possibility that the spectra obtained from such small particles may not be representative of the bulk sample, and could provide misleading information. The second, stems from the fact that rotational Brownian motion is known to be more effective for larger rather than smaller suspended particles. It is surprising, therefore, that nanometer size particles produce such narrow resonances as those reported by Kimura [22]. In this context a preliminary reinvestigation of hydroxy apatite, as studied by Yesinowski, suggests that the narrow lines attributed by him to the solid suspension may have in fact arisen from dissolved species.

4.3 Sonically Induced Narrowing Nuclear Magnetic Resonance (SINNMR) [24,25]

Sonically induced narrowing nuclear magnetic resonance (SINNMR—pronounced cinema by the originator) is a new technique which appears to offer exciting new possibilities for NMR studies of the solid state.

SINNMR is based upon the ultrasonic irradiation (originally the 20-kHz region of power ultrasound) of a suspension of solid particles in a suitable support medium. The underlying principle is that under such conditions cavitational and interparticle effects cause the solid particles to take on the motional characteristics of large molecules in solution. The resulting incoherent motion produces effective narrowing of the solid resonances by removing line-broadening normally found in solids. The equipment necessary to implement the technique is much cheaper than that involved in the MAS techniques. Moreover, SINNMR produces spectra without complicating spinning side bands due to the averaging motion being incoherent, as opposed to the coherent motion imposed by MAS. Naturally, considerable effort is being devoted to the development of this technique for the routine study of a wide range of samples.

The Role of Ultrasound in SINNMR

Ultrasonic manipulation of particles in suspension lies at the heart of SINNMR. The incoherent motion necessary for line-narrowing is thought to be generated from a number of ultrasonic effects.

In SINNMR experiments the acoustic field plays several parts. First, it is used to hold the particles in the "active region" of the NMR probe. Second, it induces the required incoherent rotational motion of the particles through several mechanisms.

The underlying physics of the process was first characterized by Dysthe [26]. He considered single particles, with dimensions much smaller than the acoustic wavelength, in a standing wave. He showed that a small anisotropic body can assume three stable orientations relative to the direction of the acoustic field. For an anisotropic particle these three orientations correspond to each of the three principal axes of the particle being parallel to the direction of the acoustic field, with the most stable orientation being when the long axis of the particle is parallel with the field. If a particle does not have the most stable orientation there is a restoring force, perpendicular to the direction of propagation of the acoustic wave, that causes very rapid rotation of the particle to its most stable orientation. When in this state the motion of the particle is purely translational, and by balancing the forces due to the acoustic field (progressive), buoyancy, and gravity it is possible to bring the particle to spatial equilibrium, e.g., in the detector region of an NMR spectrometer.

Evidently, in order to achieve the necessary rapid incoherent rotational motion to produce line-narrowing, orientational perturbations of a particle are required. A variety of mechanisms whereby these can be achieved are possible.

In heterogeneous solid–liquid systems cavitation often involves the asymmetrical collapse of cavitation bubbles near a solid surface to produce localized microjets of high-velocity liquids that impinge on the solid surface. When directed at the surface of an individual particle in a SINNMR experiment, the microjets are able to produce the required rotation as well as translation of the particle. There is strong evidence that cavitational effects provide the largest contribution to the incoherent motion of particles used in SINNMR.

An unfortunate consequence of the significant dependence of SINNMR on cavitational microjets is that they are capable of causing severe surface damage, local heating, and chemical reaction. Since chemical reaction is undesirable in the SINNMR experiment it has proved necessary to encapsulate some samples in a nonreactive matrix.

Sonicated particles can also have their motion influenced by the presence of cavities remote from their surfaces. This may arise through the action of shock waves propagated from the collapse of unstable cavities. Since these shock waves have been suggested to be capable of causing metal particles to fuse, it is not surprising that they are capable of increasing the number of interparticle collisions. In turn, these, like microjets, can result in the oscillation of the particles about their Dysthe equilibrium positions with such rapidity that NMR line-narrowing can be achieved.

The bulk flow of fluid that results whenever a sound wave is present in the medium is given the term "microstreaming." Microstreaming, enhances mass transfer and as such aids interparticle collisions, again resulting in the particles achieving a non-Dysthe equilibrium orientation.

SINNMR Equipment

The spectrometer used to perform the preliminary SINNMR experiments was a JEOL FX-90Q multinuclear iron-core magnet NMR spectrometer. Although equipped with a variable temperature system to enable the probe temperature to be set between –100 and +180 °C, normal operating conditions were such that the probe and magnet were maintained at a constant 28 °C.

The acoustic field was derived from a titanium alloy ultrasonic horn attached to a piezoelectric transducer unit that was driven by a Kerry Ultrasonics 20-kHz power generator. Two horns were used: one a short horn 27.5 cm long for visual characterization of particle behavior, and a long horn 77 cm long for the actual SINNMR experiments. Each horn had a 19 mm diameter coupling surface and was machined exponentially to provide mechanical amplification at a 5 mm diameter probe tip. The ultrasonic transducer and horn unit were mounted on a rig that permitted its complete orientation in the laboratory frame. The equipment was calibrated calorimetrically and shown to deliver 6.5 W cm^{-2} to water for a 1-μm tip displacement. As an aside it should be noted that the authors have misgivings regarding the calorimetric calibration of the amount of acoustic energy that is actually generated by an acoustic horn and absorbed by a given system. Inevitably there will be significant acoustic losses. What is required is a "molecular thermometer" to measure the energy actually absorbed by a suitable system, even though this does not, of course, guarantee that another system will absorb the same amount of energy under equivalent acoustic conditions. An indication of a possible way forward in this general connection is that Homer et al. [25] have observed that the NMR chemical shifts of some materials differ under sonicated and nonsonicated conditions. In the case of potassium hexacyanocobaltate(III), Sutcliffe et al. [27] have shown that the spin lattice relaxation times are temperature- but not pressure-dependent. It may be possible, therefore, that for this compound measurements of the T_1 dependence on varying notional acoustic intensities may be used to calibrate accurately the energy absorbed.

4.4 The Origin of Line-Narrowing in SINNMR

Trisodium Phosphate Dodecahydrate [25]

The parameters that are most likely to govern the efficiency of the SINNMR experiment are particle size and shape, support liquid density and viscosity, and the intensity of the applied ultrasound. Trisodium phosphate dodecahydrate (TSP), containing both dipolar (^{31}P) and quadrupolar (^{23}Na) nuclei, was the principal material used to assess the relative importance of the above parameters and, therefore, the optimization of SINNMR experimental conditions. The support media used were obtained from mixtures of chloroform and bromoform. In general terms it was found that an optimum concentration (ca. 1.2 g in 1.5 cm^3) of particles with sizes larger than ca. 100 μm supported in a liquid medium of higher density (by ca. 0.8 g cm^{-3}) than the solid were necessary to achieve significant line narrowing. Using these conditions the static solid ^{23}Na FWHM of about 11,500 Hz could be reduced to ca. 100 Hz with a corresponding reduction in the ^{31}P FWHM from ca. 6500 to 30 Hz (Figure 2). The fact that such line narrowing could only be achieved using ultrasonic intensities above the cavitational threshold proves inter-

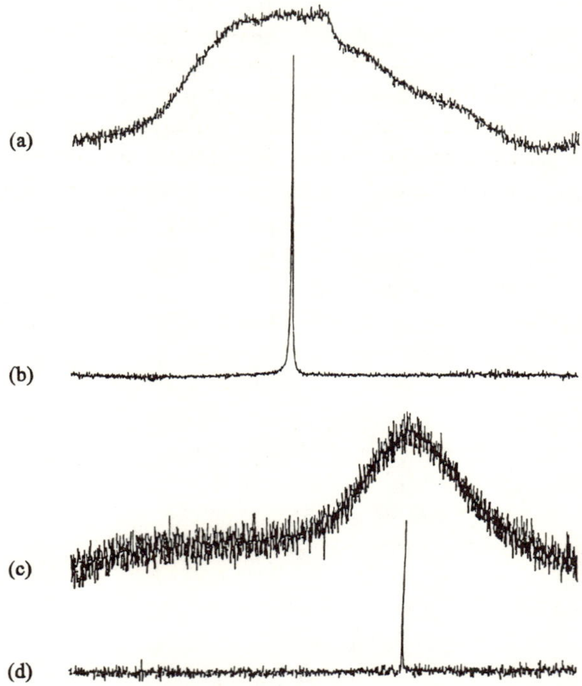

Figure 2. ^{23}Na spectra of TSP (**a**) static solid and (**b**) during the SINNMR experiment; (**c**) and (**d**) are the corresponding ^{31}P spectra. (The spectral width is 30,030 Hz for all spectra).

esting. It is known that for microjets to be produced at a solid surface the dimensions of the latter must be greater than the resonant bubble size. Applying the approximate Eq. (10) for the resonant bubble radius, R_r, [28] for the support media used reveals that the relevant dimension for the conditions used is ca. 95 μm.

$$R_r = \frac{1}{2\pi\nu}\sqrt{\frac{3\kappa P_0}{\rho}} \tag{10}$$

In Eq. (10) κ and ν are the polytropic constant and ultrasonic frequency, respectively, P_0 is the liquid ambient pressure, and ρ is the density of the liquid.

The fact that SINNMR narrowing could not be achieved with particles smaller than ca. 100 μm was taken to reflect the importance of microjetting to the SINNMR phenomenon.

It is important to acknowledge that there are a number of ways in which a narrowed resonance could be generated from experiments of the type described above, but which might not result from the incoherent motion of the solid particles used.

The narrowing effects could be due to particle fragmentation and the formation of ultrafine particles, with the observed spectral line narrowing being due to UFPNMR. This possibility was discounted as a narrowed spectrum could not be detected when studying the filtrate of a coarsely filtered (one which would allow only submicron particles through into the filtrate) SINNMR suspension.

Line narrowing could be due to dissolution of the sample or, in principle, reaction of the solid particles with the support medium: these possibilities were discounted also by the inability to observe either ^{23}Na or ^{31}P signals from the filtered support liquor after the SINNMR experiment.

As the temperature of the support medium rises due to the prolonged sonication of the sample, thermally induced changes which can result in increased molecular motion also had to be considered. Consequently, TSP was heated and it was observed that thermally induced narrowing does occur and appears to begin at 40 °C for ^{23}Na and 50 °C for ^{31}P. At 55 °C a phase change occurs resulting in the formation of the octahydrate allowing the possibility of TSP dissolving in the released water of crystallization. Bench experiments have shown that the maximum temperature attained after prolonged sonication was 60 °C. Corresponding experiments in the NMR spectrometer have shown that the maximum temperatures reached were significantly less than the bench temperatures, due no doubt to the cooling effects of various air flows in the NMR spectrometer.

The temperature dependence of the nonsonicated narrowing of the ^{23}Na signal was measured. The signal/noise ratio was a maximum at 70 °C and FWHM a minimum at 62 °C, but still 15 Hz greater than the corresponding SINNMR result.

These results indicate that the phase changes to octahydrate from the dodecahydrate and corresponding dissolution effects could not account for the SINNMR effect because of the much greater signal/noise ratios and a smaller FWHM found in the SINNMR experiment.

The time-dependent decay of the signal/noise ratios for the SINNMR narrowed ^{23}Na spectra of TSP after the cessation of sonication was compared to that for a similar sample preheated to 55 °C in the NMR spectrometer and allowed to cool. Regressions of the signal/noise ratio on time for both the SINNMR data (after sonication had been terminated) and the thermal data alone paralleled each other, indicating that each system was relaxing by similar mechanisms. It is thought that the common mechanism depends on ultrasound promoting molecules within a solid lattice to free rotor states in a similar way to thermal excitation: in both cases the excited states then relax back to their initial "state" by a purely thermal mechanism. However, the initial sections of the signal/noise "regressions" were considerably different, a dramatic decrease in signal/noise ratio being observed with the SINNMR spectra immediately after switching off the source of sonication. This is thought to be due to the cessation of incoherent molecular averaging in the SINNMR experiment, which provides the major source of line narrowing.

Detailed T_1 and T_2 measurements for the ^{31}P resonance of TSP [25] provided considerable information regarding the rotational motion of the particles studied.

Table 4. Correlation Time, τ_c, Estimated from ^{31}P T_1 and T_2 Studies of Trisodium Orthophosphate

Particle Size Limit in NMR Probe Region	τ_c / s
280–210 μm	6.6×10^{-7}
346–210 μm	5.6×10^{-7}
743–407 μm	4.9×10^{-7}
1000–534 μm	4.1×10^{-7}
1000–834 μm	4.1×10^{-7}

It was found that the ^{31}P T_1's measured for particles constrained by an open mesh nylon bag and those of the same size but unconstrained in the SINNMR experiment, under otherwise identical sonication conditions, were different. The former was found to be 2.5 s while the latter was 0.5 s. As the only difference between the two experimental configurations related to the freedom of particle motion, the difference between the two T_1 values was shown to be due to rotational motion of the particles in the SINNMR experiment, with a correlation time of ca. 10^{-7} s.

Combined T_1 and T_2 studies revealed the superficially surprising fact that it is the larger particles, in the range 250 to 1000 μm, that have the shorter correlation times (see Table 4).

Nevertheless, this observation is consistent with microjetting being the primary source of incoherent motion. This follows from the fact that the smaller the particle the more likely is the microjet to strike a particle along a line through its center of mass and result mainly in translational motion. As the size of the particles increases the chance of the microjet (of constant dimensions) striking the particle away from a line through its center of mass increases so that there is a greater probability of inducing rotational motion with the shorter correlation times.

Polytetrafluoroethylene [25]

Both amorphous and crystalline regions of polytetrafluoroethylene (PTFE) may be distinguished by observing its ^{19}F NMR spectrum. However, due to the strong dipolar coupling between ^{19}F nuclei MAS NMR techniques cannot be applied effectively and pulse sequences such as M-REV8 [29] are normally employed to effect line narrowing. PTFE, therefore, presents a particularly demanding test for the demonstration of ^{19}F SINNMR.

^{19}F SINNMR studies were conducted on a sample consisting of ca. $2 \times 2 \times 0.75$ mm particles of highly crystalline PTFE, using bromoform as the support medium. Examination of the major broad ^{19}F resonance from PTFE showed that under optimized conditions the FWHM were reduced significantly to 1600 Hz from the value of 3750 Hz obtained for a static sample.

As PTFE is unlikely to melt or dissolve under the experimental conditions employed, the resonance must be due to an averaging effect either of the particles themselves or of molecules within the solid. As the solid undergoes phase changes at 19, 30, and 130 °C, the narrowing of linewidths due to thermally induced temperature changes was examined. To achieve linewidths of the alleged SINNMR results, temperatures of 90 °C would have to be reached, whereas a temperature of only 30 °C was measured immediately after the experiment. Line narrowing due to the presence of ultrafine particles, swelling, or reaction with the support medium were also discounted. It, nevertheless, remains a possibility that ultrasound could stimulate sufficient motion of the molecular chains in the solid matrix to enable line narrowing.

Aluminum and Alloys [25]

A study of SINNMR using metal samples provides an ideal method of determining whether SINNMR line narrowing effects can be induced via incoherent molecular motion alone. Motions within a metal lattice, similar to that suggested for TSP or PTFE, can be discounted due to the strong interatomic bonding present within metals and their alloys. A particularly important and characteristic feature of metals is their ability to produce large Knight shifts. Knight shifted resonances from the metallic state are usually highly deshielded when compared to the normal chemical shifts of the same nuclide when found in a nonmetallic state. Consequently, if narrowed and Knight shifted SINNMR resonances can be observed they can only be attributed to the sonically induced incoherent motion of metallic particles.

^{27}Al SINNMR studies of aluminum and several of its alloys have been undertaken. Since the sonication of aluminum in haloform support media results in a violent reaction occurring, the metal particles were encased in resin. Static ^{27}Al NMR studies of the pure metal yielded resonances with FWHM in excess of 9 kHz. These can be reduced to 700 Hz FWHM by using MAS NMR, but SINNMR revealed even narrower Knight shifted resonances with FWHM of ca. 500 Hz.

Further, ^{27}Al SINNMR studies of an aluminum alloy containing 3% lithium revealed a Knight shifted resonance of 353 Hz FWHM at 1369 ppm deshielding from the reference aqueous $AlCl_3$. A corresponding broader MAS NMR Knight shifted resonance was observed at 1370 ppm. In a limited number of experiments an additional resonance was detected at some 1000 ppm to higher shielding than the Knight shifted resonance as a consequence of the ultrasound removing part of the protective matrix, and reactions occurring.

The observation of sonically induced narrowed Knight shifted resonances from metallic aluminum species may be taken as definitive proof that the SINNMR phenomenon is genuine.

SINNMR at High Ultrasonic Frequencies [30]

The apparent total dependency of SINNMR on cavitational processes restricts studies using 20-kHz ultrasound to particles with dimensions above about 100 μm. Reference to Eq. (10) suggests that in order to conduct SINNMR studies of smaller particles it is necessary to use higher frequency ultrasound. It is more difficult to produce sufficiently high intensities of high-frequency ultrasound to cause cavitation, than at say 20-kHz. It is similarly difficult to produce sufficient acoustic intensity to drive particles downwards to spatial equilibrium in high-density support media, as done in the original 20-kHz SINNMR experiments. Consequently, for successful high-frequency SINNMR experiments it would appear desirable to irradiate particles in lower density support media from the bottom rather than the top of the samples. In this way the solids might be both levitated to the correct location and subject to cavitational processes. The appropriate experimental configuration to enable this is difficult to achieve in iron magnet NMR spectrometers, since the supply cables to the transducers cannot be allowed to pass the NMR detector coil region, or ringing effects will result in excessively reduced spectral signal-to-noise ratios. These problems can be avoided in cryomagnets where access to the NMR detector coil region is readily possible from below.

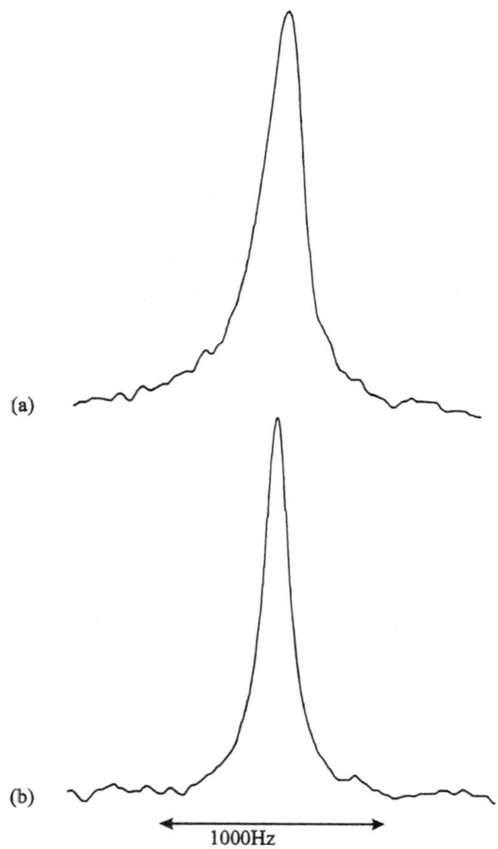

Figure 3. ^{11}B spectra of a 20%/70%/10% : $Na_2O/AlO_3/B_2O_3$ glass: (**a**) as a suspension in a chloroform/bromoform mixture and (**b**) the same suspension subjected to 2 MHz ultrasound.

The use of cryomagnet spectrometers for SINNMR studies raises the crucial question as to whether the introduction of ultrasound into the bore of such magnets will cause them to quench. Preliminary studies using 1.5- and 3.0-MHz transducers with different orientations and operating at intensities up to ca. 250 W cm^{-2} have shown that quenching does not occur under these conditions (most recently, devices delivering ca. 650 W cm^{-2} at 5 MHz have been fabricated and no quenching of a cryomagnet has been stimulated using these). Although successful high-frequency SINNMR studies in cryomagnet spectrometers have yet to be completed, encouraging high-frequency results have been obtained using top-mounted transducers in an iron magnet spectrometer. In order to illustrate these some comments now follow on ^{29}Si SINNMR studies of glasses dispersed in bromoform.

Of particular interest is the fact that significantly narrowed resonances have been observed for nonsonicated suspensions of $Na_2O{\cdot}B_2O{\cdot}Al_2O_3$ glasses, due presumably to the effects of Brownian motion. When sonicated at 2 MHz the FWHM of the Brownian narrowed resonances were further narrowed by a factor of about 2 from 240 to 132 Hz (see Figure 3). In order to achieve this by enhancing the effect of Brownian motion by raising the temperature it has been shown that it is necessary for the *in situ* sonication to have raised the sample temperature by some 50 °C above the bromoform boiling point. It does, therefore, appear that the use of high-frequency ultrasound in SINNMR can cause line narrowing directly.

5. ULTRASOUND COMBINED WITH ELECTRON SPIN RESONANCE SPECTROSCOPY

It is probably now widely accepted that the high temperatures and pressures generated by cavitation are sufficiently extreme to initiate radical formation and reaction. Indeed, spin trap electron spin resonance techniques have been employed to confirm the presence of radicals in some sonochemical experiments. Hydrogen and hydroxyl radical formation due to ultrasound has been positively identified by ESR measurements, and in the relevant reactions are now believed to be formed through thermal dissociation of water molecules at the temperatures generated within the cavitating bubble [31].

Zhang et al. [32] have reported the formation of a series of radical cations via the sonolysis of aqueous *N*-tetraalkyl-*p*-phenylenediamines. It is suggested that on sonolysis the aqueous solution forms hydroxyl radicals and via a single electron transfer the corresponding alkyl radical cation is produced. In this study both hydroxyl and the *N*-tetraalkyl-*p*-phenylenediamine radicals were identified by ESR techniques. Christman et al. [33] obtained evidence for the production of free radicals in aqueous solutions due to microsecond pulsed ultrasound. Employing spin traps such as 5,5-dimethyl-1-pyrrolidine-*N*-oxide (DMPO) and 4-pyridyl-1-oxide-*N*-*tert*-butylnitrone (4-POBN) the ESR spectra obtained provided evidence for the formation of the free radicals OH· and H·.

Observations such as those referred to above have been made through the use of spin traps and the subsequent introduction of the samples to ESR spectrometers. An exciting alternative would be to conduct ESR studies of radicals produced by ultrasound in the spectrometer resonant cavities. It has been shown that this is possible by irradiating 2,2,6,6-tetramethyl-1-piperidinyloxy (TEMPO) in CCl_4 with 20 kHz ultrasound introduced via a fine horn into the sample contained within a glass tube situated in the resonant cavity of a continuous wave spectrometer [34]. As the intensity of the ultrasound was increased the TEMPO ESR signal was observed to decrease in intensity with time. When the ultrasound was switched off the signal slowly reappeared. The possibility of observing directly signals from radicals produced by ultrasound in a pulsed spectrometer remains to be examined.

ACKNOWLEDGMENTS

The authors are grateful to the Royal Society of Chemistry for permission to reproduce the material contained in Tables 1, 2 and 3, and Figures 1 and 2, and acknowledge the origin of the relevant data in the Ph.D. theses of Drs. S.U. Patel and M.J. Howard, which in the latter case is the source of Table 4. The authors also thank S.A. Reynolds for providing Figure 3.

REFERENCES

[1] Broeckaert, L., Caulier, T., Fabre, O., Maershalk, C., Reisse, J., Vandereammen, J., Yang, D.H., Lepoint, Th., and Mullie, F. In Price, G.J. (ed.), *Current Trends in Sonochemistry*. Royal Society of Chemistry, Cambridge, 1992, pp. 8–25.

[2] Akitt, J.W. *NMR and Chemistry: An Introduction to Modern NMR Spectroscopy*, Third Edition. Chapman and Hall, London, 1992.

[3] Abragam, A. *Principles of Nuclear Magnetism*. Oxford University Press, New York, 1989.

[4] Kastler, A. *Experimentia*, 8 (1952) 1–9.

[5] Altshulter, S.A. *Zhur. Eksptl. i. Theoret. Fiz.*, 28 (1955) 38–48 (translated in *Soviet Phys. JETP*, 1 (1955) 29–36).

[6] Altshulter, S.A. *Zhur. Eksptl. i. Theoret. Fiz.*, 28 (1955) 49–60 (translated in *Soviet Phys. JETP*, 1 (1955) 37–44).

[7] Altshulter, S.A., Kockelaev, B.I., and Leuslin, A.M. *Usp. Fiz. Nank. (USSR)*, 75 (1961) 459–499 (translated in *Soviet Phys. Uspekhi*, 4 (1962) 880–903).

[8] Bolef, D.I. and Menes, M. *Phys. Rev.*, 109 (1958) 218–219.

[9] Bolef, D.I. and Menes, M. *Phys. Rev.*, 114 (1958) 1441–1451.

[10] Proctor, W.G. and Tanttila, W.H. *Phys. Rev.*, 101 (1956) 1757–1763.

[11] Proctor, W.G. and Robinson, W.A. *Phys. Rev.*, 104 (1956) 1344–1352.

[12] Patel, S.U. Nuclear Magnetic Resonance Spectroscopy and Ultrasound. Ph.D. Thesis, 1989.

[13] Homer, J. Ultrasonic irradiation and NMR. In Grant, D.M. and Harris, R.K. (eds.), *Encyclopaedia of Nuclear Magnetic Resonance*. John Wiley and Sons (1996) 4882–4891.

[14] Bowen, L.O. *Brit. J. App. Phys.*, 15 (1964) 1451–1453.

[15] Bowen, L.O. *Proc. Phys. Soc.*, 87 (1966) 717–719.

[16] Homer, J. and Patel, S.U. *J. Chem. Soc. Faraday Trans.*, 86 (1990) 215–216.

[17] Engelhart, G. and Michel, D. *High-Resolution Solid-State NMR of Silicates and Zeolites*. John Wiley and Sons, 1987.

[18] Andrew, E.R., Bradbury, A., and Eades, R.G. *Nature*, 182 (1958) 1659.

[19] Lowe, I.J. *Phys. Rev. Lett.*, 2 (1959) 285–287.

[20] Samoson, A., Lipmaa, E., and Pines, A. *Mol. Phys.*, 65 (1988) 1013–1018.

[21] Yesinowski, J.P. *J. Am. Chem. Soc.*, 103 (1981) 6266–6267.

[22] Kimura, K. and Satoh, N. *Chem. Lett.*, 2 (1989) 271–274.

[23] Kimura, K. and Satoh, N. *Chem. Lett.*, 7 (1989) 1317.

[24] Homer, J., McKeown, P., McWhinnie, W.R., Patel, S.U., and Tilstone G.J. *J. Chem. Soc. Faraday. Trans.*, 87 (1991) 2253–2254.

[25] Homer, J. and Howard, M.J. *J. Chem. Soc. Faraday Trans.*, 89 (16), (1993) 3029–3038.

[26] Dysthe, K.B. *J. Sound, Vib.*, 10 (1969) 331–339.

[27] Sutcliffe, L.H., private communication.

[28] Aptel, R.E. In Edmonds, P.D. (ed)., *Methods of Experimental Physics*. Academic Press, New York, 1981, Vol. 19, Chapter 7.

[29] Mansfield, P. *J. Phys. Chem.*, 4 (1971) 1444; Rhim, W.K. and Elleman, D.D. *J. Chem. Phys.*, 59 (1973) 3740–3749.

[30] Homer, J. and Howard, M.J., unpublished work.

[31] Makino, K., Mossoba, M.M., and Riesz, P. *J. Phys. Chem.*, 87 (1983) 1369–1377.

[32] Zhang, F., Yang, W.P., Liu, Z.L., and Liu, Y.C. *Chinese Science Bulletin*, 35 (1990) 25–27.

[33] Christman, C.L., Carmichael, A.J., Mossoba, M.M., and Riesz, P. *Ultrasonics*, 25 (1987) 31–34.

[34] Homer, J., Palfreyman, S.A., and Lee, J., unpublished work.

DEGASSING, FILTRATION, AND GRAIN REFINEMENT PROCESSES OF LIGHT ALLOYS IN A FIELD OF ACOUSTIC CAVITATION

Georgy I. Eskin

OUTLINE

Advances in Sonochemistry
Volume 4, pages 101–159

ISBN: 1-55938-793-9

ABSTRACT

Theoretical and technological problems regarding the ultrasonic treatment of liquid and solidifying melts of light alloys are under consideration. It is established that efficiency of the processes of degassing and fine filtration of a liquid metal through multilayer fiberglass filters is determined by a degree of development of cavitation processes in a melt. The mechanism of nondendritic solidification in a field of acoustic cavitation, when the ultimate refinement of the as-cast grain may be obtained during the ultrasonic treatment of a melt in processes of shape and continuous casting, is discussed. Practical examples are given on the profitable use of the ultrasonic treatment of melts in processes of light alloy metallurgy.

1. INTRODUCTION

Conventional methods of melting and casting cannot provide the quality of metal needed for modern requirements to the properties of products from nonferrous alloys. Accordingly, novel technological processes using vacuum and electromagnetic treatments of melt are under wide use in the metallurgy of light and special alloys. Among the newer physical methods for treating liquid or solidifying metals, one of the most promising is ultrasonic treatment (UST) at high intensity. The activation of physical processes due to cavitation contacts at interfaces within metal liquid provides a powerful source of nuclei of degassing and solidification. This peculiarity of the ultrasonic treatment opens up fresh opportunities for intensification of the metallurgical processes with achievement of principally new results in cleaning of metal from hydrogen and oxides and in ultrafine refinement of an as-cast structure transferred to deformed semifinished products.

For the first time an idea for improvement of quality of as-cast metal with the superimposition of elastic oscillations from mechanical vibration or shaking on a process of steel solidification has been proposed by the prominent Russian metallurgist, D.K. Chernov, in the middle of the nineteenth century. But investigations on the influence of ultrasonic oscillations on structure and properties of substances started only in the 1920s. The detailed historical review on the development of such investigations is given in a number of works [1–7]; we shall restrict our consideration to the brief review of some studies.

In 1926, Boyle and Taylor reported the possibility of degassing of light alloys melts by ultrasonic oscillations. A year later Wood and Loomis published the results of their study on powerful ultrasonic oscillations. They excited a quartz plate using a 2-kW generator in the frequency range of 200 to 500 kHz and investigated an effect of oscillations on processes of dispersion, emulsification, degassing, etc.

The first study of the influence of ultrasound on processes of metal solidification was referred to 1935 when the work by Sokolov concerning the effect of ultrasound on molten zinc, tin, and aluminum was published. A year later, Seeman as well as Schmid and Ehret repeated the Sokolov's experiment using antimony, cadmium, duralumin, and silumin.

From the beginning of the 1930s, studies on a melt treatment by elastic oscillations were carried out in three main directions: (1) the study of an effect of elastic oscillations of various frequencies with the aim to establish a mechanism of nucleation and growth of solidification nuclei in supercooled liquids, i.e. melts and solutions; (2) the study of structure and properties of metals and alloys subjected to low-frequency vibration; and (3) the study of an ultrasonic oscillation effect on molten metals. Significant research in this area was performed in the 1950s by Danilov, Kapustin, Polotskii, Sirota, and their associates on solidification of organic substances and a number of metals in ultrasonic field.

Numerous investigations on the ultrasonic treatment of molten ferrous and nonferrous metals and alloys were carried out in the 1960s. We should mention the works by Teumin and Abramov [6, 8] on the basics of the ultrasonic treatment of a melt during solidification of ferrous metals and alloys, as well as the investigations by Balandin [9] on solidification of aluminum alloys. Further studies by Rostoker and Richards [10] and Seemann [11], and by a number of other scientists were devoted to the effect of UST on structure formation in nonferrous alloys.

The author started his own investigations on the influence of the ultrasonic treatment on metallurgical processes of melting and shape and continuous casting of light alloys in 1956, and a number of these works are devoted to the problem [3–5, 12–14].

An interesting set of investigations on an effect of ultrasound with pulsed and continuous initiation of cavitation on solidification of low-melting metals and alloys has been performed by Buxmann [15] in Switzerland. The results of studies on the ultrasonic treatment of melts of aluminum and its alloys have been published by Angelov in Bulgaria [16], Kratky in Czechoslovakia [17], and Bondarek in Poland [18].

A number of investigations on ultrasound effects have been carried out in the U.S.; these include studies on vacuum-arc melting of nickel-containing alloys and steels under an ultrasonic field [19], on a mechanism of grain refinement in as-cast alloys [20, 23], on amorphization of metals (Suslick [21]), on ultrasonic atomization (Grant [22]), and on production of composite materials [23].

In Great Britain, Crowford [2], Notlingk and Neppiras [24], Chalmers [25], and a number of other scientists have devoted their work to the ultrasound effect on metallurgical processes. Recently the investigations on intensification of various chemical and metallurgical processes under acoustic cavitation field are centered at Coventry University (Mason [26]).

For a number of years, investigations on ultrasonic methods of atomization of metals have been carried out in the Leybold–Heraews Gmbh. and Rheinish Westfalishe Technische Hochschule (Polmann).

Some studies have been performed on the continuous casting of aluminum alloys with UST in Japan (Furucawa Co.) and on an intensification of various metallurgical processes using ultrasound in China [27].

By and large we can consider the interest in the problem of the ultrasonic treatment of melts of metals and alloys by the quantity of published works. Thus, the well-known review by Hiedemann (1954) [1] contains 120 references, mainly reflecting the investigations performed before World War II, our monograph [5] written in 1965 reveals about 260 original works and patents on UST, while in 1988 when we prepared the second edition of the monograph [5] the quantity of published works was even higher.

This chapter sums up the investigations performed by the author for years on establishing the scientific basics of the ultrasonic treatment of melts of light alloys with the aim to improve purity of the melt in nonmetallic inclusions and to control the structure of as-cast and deformed metal.

2. ACOUSTIC CAVITATION IN LIQUID METALS

2.1 Nature of Cavitation Strength of Metallic Liquid

The action of alternating pressure on a liquid as it takes place with propagation of waves of powerful ultrasound results in discontinuity of the liquid. Small cavities formed in discontinuities behave differently under the sound field action. Some may pulse without changing the amount of vapor-gaseous mixture within the volume; others grow intensively due to the action of tensile stresses caused by the sound wave and one-direction (rectified) diffusion from liquid to the cavity, and thirdly others begin to collapse under compressing stresses from the sound wave forming extremely fine "fragments" of the bubbles and developing high local pressures near the sites of the collapse.

All the effects reflect the physical phenomenon of acoustic or ultrasonic cavitation. With this, to initiate the cavitation one must apply a certain threshold sound pressure designated as a cavitation threshold and determine the cavitation strength of a liquid.

According to the kinetic theory of liquids, liquid substances of high purity should break under sufficiently high pressures (ca. tens of GPa). Nevertheless, the actual strength of liquids under alternating loading is significantly lower than that calculated. This is caused by the fact that, according to the up-to-date view on the cavitation strength of liquids, the break under the action of tensile stresses occurs at nuclei of a new phase.

As the cavitation strength of an actual liquid is determined by weak sites in a form of impurities, we should consider main models of cavitation nuclei to choose the most suitable one for molten metals of commercial purity.

Obviously, we should exclude liquid soluble impurities from consideration because they, changing the surface tension, may influence the process of liquid break only by the indirect route. For example, alloying of solid solution-type alloys, e.g. of the Al–Mg system, with magnesium or surface-active substances such as sodium or bismuth may result in decreasing surface tension of aluminum by up to 30%. Consequently, insoluble impurities existing in all of the three aggregation states contribute significantly to the cavitation strength of a liquid. Apparently, insoluble liquid impurities can not decrease the liquid strength to significant extent due to considerably high molecular forces of bonding between the matrix liquid and the impurity. That is also true for solid impurities with surface well wetted by the matrix liquid.

As it has been shown by Frenkel' [28] and Harvey [29], a decrease in the resistance of a liquid at the cavitation break is most likely to be determined by nonwettable hydrophobic solid particles having small cracks filled with gas insoluble in the liquid (Figure 1).

The adhesion of a liquid to a solid surface and wetting of the latter is determined by the value of $\Delta\sigma$ which can be written in the following form,

$$\Delta\sigma = \sigma_{1,0} + \sigma_{2,0} - \sigma_{1,2}, \tag{1}$$

where $\sigma_{1,2}$ is the surface energy at the solid–vapor (gas) interface; $\sigma_{2,0}$ is that at the vapor (gas)–liquid interface; and $\sigma_{1,0}$ is that at the solid–liquid interface.

Considering a small crack in a nonwettable solid particle where a gas bubble can be located and designating the contact angle of the bubble as θ, we can write,

$$\sigma_{1,2} = \sigma_{1,0} + \sigma_{2,0} \cos\theta \tag{2}$$

thus we can write Eq. (1) as:

$$\Delta\sigma = \sigma_{2,0}\,(1 - \cos\theta) \tag{3}$$

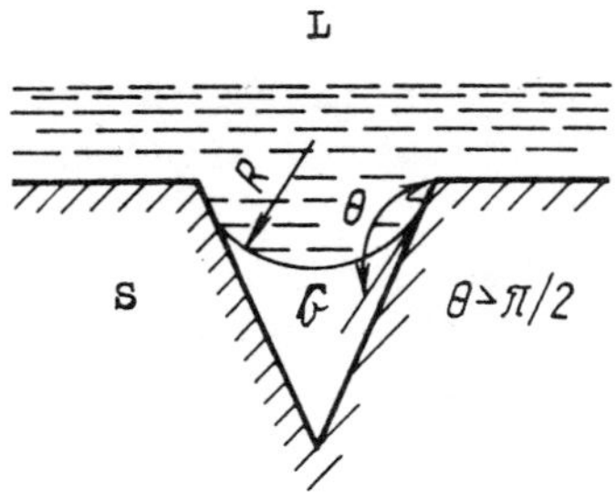

Figure 1. A model of a cavitation nucleus (after Frenkel′–Harvey) as an unwettable (hydrophobic) solid particle in liquid [3].

The case when θ = 0 corresponds to the absolutely flat bubble and, so, to fully nonwettability of the solid surface by the liquid.

Frenkel' particularly stresses that only physico-chemical properties of a liquid and primarily the value of adhesion or wettability, $\Delta\sigma$, determines the ability of the particles to become nuclei of cavitation, but not the sizes of the impurity particles.

Thus, if a liquid contains suspended particles with complex microrelief, a vapor–gas nucleus often remains in small cracks in such particles. With poor wettability by the liquid, nucleation of a cavity under action of tensile acoustic stresses should always begin from the vapor–gas state in the entrance of the crack.

But an actual metal is far from the ideal one. In the opinion of most expert metallurgists of our day, a liquid metal contains a significant amount of insoluble impurities. The cavitation strength in the liquid state and its structure in the solid state after solidification are mainly determined by the purity of the metal in such solid nonmetallic impurities.

Gas dissolved in a liquid metal offers problems in an analysis of the dependence of the cavitation strength of a melt on the amount of insoluble impurities. Accordingly, discussing the conditions of cavity nucleation in a liquid metal, it is appropriate to consider the system as a whole, i.e. melt—nonmetallic impurity—gas.

As we will mainly consider the cavitation strength of light metal melts, let us discuss the problem by the example of the aluminum melt by taking into account the following system: liquid aluminum–alumina–hydrogen.

Aluminum interacts virtually only with hydrogen and oxygen, the latter existing in a form of aluminum oxide Al_2O_3—a strong chemical compound with a low degree of thermal dissociation. Usually, alumina is present in the metal in the form of fine-dispersed suspension of less than 1.0 μm [30].

Commercially pure aluminum does not contain pure alumina but rather alumina contaminated with oxides of associated impurities, e.g. Fe, Mg, Cu, Ti, and Si, as well as oxides of alloying and modifying elements, mainly transition metals with imperfect *d*-shell.

It is confirmed that hydrogen forms in aluminum at above 500 °C as a result of the interaction of water vapor absorbed at the surface of aluminum; with this, the essence of the interaction is in vapor dissociation, formation of alumina, and precipitation of hydrogen.

Hydrogen is mainly dissolved in molten aluminum, its equilibrium presence in a form of free bubbles is under discussion and is not confirmed experimentally. At the same time, a number of scientists [30] are developing hypotheses concerning the bonding of hydrogen with oxide inclusions. Experimentally it has been shown that alumina suspended in a melt hinders the process of diffusion removal of hydrogen from the melt. Under these conditions, the cause of the reactivity of alumina to hydrogen may be in the possible change of the nature of the bonding and in capillary effects, i.e. formation of a large number of absorbed layers, which realize the step-by-step transition to a capillary condensation, may occur in capillaries. Since an actual oxide particle suspended in a melt has a widely spread system of capillaries, we may imagine that the interaction of hydrogen with alumina will be sufficiently active.

The majority of experimental quantitative evaluations of the cavitation strength of liquids were carried out using water and its solutions as model systems. This is because of the reasonable simplicity of such experiments in this easier-to-handle low-temperature fluid. Measurements in a liquid metal, particularly in molten aluminum and its alloys which react and dissolve virtually all known substances, result in significant difficulties. These are connected with the methods of introduction of ultrasound into the melt as well as with the methods of control of the experimental conditions during the development of cavitation.

The technique of a cavitation noise measurement is the basis for the methods of determination of the cavitation threshold [31]. It is well known that the occurrence of cavitation phenomena in liquid—formation and collapse of a large quantity of bubbles of different sizes which have their own resonance frequencies—results in radiation of white acoustic noise. Separate frequency peaks corresponding to resonance frequencies and subharmonics of cavitation bubbles being added together with frequency tones from other cavitation bubbles produce a spectrogram of noises quantitatively different from that of the base tone of the carrier ultrasonic frequency which is used during the current US treatment.

To study the cavitation strength of light alloys melts [32], we used a setup the block diagram of which is given in Figure 2.

A direct measurement of acoustic power introduced in the melt was performed calorimetrically for the qualitative evaluation of the cavitation threshold in the melt.

The values of acoustic power obtained were correlated with the values of oscillation amplitude of an ultrasonic radiator. The latter were obtained using feedback detectors and valve voltmeter. In the following experiments, the values of feedback voltage give an opportunity to indicate the values of acoustic power at which the melt starts to cavitate. A decrease in acoustic power introduced in the

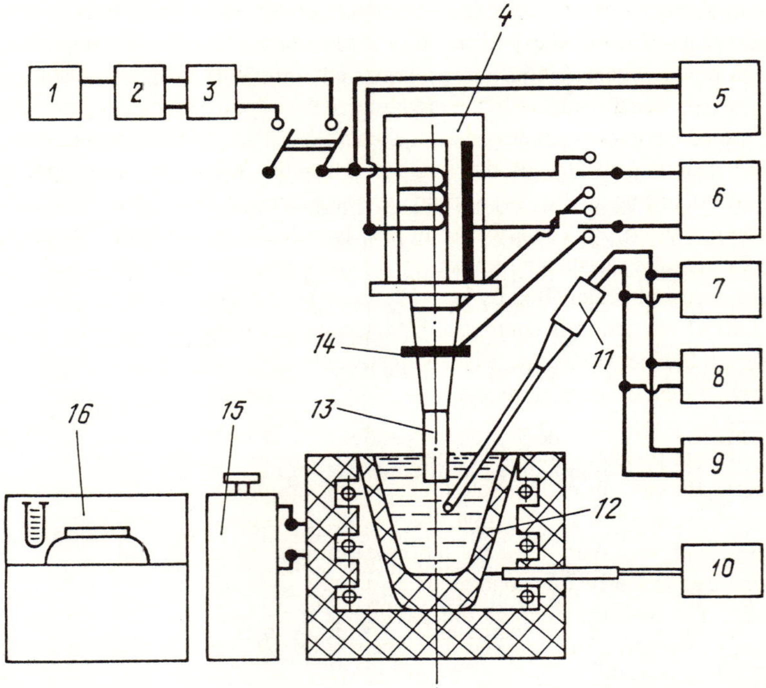

Figure 2. Experimental apparatus for the investigations of acoustic cavitation in a liquid metal. (**1**) Signal-Generator; (**2**) Amplifier; (**3**) Ultrasonic Generator; (**4**) Transducer; (**5**) Frequency meter; (**6**) Valve voltmeter; (**7**) Sensor of a waveguide stick; (**8**) Recorder; (**9**) Cavitometer; (**10**) Potentiometer; (**11**) Probe; (**12**) Crucible with a melt; (**13**) Source of ultrasound; (**14**) Receiving stick; (**15**) Electric furnace; (**16**) Detector of the first bubble.

melt was performed by 0.5–1.0-W steps in all the experiments, and the latter were carried out not less than 3 times.

The amplitude of sound pressure P_A in a mode without cavitation may be evaluated by the following evident relationship for a plane running wave,

$$P_A = [(2W_a\rho_o c_o)/S]^{1/2}, \tag{4}$$

where W_a is the acoustic power obtained experimentally; $\rho_o c_o$ is the wave resistance of a noncavitating liquid; ρ is the density of a melt; c is the sound velocity in a melt; and S is the area of a radiator.

Acoustic powers and amplitudes of sound pressure in the melt of commercial aluminum in a mode without cavitation are given below

W_a (W):	1.5	3.3	5.0	6.5	7.7	8.4
P_A (MPa):	0.37	0.55	0.68	0.77	0.88	0.96

The calculation is based on the experimentally obtained values of the oscillation amplitude of a radiator and experimentally registered by the noise actual picture of cavitation development. Under experimental conditions, there was a distance of 10–15 mm between the surface of the radiator and the edge of the stick. Accordingly, the picture of cavitation development is somewhat different near the edge of the stick and near the surface of the radiator. Thus the values of sound pressure calculated according to Eq. (4) should be considered as the upper limit of the cavitation threshold, P_c. The lower limit may be calculated by taking into account the values of wave resistance reduced by 3 to 4 times. With this, the value of the cavitation threshold calculated by Eq. (4) should be reduced by 1.4 to 2.

The start of cavitation is controlled by a waveguide stick **14** and a C1–13 type oscillograph **8**. The waveguide with the resonance frequency of 200 kHz (significantly higher than the carrier frequency—18 kHz) is a rod of conic form made from commercially pure titanium of the VT-1 grade. A cylindrical receiver made from barium titanate piezoceramics is cemented onto the edge of the waveguide. One can observe the cavitation threshold on the oscillograph screen by the distortion of the base frequency (18 kHz) signal, the intensive cavitation mode being significantly different from that of the cavitation start mode (Figure 3).

Varying sequentially the content of hydrogen and alumina in an A7 grade aluminum, the measurement of the cavitation threshold has been carried out according to the technique worked out. The results obtained are given in Table 1. One can see that the increased amount of hydrogen dissolved in an aluminum melt affects the decrease in the cavitation threshold only at low content of alumina of 0.001 to 0.005%. Further decrease in the cavitation threshold is due to the purity of the melt in solid nonmetallic inclusions of alumina. Increasing alumina content from 0.005 to 0.1% results in decreasing cavitation threshold from 0.85 to 0.55 MPa or by 37%.

Table 1. Effect of Content of Hydrogen (H_2) and Nonmetallic Inclusions (Al_2O_3) on Cavitation Threshold P_c in a Melt of Commercially Pure Aluminum[a]

	P_c (MPa)		
	Al_2O_3 (%)		
H_2 [cm^3 $(100\ g)^{-1}$]	*0.005*	*0.05*	*0.1*
0.1	0.85	0.70	0.55
0.2	0.80	0.70	0.54
0.3	0.76	0.68	0.54
0.4	0.74	0.65	0.54

[a]Ultrasonic Frequency of 18 kHz

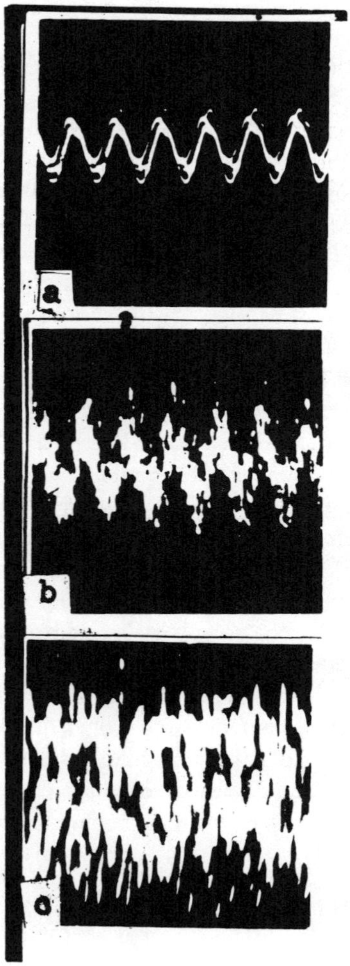

Figure 3. Distortion of the base frequency signal with cavitation. (**a**) No cavitation; (**b**) the cavitation threshold; (**c**) developed cavitation mode.

Table 2.

Alloy	*Temperature (°C)*		
	680	*700*	*750*
Al A99	0.9	1.0	1.3
Al–6% Mg	0.7	0.8	0.9
Al–0.2% Zr	0.6	0.7	0.8

The cavitation threshold (MPa) in aluminum and its alloy melts depends on the temperature of the melt, composition, and on presence of impurities of transition metals (Ti, Zr etc.) increasing hydrogen absorption on oxides as shown in Table 2.

The measurement of the cavitation threshold in magnesium melts of various purity with surface protection from a flux shows the values of 0.6 to 0.8 MPa at 700 °C, which correspond well to the values of the cavitation strength of commercially pure aluminum.

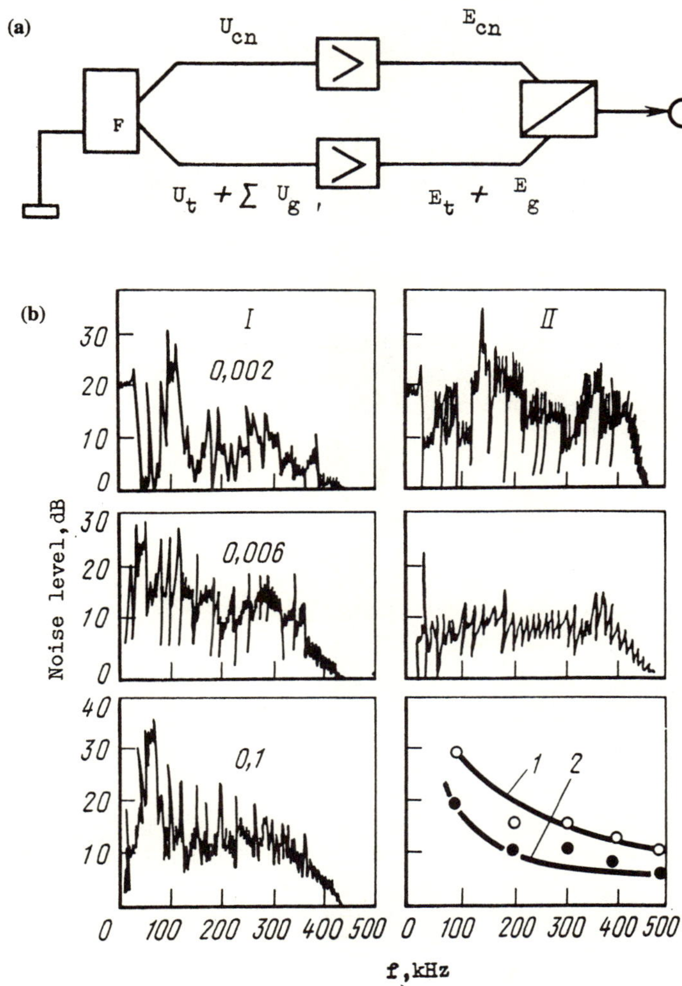

Figure 4. A block diagram of cavitation indicators (**a**) and spectrum patterns of cavitation noises of aluminum melts of different purities with the ultrasonic treatment at 18 kHz with A = 20 μm at 720 °C (**b**). I. Noise level at different Al_2O_3 content (mass %); II. A melt of a V95pch grade alloy before (**1**) and after (**2**) fine filtration.

Abramov et al. [33] studied the cavitation threshold in melts of low-melting metals (Bi, Sn, In, Pb, Cd). They calculated that the cavitation threshold is from 0.5 to 1.0 MPa at 20 kHz and bubble radii R_0 are from 0.01 to 100 μm.

Important information may be obtained by spectrum cavitometers allowing one to express the relationship between components of acoustic signals in the form of dimensionless (normalized) values which are received from the cavitation region by the stick. These are the base tone signal (E_T), harmonic and subharmonic signals (E_h), and incoherent noise from closing cavitation bubbles (E_{cn}).

The use of special crested filters allows one to distinguish the cavitation noise from all other components of total acoustic radiation [5] in the following form:

$$\gamma = (E_{cn}/(E_T + E_h)) \times 100\% \tag{5}$$

Figure 4 displays the block diagram of such a setup and an example of the record of cavitation development and noise level in the frequency range from 0 to 500 kHz for aluminum melts of various purity (I) and for V95pch grade alloy (II).

Margulis [34] has shown that the cavitation threshold may be confidently determined by the method of liquid sonoluminescence. Although the mechanism of liquid luminescence with cavity closing is not conclusively established—we do not know which factor is the main one in the luminescence in the moment of collapse: thermal phenomena, electric charges, or the superimposition of these processes—the method is of practical use for early observation of cavitation origination.

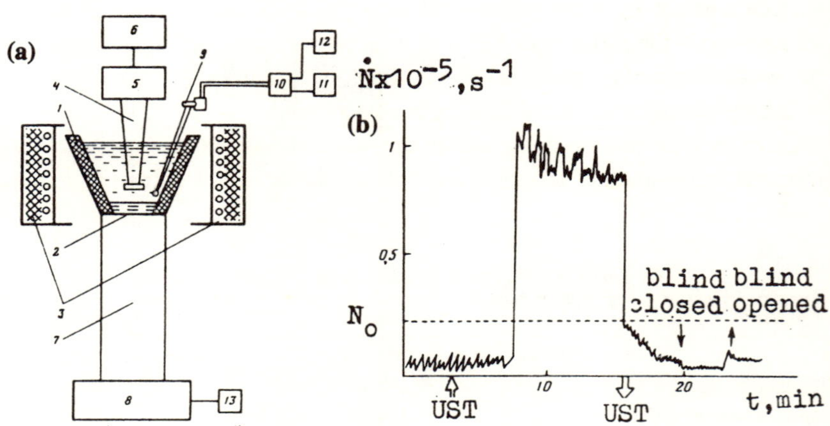

Figure 5. A scheme of an experimental setup for investigations on sonoluminescence and cavitation noise in metal melts (**a**) and dependence of light impulses count rate, N, on time and ultrasound intensity, I (I was increased by 10 mW cm^{-2}-steps at 3-min intervals) (**b**).

Most works on sonoluminescence were carried out using water or other low-temperature liquids [34]. Only ref. [35] reports the use of the method for revealing cavitation in mercury.

As it has been shown in our mutual investigations with Margulis [36], the sonoluminescence method can be used with some advancement of the technique (Figure 5) for an analysis of cavitation processes in liquid metals. Under ultrasonic action on a tin melt at 22-kHz frequency and 1.0-W-cm^{-2} intensity we managed to register the start of cavitation by this method. One can easily see in Figure 5b the increase in light radiation from the cavitating melt by 2 to 3 orders with increasing intensity of ultrasound; the radiation cancels with switching off the ultrasound source.

The future will show whether this method may be used for an analysis of cavitation phenomena in melts of aluminum and its alloys.

2.2 Dynamics of a Cavitation Bubble in Metallic Liquid

Dynamics of a separate vapor–gaseous cavity in noncompressible liquid without regard to gas diffusion into the cavity can be represented by the Notlingk–Neppiras equation [24],

$$\rho(R\ddot{R} + 3/2\dot{R}^2) + 4\mu\dot{R}/R(P_o - P_v + 2\sigma/R_o)\,(R_o/R)^3 +$$

$$+ 2\sigma/R - P_v - P_A \sin \omega t + P_o = 0 \qquad (6)$$

where R is the radius of the cavity; R_o is the initial radius of the cavity; σ is the surface tension; P_v is the vapor pressure; P_o is the static pressure; μ is the melt viscosity; ρ is the melt density; P_A is the sound pressure amplitude; and $\omega = 2\pi f$ is the round frequency where f is the frequency.

Table 3 contains the initial data for the calculation.

Cavities with initial radii R_o within the range of 1 to 100 μm were studied, the minimal initial radius, so called the critical radius, being determined from the

Table 3. Physical Constants of Liquids Studied

Constant	*Aluminum*	*Water*
	(700 °C)	(20 °C)
Density, ρ (kg m^{-3})	2350	1000
Surface tension, σ (N m^{-1})	0.860	0.079
Viscosity, μ (MPa s^{-1})	1.0	1.0
Pressure, P_o (MPa)	0.1	0.1
Frequency, f (kHz)	18	18
Vapor pressure, P_v (kPa)	0	2.2

condition of stability and the maximum initial radius, according to the Minnaert equation.

Figures 6 to 8 show the results of the calculation of relative radius R/R_0 and gas pressure in the cavity P_g at various sound pressures. These are the series of curves varying in the initial radius of the cavity and in the level of sound pressure applied and reflecting dynamics of the cavity during 1 to 3 periods, T, of a sound wave.

Considering the results presented in Figures 6 to 8, one can see that cavities pulse and do not close during the counting time at low amplitudes of sound pressure, $P_A \leq 0.6$ MPa and $P_A < P_c$ where P_c is the cavitation threshold. With this, gas pressure inside bubbles varies negligibly. With increasing amplitude of sound pressure ($P_A \geq 1$ MPa), as it approaches P_c, the majority of cavities with $R_0 > R_{cr}$ behave as typical cavitation bubbles collapsing at the end of the period. Finally, at $P_A >> 1.0$ MPa, i.e. $P_A >> P_c$, we can see a picture of developed cavitation when all the cavities considered expand during 1–2 periods of a sound wave and then collapse.

An expansion of a bubble by tens and hundreds times compared to the initial size results in significant pressure drop within the cavity which reaches the values of 100 to 133 Pa at applied pressure $P_A > 2.0$ MPa. The further increasing sound pressure leads to increasing rate of cavity collapse.

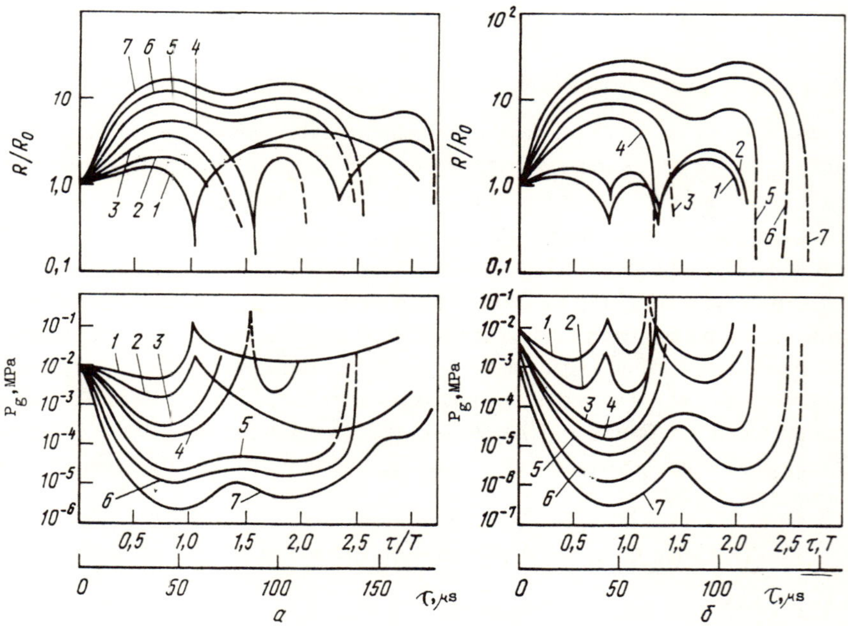

Figure 6. Dynamics of a cavitation bubble in an aluminum melt with $R_0 = 100$ µm (**a**) and $R_0 = 50$ µm (**b**) at the following values of P_A (MPa): 1. 0.1; 2. 0.2; 3. 0.6; 4. 1.0; 5. 2.0; 6. 3.0; 7. 10.

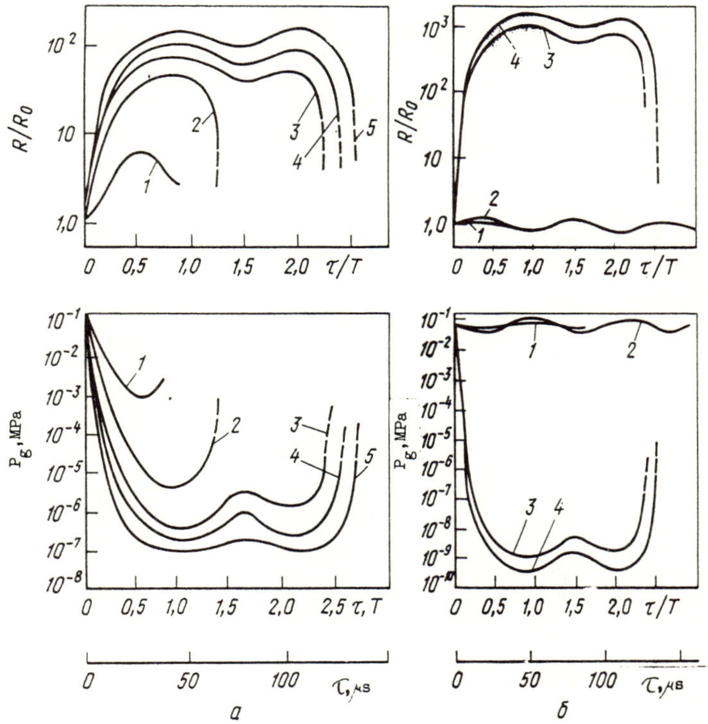

Figure 7. Dynamics of a cavitation bubble in an aluminum melt with R_0 = 10 μm (**a**) and R_0 = 1.0 μm (**b**) at the following values of P_A (MPa): 1. 0.2; 2. 1.0; 3. 3.0; 4. 5.0.

The experimental studies of high speed photography of instable cavities recording the behavior of a bubble with compression, e.g. the studies by Knapp et al. and Lauterborn et al. [38, 39], show that if the bubble can preserve the spheric form with a smooth and glossy surface during initial expansion and closing, its form is disrupted with further expansion—the sphere becomes scarred and the bubble transforms to a cloud of very small fragments after the collapse. It should be noted that the recent investigations (Margulis, Grivin, Zubrilov, etc.) show that pulsation of a cavitation bubble results in the loss of a spheric form and strong distortion of its surface [34, 40].

2.3 Diffusive Growth of a Cavitation Bubble in an Ultrasonic Field

One of the most interesting problems of ultrasonic cavitation is the existence of the diffusive growth of bubbles in a sound field. As, without any field, a gas bubble should slowly dissolute due to gas diffusion from the bubble to a liquid, directional gas diffusion from liquid to the bubble arises under conditions of the bubble surface

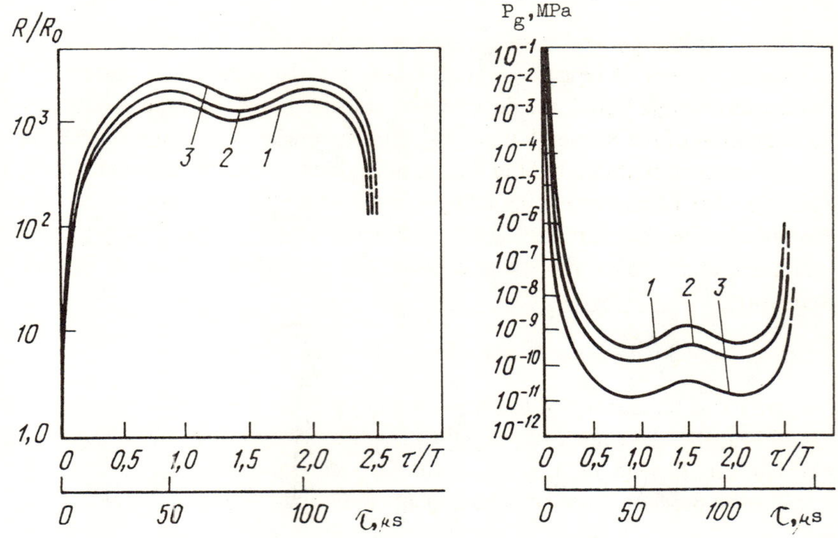

Figure 8. Effect of surface tension (σ) on dynamics of a cavitation bubble with R_0 = 1.0 μm at P_A = 10 MPa. (**1**) Aluminum melt (σ = 0.86 N m^{-1}); (**2**) Magnesium melt (σ = 0.50 N m^{-1}); (**3**) Water (σ = 0.079 N m^{-1}).

pulsation. Being added with normal steady-state diffusion from the bubble to a liquid, the process of rectified diffusion may exceed that of gas from the bubble under certain values of sound pressure.

The qualitative understanding of the process of rectified diffusion may be obtained from consideration of three effects of cavity pulsation:

1. Gas diffusion from a bubble to a liquid is directed when the bubble is compressed and gas concentration within it is decreased. The expansion of the bubble and hence decreasing gas concentration within it may be considered as the condition of gas diffusion inside the bubble.
2. The cavitation dynamics of a cavity provides the conditions when the surface area of the bubble in a phase of expansion is significantly higher than that in a phase of compression. Accordingly, an amount of gas diffusing into the cavity is considerably more than that leaving the bubble with its compression.
3. It is well known that the process of diffusion is controlled by thickness of so-called diffusive layer which locates in the layer of a liquid around a bubble. With bubble compression, this layer grows and the concentration gradient decreases; with bubble expansion, thickness of the layer decreases and the gradient increases, hence, a rate of gas flow into the bubble increases.

Thus, considering the behavior of bubbles under cavitation we must take into account the variation in gas amount within a cavity during its expansion and collapse. In the general case the bubble behavior is described by a sufficiently complex system of equations. With taking into account that a bubble pulses in noncompressible liquid, the dependence of its radius, R, on time is represented by an equation of the Rayleigh type [44]. But if we want to account for the variation of gas amount within a bubble during pulsations, the value of gas pressure in the bubble, $P(t)$, cannot be expressed by a simple analytical dependence on the radius.

The change of gas amount is $M = 4\pi R^3 m\gamma/3$, where γ is the density of gas within a bubble and m is the molecular mass of gas, may be expressed by the following equation,

$$4/3\pi\ \mathrm{d}(\gamma R^3)/\mathrm{d}t = 4\pi R^2 i(t) \tag{7}$$

where $i(t)$ is the density of gas flow through the surface. Considering gas within a bubble as the ideal one, $P = \gamma kT$, where T is the temperature of gas within a bubble which can depend on time, k is the Boltzmann constant. Accepting that the cavitation process in a melt occurs in an isothermal mode, Eq. (7) takes the form:

$$R/3kT\ \mathrm{d}P/\mathrm{d}t + P/\mathrm{k}\mathrm{T}\ \mathrm{dR}/\mathrm{d}t = i(t) \tag{8}$$

The gas flow may be caused by different processes, the major one of which is diffusion. The solution may be simplified by taking into account the results from ref. [41] where the expression for the density of the gas flow into a cavitating bubble is given.

If the equations of bubble pulsation and gas diffusion into the pulsing bubble are written in the way when the unknown item is an area of bubble surface $S = 4\pi R^2$ and if we use the expression, [41]

$$I = 8/3\ \sqrt{5\pi}\ Z_o/\rho\ D^{1/2}\ C_o t^{3/2}$$

as the value of the total gas flow into the bubble (where $Z = 0.8\ P_A$), then the system of equations describing the bubble dynamics with regard to gas diffusion into the pulsing bubble takes the following form:

$$\begin{cases} \mathrm{d}^2S/\mathrm{d}t^2 = -1/4S(\mathrm{d}S/\mathrm{d}t)^2 + 2\pi/\rho[P(t) - P_o - P_A \sin \omega t] - \\ -4\sigma\pi^{3/2}/S^{1/2} - 4\mu\pi\dot{S}/S \\ (3S^{3/2}/\sqrt{\pi}kT)\ \mathrm{d}P/\mathrm{d}t + (PS^{1/2}/2\sqrt{\pi}kT)\ \mathrm{d}S/\mathrm{d}t = I \end{cases} \tag{9}$$

Here σ is the surface tension of the melt, C_o is the gas (hydrogen) content in the melt, and D is the diffusion coefficient of hydrogen in the melt. The system (9) may be solved by numerical methods using a computer by taking into account the starting conditions: $S(0) = S_o$; $P(0) = P_o$; and $\dot{S}(0) = 0$.

The numerical calculations were performed for aluminum melts with hydrogen content of $C_o = 0.2\ \mathrm{cm}^3\ (100\ \mathrm{g})^{-1}$ and the diffusion coefficient $D = 1.0\ \mathrm{cm}^2\ \mathrm{s}^{-1}$ [30] and for different initial radii of cavitation bubbles, R_o. The latter were selected from the conditions of the actual outline of existence of cavitation nuclei in the forms of

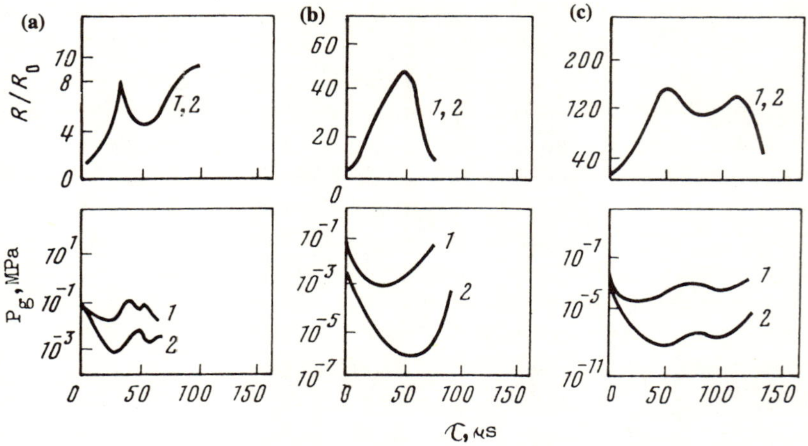

Figure 9. Dynamics of a cavitation bubble with R_o = 10 μm with the ultrasonic treatment of an aluminum melt at P_A: (**a**) 0.2 MPa; (**b**) 1.0 MPa; (**c**) 10 MPa. The calculations were performed by taking into account hydrogen diffusion into a cavity (**1**) and without taking the latter into account (**2**).

nonwettable particles of fine-dispersed alumina of 0.3–1.5 μm in size and hydrogen bubbles of about 1.0 μm in size formed on these particles [30].

Figure 9 displays time dependencies of variation in a relative radius, R/R_o, and in hydrogen pressure inside a cavitation bubble, P_g, at three values of sound pressure for R_o = 10 μm; these are P_A = 0.2 MPa (without cavitation), P_A = 1.0 MPa (cavitation threshold), and P_A = 10 MPa (developed cavitation).

As a result of graphical integration of the dependencies $P_g = f(\tau)$ (Figure 9) the values of ΔM, 10^{-9} g, given in Table 4 were calculated.

The character of cavitation bubbles pulsation as well as the conditions of rectified diffusion of hydrogen inside a cavity allow us to confirm the supposition that the cavitation threshold for liquid aluminum and magnesium is 0.65 to 1.0 MPa at a frequency of 18 kHz.

2.4 Mechanism of Compression and Splitting of Cavitation Bubbles

The behavior of bubbles with compression at half-period of compression of a sound wave is the least known feature of cavitation in liquid metals. Ref. [42]

Table 4. ΔM Values

P_A(MPa)	R_o = 1.0 μm	R_o = 10 μm	R_o = 100 μm
0.2	4×10^{-3}	4×10^{-4}	4×10^{-7}
1.0	4.6×10^{-3}	4.6×10^{-3}	7×10^{-4}
10.0	2.0	1.0	0.65

reports a quantitative estimation of compression pressure, P_{max}, in a shock wave for a number of low-melting metals. According to this work, P_{max} at a frequency of 20 kHz is 107 to 110 GPa for Sn and Bi and 95 to 99 GPa for Cd and Pb, which is about 10^5 times higher than the cavitation threshold, i.e. the amplitude of sound pressure applied to the melt.

Due to analytical investigations and experimental studies by rapid filming and holography, we know much more about the conditions of compression and splitting of cavitation bubbles in transparent liquids (water and solutions). According to the data by Margulis and Dmitrieva [43] obtained for water by taking into account heat transfer from a bubble to a liquid, the calculated value of P_{max} is from 1 to 10 GPa at P_A = 0.2–1.0 MPa at the 20-kHz frequency, the maximum temperature within the bubble being equal to 800–1500 K. These values, as it is pointed out by the authors, may play a certain role in the development of sonoluminescence and accelerating of chemical reactions under an ultrasonic field.

Experimental studies of a process of cavitation bubbles pulsation, involving final stages of their compression, by means of rapid filming and holography [31, 38–40] show that the loss of bubble stability, deformation of its surface, and splitting to new nuclei ("fragments") of cavitation bubbles should be taken into consideration. Depending on the conditions of interaction with a solid sublayer (a wall) on which a cavitation bubble has appeared and is developing, its splitting occurs with formation of a cumulative jet (or jets) with its speed of tens and hundreds ms^{-1}.

Along with the thermal theory of cavitation bubbles collapse [24], a number of electrical theories of cavitation are evolving which allow one to explain many energy-intensive chemical and physicochemical phenomena as a result of the production of plasma states within a cavitation bubble. The Frenkel' electrical theory was one of the most well known [28]. But it has been shown [34] that this theory and a number of later electrical theories are insufficient. Since there were much experimental data contradicting the thermal theory, a new electrical theory of cavitation phenomena was developed in 1985 [34]. According to the Margulis theory [34], a double electrical layer forms at a cavitation bubble–liquid interface in any liquid. A pulsing bubble gradually grows, loses its stability when it achieves the resonance size, and begins to split. As a result of removing the diffuse part of the double electrical layer, a noncompensated charge accumulates at the neck between the main bubble and fragmentation bubbles. After splitting, this charge localizes in a narrow part of the cavitation bubble surface, the size of the part being virtually equal to the neck cross-section at the instant of its break. Due to the small cross-section of the neck, electric field intensity within the bubble reaches the values of 10^{10}–10^{11} V m^{-1}, which is sufficient for an electrical breakdown even at pressures of up to several thousand atmospheres.

Plasma formed as a result of the electrical breakdown at high pressure facilitates neck break, formation of a new bubble, and, probably, formation of shock waves and cumulative jets.

2.5 Actual Outline of a Cavitation Field in Melts under Ultrasonic Treatment

As mentioned, the multitude of cavitation bubbles appear in actual liquids (melts) even at relatively low-sound pressures above the cavitation threshold. These bubbles occupy the certain part of a volume which is called a cavitation region. As the value characterizing a degree of cavitation development, Rozenberg [44] coined a term of cavitation index, K, reflecting the ratio of the volume of all cavitation bubbles in the phase of their maximum expansion to the corresponding cavitation region, V:

$$K = \Delta V / V \tag{10}$$

Assuming $0 \leq K \leq 0.1$, one can estimate the drop of wave resistance with cavitation, $\overline{\rho_c c_c}$, in dependence on the cavitation index for any liquid:

K	0.001	0.06	0.1
$\rho_o c_o / \overline{\rho_c c_c}$	1.0	5.0	10.0

Rozenberg [44] has been the first to show that cavitating liquid represents a nonlinear medium, and its average wave resistance, $\overline{\rho_c c_c}$, determining by actual ultrasonic power emitting into a liquid, W_a, and by oscillation speed, $(\omega A)^2$, of the surface emitting sound, S, should gradually decrease to a given value.

According to our own data, the value of $\overline{\rho_c c_c} \cong 0.1\ \rho_o c_o$ in a cavitating aluminum melt. This data well agree with previous results obtained by Rozenberg and Sirotyuk [31, 44] ($\overline{\rho_c c_c} = 0.3\ \rho_o c_o$), Kikuchi [45] ($\overline{\rho_c c_c} = 0.25\ \rho_o c_o$) for water, Astashkin and Abramov [33] ($\overline{\rho_c c_c} = 0.25\ \rho_o c_o$), for tin melt, and Bertnik et al. [46] ($\overline{\rho_c c_c} = 0.1\ \rho_o c_o$) for a POS-1 grade solder.

The wave resistance drops by 7 times with cavitation development, being virtually unchanged with further increasing amplitude of radiator oscillations, although additional acoustic power is transferred into the melt.

To design a technological process of the UST of melts one has to know the geometrical dimensions of the cavitation region. The most reliable methods for the determination of a topography of a cavitation field are considered to be the estimation of cavitation erosion of a thin foil or measurement of mass loss of special specimens which are situated in a liquid medium treated by ultrasound. With decreasing degree of cavitation phenomena development, dimensions of a cavitation field increase occupying the volume with linear dimensions (a) which are virtually equal to $\lambda/4 \leq a \leq \lambda/2$ where λ is the wave length in the liquid. The a value is 20 to 40 mm for water, and 40 to 60 mm for an aluminum melt.

Foils from refractory metals (Ti, Mo, W, etc.) of 20 to 500 μm in thickness were used for the cavitation study in melts of aluminum and its alloys. Specimens of the foil in stiff frames were situated into the melt under a radiator in such a way that the axis of the waveguide–radiating system passes through the plane of the frame

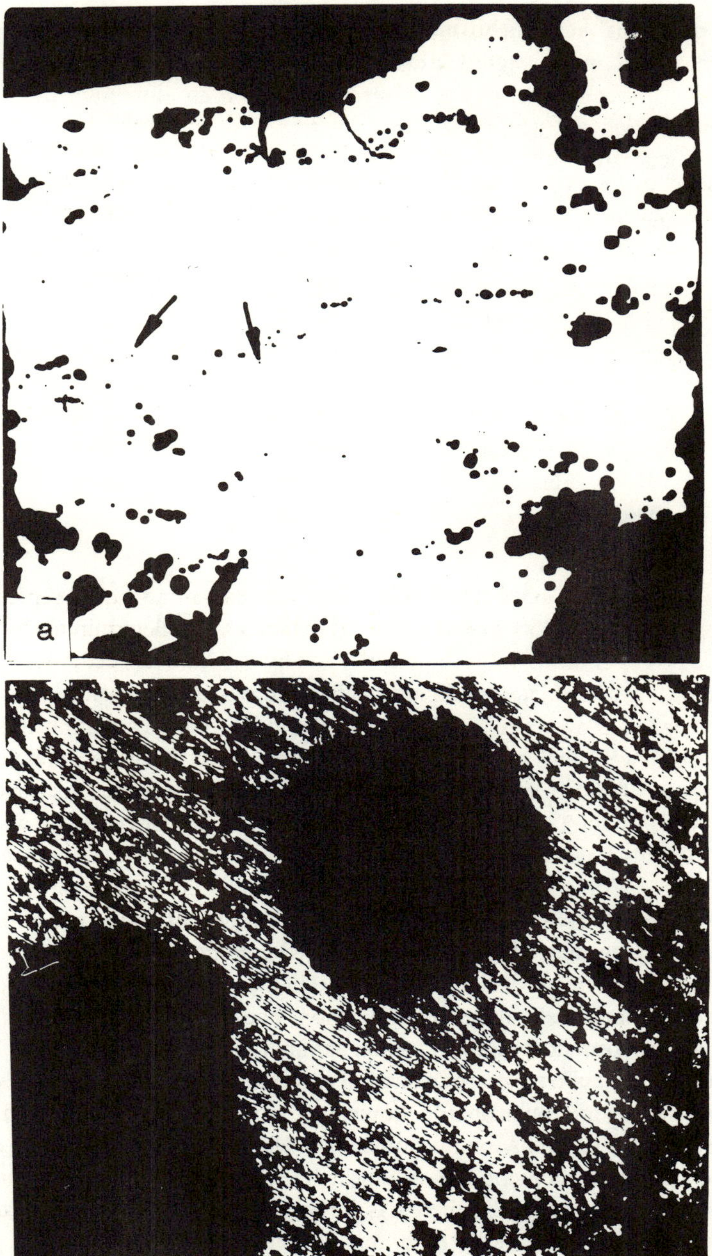

b

Figure 10. Appearance of a titanium foil of 100 μm in thickness (**a**) with cavitation erosion and a form of cavitation failure (**b**, ×300) after 15-min period of ultrasonic treatment in the developed cavitation mode in an aluminum melt.

[5]. Some of the experiments were performed with the foil location in the plane perpendicular to the waveguide axis.

Usually, holding a foil (for example, from Ti) in an aluminum melt without ultrasonic action for up to 1 h does not result in its wetting and subsequent metallization. Under ultrasonic treatment, a strong diffusive layer forms at the foil surface after 1–3 min of exposure, and typical punctures—traces of cavitation erosion—appear on the foil after 5–15 min (Figure 10).

The degree of cavitation development in the melt can be judged from the quantity of the punctures. In the case when the punctures in the foil are located close to each other and virtually coincide forming holes, the quantitative estimation of the cavitation field is somewhat corrupted. It should be noted that we managed to reveal the punctures only after etching of a surface layer of the foil by holding it in a 10% solution of NaOH for 2–3 h.

The investigations show that the stability to cavitation erosion reduces in the series: Ti, Mo, W. This data on cavitation erosion of various materials in liquid aluminum were obtained during the selection of a material for a radiator [5].

3. MAIN REGULARITIES OF DEGASSING OF LIQUID METALS IN A FIELD OF ACOUSTIC CAVITATION

3.1 Thresholds of Cavitation and Degassing

It is well known that a process of gas dissolution and precipitation from a melt has its origins in diffusion of atoms to the gas–liquid interface. So we should take into consideration not only the thermodynamic conditions of gas transition from a solution but the kinetic conditions, particularly an amount of degassing centers, as well.

To better appreciate the role of ultrasonic treatment of aluminum melts, it is necessary to consider the presence of solid nonmetallic inclusions within liquid aluminum (nonwettable by the melt).

If homogeneous melts (containing no oxides) are able to form supersaturated hydrogen solutions during solidification, then heterogeneous formation of bubble nuclei can easily occur onto oxides within actual liquid metals containing oxides.

The analysis of hydrogen distribution within the volume of liquid metal, containing dispersed particles of Al_2O_3, shows the possibility of precipitation of molecular hydrogen on the oxide particles.

The calculations performed by Makarov [47] testify that despite the low content of molecular hydrogen (about 0.01 to 0.1 vol.%) in the melt, its role is quite important because it serves as nuclei of cavitation and degassing and determines the behavior of the melt during cleaning and solidification. The amount of molecular hydrogen absorbed on particles of alumina is given in Table 5 according to ref. [47].

Table 5.

Content of oxides (mass%):	0.01	0.005	0.001
Volume of the oxide per 100 g Al (cm^3):	0.0028	0.0014	0.00028
Volume of hydrogen on an oxide particle at 700 °C (cm^3):	0.00143	0.00072	0.00014
Volume of hydrogen on an oxide particle reduced to normal conditions (cm^3):	0.004	0.002	0.00004
Hydrogen proportion in an aggregation (mass%):	0.16	0.08	0.016

These data may give a certain answer to the repeatedly discussed question of the amount of hydrogen absorbed in the molecular form on suspended particles of nonmetallic impurities in the aluminum melt.

A liquid metal contains, as a rule, no free gas bubbles: large bubbles rise to the surface and small ones dissolute. With propagation in a liquid metal of a sound wave the pressure of which is over the cavitation threshold, cavitation bubbles appear in the melt and the kinetics of mass transport of gas from the solution to a bubble changes significantly.

The improvement of the contribution of diffusion processes to the growth of cavitation bubbles as well as the concepts on the nucleation of cavitation centers in a metallic liquid allow one to consider the cavitation nature of a process of ultrasonic degassing of melts of aluminum and other nonferrous alloys.

Considering the results of the ultrasonic degassing action on the melt of commercial aluminum in a wide range of ultrasonic intensity, one can note that the dependence given in Figure 11 is characterized by three specific regions. The formation of these regions may be interpreted as the initiation and development of ultrasonic cavitation in a liquid metal [48].

Region I, where no ultrasonic degassing occurs, may be called the region of precavitation modes of the treatment. Region II, where the efficiency of degassing initially sharply increases and then gradually stabilizes, refers to cavitation modes of the ultrasonic treatment.

With further increase in the ultrasonic intensity and achievement of a certain degree of cavitation phenomena development, one can reveal Region III where the efficiency of ultrasonic degassing starts to increase linearly with the increasing ultrasonic intensity. This is the region of developed cavitation with wave resistance of the melt $\overline{\rho_c c_c}/\rho_o c_o << 1.0$.

The analysis of the efficiency of ultrasonic degassing in relation to the modes of the UST allows one to conclude that the degassing threshold coincides with the cavitation threshold in liquid metals, which confirms the cavitation nature of the process of gas removal from a melt under action of powerful ultrasound [48].

Acceleration of the degassing process with the melt treatment under the conditions of developed cavitation is connected with the fact that acoustic streams

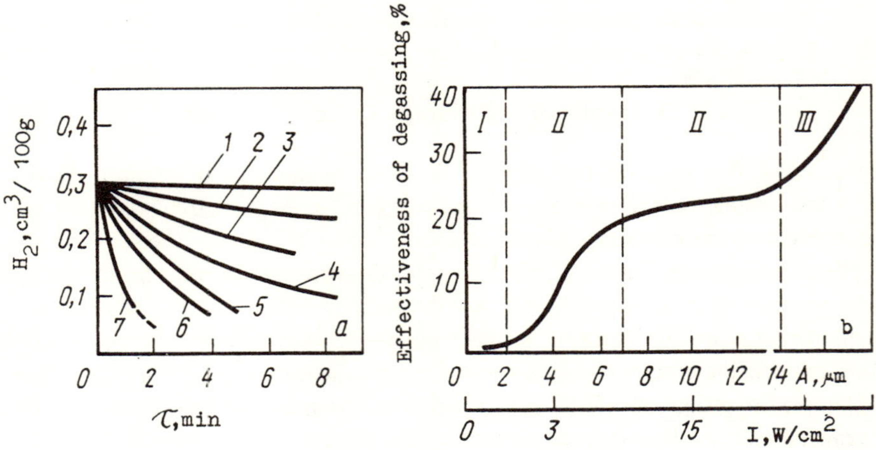

Figure 11. Kinetics of degassing of an A7 grade aluminum melt at the ultrasonic intensity of to 40 W cm^{-2} (**a**) and isochrone of the ultrasonic degassing efficiency in dependence on the UST mode (**b**). (**1**) The amplitude of a source displacement of 2 μm; (**2**) 3 μm; (**3**) 5 μm; (**4**) 10 μm; (**5**) 15 μm; (**6**) 18 μm; (**7**) 22 μm. I. No cavitation; II. The cavitation start; III. Developed cavitation.

develop within the metal volume with increasing density of cavitation bubbles. These streams promote coagulation of separate hydrogen bubbles and their accelerated evolution to the surface of a bath with the melt.

Similar dependencies are inherent in processes of ultrasonic degassing of the majority of industrial wrought aluminum alloys as well as aluminum alloys for shape casting.

Accordingly, the following stages of the process of cleaning and degassing (corresponding to the peculiarities of ultrasonic cavitation of liquid metal) occur simultaneously or sequentially in the melt subjected to the ultrasonic treatment in the mode of developed cavitation:

1. Nucleation of hydrogen bubbles on the surface of solid particles of the metal oxides nonwettable by the melt at sound pressures over the threshold when disturbance of liquid–metal adhesion to the solid surface occurs.
2. Growth of the hydrogen bubbles due to gas diffusion directed inside the bubble from the solution and depending on dimensions of the initial nucleus, the initial content of hydrogen in the melt, the amplitude of sound pressure, and duration of cavitation action.
3. Coagulation of the separate pulsing bubbles as a result of acoustic streams developing near the cavitating bubbles.
4. Adsorption transport of solid particles of the oxides which serve as a base for cavitation nuclei to the surface of a coarse gas bubble.

5. Evolution of the gas bubbles to the surface of a bath containing the melt due to the action of acoustic macrostreams.

3.2 Degassing of a Stationary Volume of a Melt

One of the first successful attempts of ultrasonic degassing of aluminum alloys for shape casting was the development of the process of the ultrasonic treatment in a crucible with the stationary volume of up to 200 kg [3–5, 49].

The investigations (Figure 12) show that the ultrasonic treatment of a stationary volume of the melt is more effective than the vacuum treatment of the same melt. Even greater effect of degassing has been observed for the melt of an AL9 grade alloy subjected to combined ultrasonic and vacuum treatment.

Our own investigations [5, 49] performed using cast and wrought aluminum and magnesium alloys show that the ultrasonic treatment provides a sufficiently high degree of cleaning from hydrogen as well as from solid nonmetallic inclusions.

The floating action of oscillating gas bubbles with acoustic cavitation has been revealed in 1960 by Rozenberg [44] in studies of a process of ultrasonic cleaning.

Later, Novitskii [50] has shown that a pulsing bubble collects around itself particles regardless of their chemical composition, density, and wettability. The significant contribution of acoustic microstreams to the floatation mechanism has been revealed by means of high speed filming of the distillate water–metal particles system at the ultrasonic treatment frequency of 20 kHz, metal particles being of 5 to 100 μm in size. Evidently, ultrasonic degassing of a stationary volume of a melt increases markedly the metal purity in solid nonmetallic inclusions. The industrial experience for more than 10 years of using the method of ultrasonic degassing in

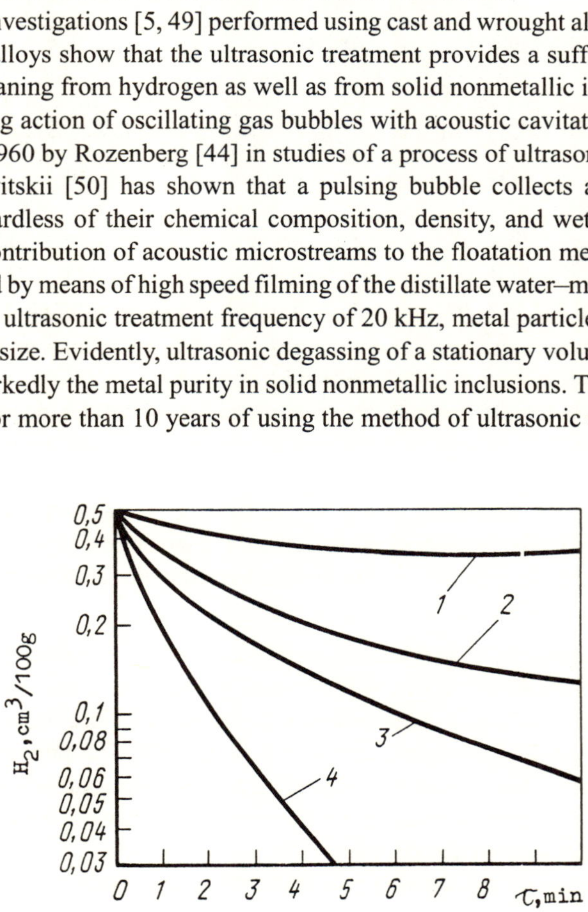

Figure 12. Kinetics of hydrogen removal at different methods of degassing of an AL9 grade alloy melt. (**1**) Chlorine salts; (**2**) Vacuum treatment; (**3**) UST; (**4**) UST in vacuum.

Table 6.

	Alloy		
	AL9	*AL3*	*ATsR-1*
Fluidity measured by the length of a spiral (mm):			
Without cleaning	500	500	600
Argon lancing	550	600	600
Ultrasonic treatment	670	670	720

the foundry workshop allow us to give data (Table 6) on melt fluidity after the ultrasonic treatment [49].

The data show that melt fluidity increases by 30% due to ultrasonic degassing of the melt, which is very important for a shape-casting production. Bondarek [18] has confirmed these results, according to his data the ultrasonic treatment in a 45-kg crucible of an AL9 grade alloy not only decreases the hydrogen content and volume proportion of nonmetallic inclusions but disperses the latter (Table 7).

To increase the efficiency of melt degassing in a stationary volume, it is necessary to reduce the height of the liquid bath. As it will be shown below, the treatment of a melt in a flow with a low height of liquid bath ($H \leq 100$ mm) results in increased rate of ultrasonic degassing. Contrary, the increasing height of a crucible with a liquid metal to $H \geq 500$ mm, as it takes place in industrial production of accurate castings in crucible furnaces, leads to increased duration of the degassing.

The temperature of a molten metal is of significant importance for the active occurrence of ultrasonic degassing. The higher is the melt temperature and the lower is its viscosity, μ, the higher is the rate of acoustic streams and the easier is the process of gas bubble evolution. However, there is the optimum temperature—temperature increase above 750 °C adversely affects the efficiency of the process due

Table 7. Effect of the UST Duration on Hydrogen Content and Inclusions Parameters in an AL9 Alloy Melt

		Inclusions Parameters		
UST Duration (min)	*H_2 Content [cm (100 g)$^{-1}$]*	*Vol. Prop. (%)*	*Mean Size (μm)*	*Amount (per 1 cm^2)*
0	0.2	1.18	2.95	763
3	0.18	0.60	2.85	550
6	0.15	0.33	2.35	672
9	0.14	0.34	2.30	667
20	0.11	0.11	2.0	369

to increasing solubility of hydrogen from atmospheric moisture. Decreasing the temperature below 700 °C results in decreased efficiency of degassing due to increased viscosity of the melt, corresponding hampering of bubble pulsation, coagulation and floating up of the bubbles as well as to the decreased diffusion coefficient of hydrogen in the melt and corresponding hampering of directed diffusion of hydrogen from the melt into a bubble under the action of alternating sound pressure.

With a sufficiently high level of ultrasonic treatment intensity, the main parameter of the efficiency of the degassing process is its duration. From the industrial experience of operation of ultrasonic degassing setups of the UZD-200 and UZD-200M types, one can reveal the optimum period of the ultrasonic treatment of 50 to 250 kg of the melt ensuring 50%-effective degassing,

$$\tau = 2k \times 1.2^{(m/50-1)} \times m^{0.5}(1 + 740/T)/n \tag{11}$$

where τ is the period of degassing (s); m is the mass of the melt (kg); n is the number of operating radiators; T is the temperature of the melt (700 to 770 °C); k is the correction coefficient ($k = 1$ at $A = 15$–20 μm; $k = 1.8$ at $A = 10$–12 μm).

The numeric multiplier 2 in Eq. (11) reflects the fact that the total period of operation consists of two successive treatments with an interruption. Such mode of the treatment is due to the kinetics of the degassing process when bubbles formed in a cavitation field must have time to coagulate and rise to the melt surface. From this point of view, continuous ultrasonic radiation may hamper the process and decrease its efficiency.

3.3 Degassing of a Melt Flow during Continuous Casting of Ingots

The development of modern techniques for the cleaning of a melt allows us to recommend—for continuous casting of aluminum alloy ingots using high-capacity furnaces and high-efficiency casting equipment—the change from melt degassing in a holding furnace to the degassing of the melt as it flows from a holding furnace to a mold.

Three basic schemes for the ultrasonic cleaning of the melt can be proposed: (1) in a liquid bath of an ingot (the melt surface is in a contact with a waveguide-radiating ultrasonic system); (2) in a mold (oscillations transmit to the melt through mold walls); and (3) in an intermediate vessel placed in the flow of melt from a holding furnace to a mold (oscillations transmits to the melt according to the first scheme).

The studies on the efficiency of the operating schemes mentioned show that the treatment of the melt in the liquid bath of an ingot at metal temperatures close to the liquidus (when the viscosity of the metal is sufficiently high and hydrogen bubbles are evolved in the opposite direction to the acoustic energy flow) is less efficient than the treatment performed on the melt flow from a holding furnace to a mold. The degassing process is incidental with the ultrasonic treatment of the melt in a mold; the main results of the treatment is grain refinement of an ingot structure (see Section 4). As for the second scheme of the treatment when oscillations transmit through mold walls, low frequencies of mechanical vibration are needed for the efficient transmission of the oscillations to a liquid metal as it is proposed in ref. [51]. With transmission of the oscillations of ultrasonic frequencies to a liquid metal through mold walls, the consumption of acoustic power should be so high (to ensure the effect of the treatment) that stresses occurring in a mold exceed the endurance limit of its material, and the mold rapidly breaks down. Accordingly, the most sufficient scheme of degassing for industrial applications is the degassing in a trough between a holding furnace and a mold. The investigations performed using a wide range of aluminum wrought alloys (AMg2, AMg5, AMg6, 1960, AK8, AK6 etc.) show high efficiency of the ultrasonic degassing.

The efficiency of ultrasonic degassing in a melt flow corresponds to the completeness of the processes of cavitation nucleation, growth and evolution of hydrogen bubbles. So with increasing flow of a melt through an ultrasonic setup, the increase in a number of working ultrasonic sources and in the duration of the residence time of the melt in the cavitation region are required.

Figure 13 displays the results of the ultrasonic degassing of a melt flow with a continuous casting of a large-scale ingot from an AMg6 grade alloy (cross-section is 1700 × 300 mm, melt consumption is to 80 kg min^{-1}). One can see that increasing from 9 to 11 kW acoustic power, W_a, inputting in a flow results in decreasing

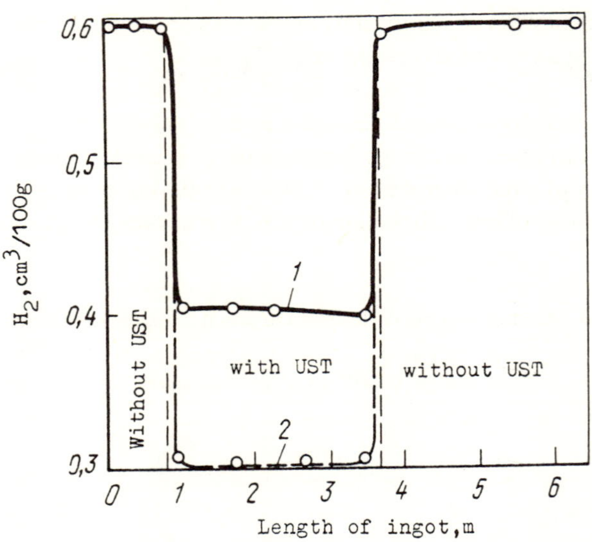

Figure 13. Effect of ultrasound on the efficiency of hydrogen removal from a melt flow during casting of large-scale flat-shaped ingots (1700 × 300 mm) from an AMg6 grade alloy. (**1**) W_a = 9.0 kW; (**2**) 11 kW.

hydrogen content from the initial one of 0.6 to 0.3 cm^3 $(100\ g)^{-1}$ instead of 0.4 cm^3 $(100\ g)^{-1}$. Accordingly, the efficiency of the degassing increases from 33 to 50% [52]. Simultaneously with the removal of hydrogen bubbles from the melt, the ultrasonic treatment favors the floatation of solid particles of nonmetallic inclusions and decrease in their content in the ingot by 10–20%.

3.4 Effect of Ultrasonic Degassing of a Melt on Properties of Shape Castings, Ingots, and Deformed Semiproducts from Aluminum Alloys

The studies on the ultrasonic degassing of aluminum and its alloys melts [3–5, 48, 49, 52] show that the significant decrease in the content of hydrogen and oxides in the melt substantially influences basic physico-mechanical properties of the as-cast metal.

As a result of the ultrasonic degassing of the melt, metal density increases and mechanical properties of castings and ingots are improved—mainly ductility, which is particularly important for loaded castings and for further plastic deformation of ingots.

The comparison of a number of industrial methods of degassing [5] show that ultrasonic degassing is distinguished for its simplicity and efficiency of the improvement of casting quality (Table 8).

The purity in nonmetallic gaseous and solid inclusions is of particular importance for the production of semiproducts from aluminum alloy ingots.

Thus, mechanical properties of a round ingot of 460 mm in diameter from an AK6 grade alloy can be significantly improved with ultrasonic treatment of the melt flow feeding a continuous casting mold (Table 9).

The effect of ultrasonic degassing of liquid metal on the quality of ingots manifests itself by increased density, decreased coefficient of ultrasonic attenuation, and increased ductility at temperatures of plastic deformation. The data on the ductility of a flat-shaped ingot (1700 × 300 mm) from an AMg6 grade alloy at the temperature of hot deformation of 400 °C are given in Table 10.

Table 8. Comparison of Industrial Degassing Techniques for an AL4 Alloy[a]

Technique	*H_2 Content [cm^3 $(100\ g)^{-1}$]*	*Density (10^3 kg m^{-3})*	*Porosity (number)*	σ_b *(MPa)*	δ *(%)*
UST	0.17	2.706	1–2	245	5.1
Vacuuming	0.2	2.681	1–2	228	4.2
Ar blasting	0.26	2.667	2–3	233	4.0
Hexachloroethane	0.3	2.663	2–3	212	4.5
Universal flux	0.26	2.660	3–4	225	4.0
Initial state	0.35	2.665	4	200	3.8

[a]From ref. [5].

Table 9. Mechanical Properties of Round Ingots from an AK6 Alloy (Ø 460 mm)

Refining Technique	*H_2 Content, [cm^3 (100 g)$^{-1}$]*	*σ_b (MPa)*	*δ (%)*
Conventional	0.4	160 to 183	4.8 to 6.4
Ultrasonic	0.21	182 to 210	6.5 to 9.5

Table 10. Effect of Ultrasonic Degassing on Mechanical Properties of an AMg6 Alloy Flat Ingots at 400 °C[a]

Efficiency of Degassing (%)	*24*	*33*	*48*
σ_b (MPa)	61/73	66/68	60/63
$\sigma_{0.2}$ (MPa)	55/67	59/62	56/57
δ (%)	54/69	58/68	55/67

[a]Numerator - the conventional technique; denominator - the ultrasonic degassing.

The decreased content of hydrogen as a result of the ultrasonic treatment of a melt flow and increased density and ductility of ingots are transferred to the final deformed product. Our investigations [5, 53] carried out on the ultrasonic degassing of ingots from an AMg6 grade alloy show that various semiproducts (sheets, plates, rods, and wires) have no flaws of the "lamination" type (according to the hot test) and have enhanced mechanical properties (Table 11).

Table 11. Effect of the Ultrasonic Treatment of a Melt Flow on Delamination of Sheets (10 mm) from an AMg6 Alloy

Parameter	*Degassing Technique*	
	Conventional	*UST*
H_2 content [cm^3 (100 g)$^{-1}$]		
melt	0.60	0.30–0.33
ingot	0.33–0.37	0.20–0.25
sheet	0.30–0.34	0.18–0.22
Number of hot probes	2–3	1
Break-off probe as-rolled (MPa)	193–235	239–247
After heating	69–80	170–206
Number of cycles to failure, 10^5 cycles (σ_{max} = 160 MPa)	0.07–2.61	1.96–6.45

3.5 Mechanism of Fine Filtration of a Melt in a Field of Acoustic Cavitation

It is well known that screen fiberglass filters with cell sizes from 0.6 to 1.3 mm which are of a wide use in industry do not retain dispersed particles, and fine filters (microfilters) or multilayer fiberglass filters with the standard size of cell are required. But the use of such filters is hampered due to poor wettability with a melt with high surface tension. To ensure the transfer of a melt flow through a cell of a single-layer screen filter, the force equivalent to the action of a column of liquid metal with the following height is needed;

$$H = 4\sigma \cos(\theta - 90°)/\rho g a \qquad (12)$$

where a is the cell size; θ is the wetting angle; and g is the acceleration of gravity. The experiments show that for capillary penetration of a melt of aluminum alloys through a filter, containing 1 to 9 layers of fiberglass with a cell of 0.6 × 0.6 mm in size, the height of the column should be 100 cm and more [54] (Figure 14, curve 1), which virtually eliminates the possibility of application of this filtration method.

The ultrasonic treatment of a melt involving the production of acoustic cavitation near to the surface of a multilayer filter offers significant possibilities for improvement in the process of fine filtration. Cavitating bubbles compressing close to filter capillaries produce impulses of high pressure ($P = 10^3$ MPa). Due to the action of these impulses, the melt overcomes easily the capillary pressure and friction of liquid metal against the surface of filter channels. This favors the transport of the melt through a capillary channel of a filter, and the required column of liquid metal reduces to the values less than 3 to 4 cm (Figure 14, curve 2):

$$H_{US} = H - \sum_{i=1}^{n} \Delta h_i \qquad (13)$$

where Δh_i is the length of melt movement through a capillary channel over one period of sound wave, T, $\Delta h_i = P_{max}(T - \tau)/\rho g n_{i-1}$; n is the number of periods. However, the authors [56] report that the maximum pressure, P_{max}, corresponds to the minimum radius of a cavity, r_{min}, which is taken equal to 9.0 μm for the calculations and experiments performed. The maximum radius of the cavity is 0.1 to 0.2 mm, which allows one to suppose that one bubble will be sufficient to ensure the liquid rise in a capillary of less than 0.5 mm in radius. Moreover, the pressure P_{max} and the generated shock wave will only have an effect in cases where the front of the wave is of the same dimension of the capillary. With this, as a result of spheric divergence of the shock wave, the pressure within it reduces, and the force affecting the liquid in a capillary will be respectively lower. Actually, the decrease in the force value, $F = PS$, will be even more significant due to the strong pressure drop which occurs with distance from the center of the cavity.

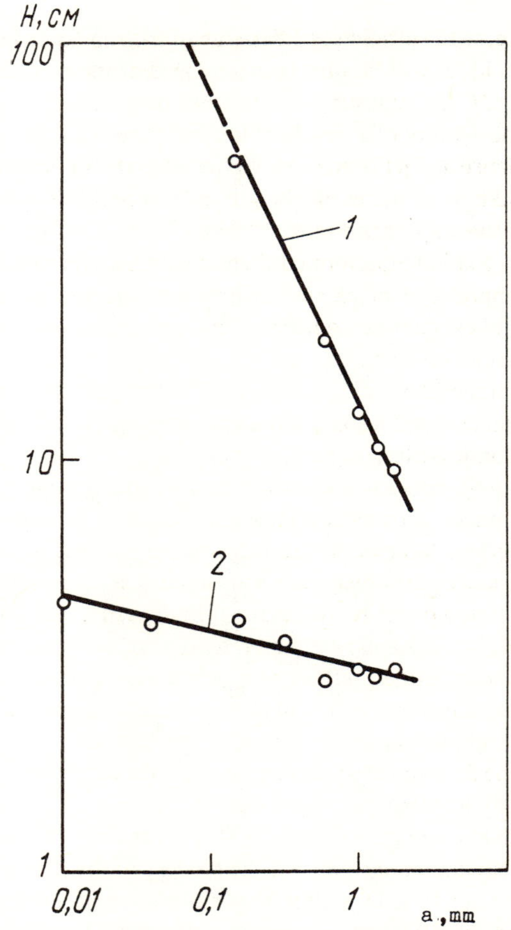

Figure 14. Effect of office size (a) in a multilayer filter with 0.6 × 0.6 mm cell on the height of liquid aluminum column required for melt transition through a filter. (**1**) Without cavitation action; (**2**) Under cavitation action.

The authors [57] have taken the disagreement in theoretical and experimental data into account and have proposed their own hypothesis of the sonocapillary effect. They have considered the asymmetry of boundary conditions of a collapsing cavitation bubble when the spheric form of the cavity becomes unstable and the cavity deforms with formation of a cumulative liquid jet. This cumulative jet is considered [57] to be responsible for the liquid rise in the capillary.

Different cases of cavitation bubbles compression with formation of a cumulative jet are considered in ref. [57]. Let us assume that a capillary locates at such large distance from a radiator that we can neglect the influence of its surface on bubble collapse. Hence, with sufficiently small size of the capillary (less than the size of a cavity), the liquid jet will be directed to the end of the capillary and result in the liquid rise of ΔH. By repeating the process over and over with the frequency reflecting the possibility of a new bubble appearance near the end of the capillary, the bubble collapse and jet formation will result in the summation of the ΔH values and, correspondingly, in increasing height and rate of the liquid rise. This phenomenon is called a sonocapillary effect.

The authors [57] have performed theoretical and experimental analyses of the process of compression of cavitation bubbles near the entrance to a capillary channel. The investigations confirmed the asymmetry of bubble collapse with formation of cumulative jets.

An appreciation of the decisive role of the hydrodynamic phenomena associated with cavitational collapse leads to a better understanding of these phenomena with respect to the sonocapillary effect.

Irrespective of the conditions ensuring the abnormally rapid movement of a liquid in a capillary under acoustic cavitation effect, it is important to note that the sonocapillary effect follows all the major effects of the ultrasonic treatment of melts. Among such phenomena are wetting and activation of solid nonmetallic impurities in a liquid metal as well as fine filtration of a melt through porous filters under action of the ultrasonic cavitation treatment. For both processes, ultrasonic cavitation and sonocapillary effect with formation of cumulative jets provide the accelerated mass transfer of a melt to slots and cracks in the surface of nonwettable solid particles and into capillary channels of fine filters.

In Eq. (13), the duration of a pressure impulse of a compressing cavitation bubble, τ, is significantly less than the sound period, i.e. $\tau << T$. The actual values for the frequency of 18 kHz are the following $T = 56$ μs and $\tau = 10^{-2}$ μs. Due to such rapid occurrence of cavitation phenomena, the process of fine filtration in an ultrasonic field starts just after 10 to 60 s after pouring (when a liquid metal enters a filter).

To determine the effect of a pressure field, resulting from cavitation bubbles compressing, on the penetration of a melt into a capillary channel of a filter, one should estimate the character of the pressure impulse propagation through the capillary. The mechanism of the flow is somewhat different under conditions of fine filtration of a melt through a porous medium (a filter from multilayer fiberglass with a cell of 0.6 × 0.6 or 0.4 × 0.4 mm in size) with rather short (about several mm) but complexly curved channels, but the main conditions remain.

For example, the determining factors for the filtration are the initial rate of liquid movement (6 to 8 cm s^{-1}) and the rate of the liquid rises to several centimeters during the first seconds of the ultrasonic action. On the other hand, due to adhesion of dispersed particles of solid nonmetallic inclusions onto the walls of a capillary, the effective cross-section of the latter continuously changes in the process of filtration. Finally, with actual filtration we deal not with the regularities of the flow through a capillary channel but with the statistics of melt movement through a porous medium with a host of channels.

Assuming that the melt flow is laminar, its flow rate through a multilayer filter does not depend on time. But in the case when a molten metal contains dispersed particles with a size less than the section of the channel, the flow rate becomes dependent on time mainly due to the adhesion of the particles to the channel walls. With this, those particles which have the size larger than that of the capillary channel section are retained at the entrance of the filter in a form of a cake which increases the apparent length of the channel and decreases the active surface of the filter. The input of intensive ultrasonic oscillations in the mode of developed cavitation results in the appearance of active acoustic streams near the filter surface and in washing-out the cake. In the ideal case, the value of the flow rate through the filter can be sustained constant for a sufficiently long period of fine filtration due to the action of acoustic cavitation and streams.

The investigations performed in ref. [56] reveal that a liquid moves through a capillary only when the cavitation region consisting of pulsing and collapsing bubbles locates just near the end of the capillary. This localization of cavitation bubbles can be disrupted with increasing ultrasonic intensity and developing acoustic streams, which result in immediate interruption of liquid supply to the capillary. The other peculiarity of the sonocapillary effect is that the maintenance of the liquid level in the capillary requires 5 to 10 times less acoustic energy input than that at the initial moment because there is no need to overcome the forces of fluid friction against the walls of the capillary.

Fine filtration with ultrasonic treatment under the conditions of developed cavitation[1] allows one to use 3- and 5-layer fiberglass filters with a cell of 0.4 × 0.4 mm in size or combined filters with 10 to 15 layers containing more coarse fiber.

Table 12 displays the results of a metallographic study of a fracture of a 5-layer filter from fiberglass with a cell of 0.4 × 0.4 mm. One can see that the first three layers do most of the work.

The efficiency of fine filtration affects the surface finish after mechanical processing of semifinished products (sheets, forgings etc.) and such service properties as fatigue endurance. For example, the use of the UZFIRALS process

Table 12. Results of Fine Filtration of Melt of an 1163 Alloy through a 5-Layer Filter with a 0.4 × 0.4 Cell

Number of Layers	*Characteristic of the Cake*	*Sizes of Particles Retained (μm)*
In front of the 1st	Agglomeration of particles	–
1st	Host of particles	5–10
2nd	Many particles	5–10
3rd	Few particles	2–5
4th	Separate particles	2.0
5th	Particles virtually absent	1.0

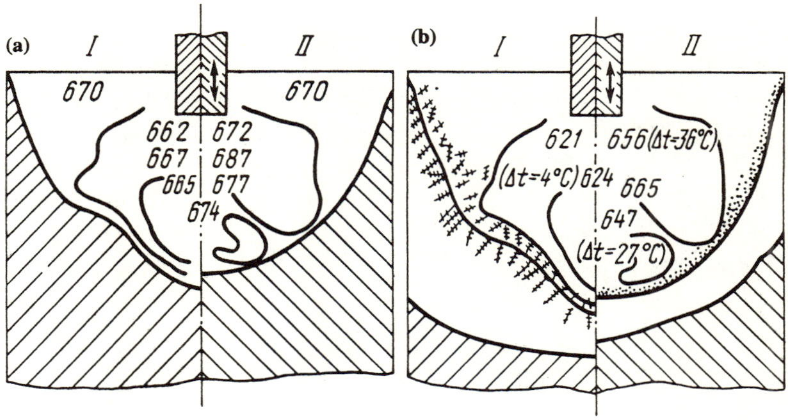

Figure 15. Topography of a liquid bath of an ingot of 270 mm in diameter from an A7 grade aluminum (**a**) and from a 1960 grade alloy (**b**). I. Quiet casting; II. Casting with UST.

increases the endurance limit of extruded plates from an 1163 grade alloy from 162 to 259 kcycles.

4. MAIN CONSIDERATIONS OF SOLIDIFICATION OF LIGHT ALLOYS IN A FIELD OF ACOUSTIC CAVITATION

4.1 Thermal Action of Cavitation on Liquid Metals

A large body of research shows that the influence of various factors on the grain size in continuously cast ingots may be primarily due to the melt temperature of the molten ingot. Variation of the temperature from the liquidus significantly changes the grain size [59]. It is also known that any methods of melt stirring in a liquid bath including those with the use of electromagnetic fields result in decreasing or equalizing temperature in the volume of the bath.

On the other hand, high-intensity ultrasonic treatment, generating cavitation, results in a significant increase in temperature of the liquid metal under conditions of continuous casting with intensive cooling of the mold with water. To quantitatively establish temperature changes during the casting of round and flat-shaped ingots from aluminum and magnesium alloys, the investigation of a temperature field was carried out by "freezing" of thermocouples [5]. According to this technique, a frame with six chromel-alumel thermocouples in an asbestos shield was located within the liquid metal in the mold. To determine the depth of the liquid bath, direct measurements of the distance to the solidification front were performed by a wire stick of a titanium alloy.

Figure 15 shows the results of the measurements of a temperature field in the liquid metal in a mold of 270 mm in diameter from an A7 grade aluminum and a 1960 grade alloy cast under silent conditions and under ultrasonic treatment using a radiator of 40 mm in diameter with acoustic power of 1.0 kW [5].

Preliminary estimations for pure aluminum show that the portion of heat input by ultrasound is about 3.0% of the total heat entering to the liquid bath. However, overheating of the melt in the liquid bath increases by 40%.

The analysis of the diagrams given in Figure 15 reveals the main peculiarities of the ultrasound effect on a temperature field in a liquid metal within a mold. Under silent conditions of casting, the liquid bath of a round ingot from aluminum has a conic form with cup curvature, and the temperature within the bath is slightly over the liquidus one, which is a typical result for pouring a liquid metal through a concentrated jet. Under action of an ultrasonic field, the topography of the bath undergoes significant changes, the profile of the solidification front becomes more flat, and the temperature of the melt increases by 15–20 °C.

The occurrence of ultrasonic cavitation and corresponding loss of acoustic power result in the appearance of intensive streams which change the direction of liquid metal movement, typical for casting under silent conditions of casting, from the bath surface to the solidification front. With this, the surface of solidification becomes more flat and the volume of the liquid metal in the bath somewhat increases.

The study of an alloy which solidifies with formation of a transition zone allows one to point out the additional effect of the ultrasonic treatment on characteristics of the zone.

Due to formation of a shrinkage gap between an ingot and the bottom of a mold, under conditions of quiet casting the transition zone broadens with approaching the periphery of the ingot. With the ultrasonic treatment of a melt, despite the large amount of heat due to overheating, dimensions of the zone decrease because of lower shrinkage and the corresponding gap.

The investigations [59] reveal that the two-phase region may be conventionally divided into two zones: solid–liquid and liquid–solid. The continuous front of solidification is the border between these zones. The analysis of the temperature within the liquid bath (see Figure 15) shows that the ultrasonic treatment and overheating of a melt narrow the transition zone from the liquid side, i.e. in the liquid–solid region.

To determine the amount of overheating in the bath due to acoustic power input, measurements were performed in the middle part of the liquid bath (at a distance of R/2 from the center and at a depth of 65–70 mm) of an ingot from a 1960 grade alloy (Figure 16). The direct dependence of overheating degree on the acoustic power input was revealed. Measurements of the temperature near the solidification surface are of great interest. It was shown that, with ultrasonic treatment, overheating reaches values of several degrees (to 8 °C) at a 2- to 3-mm distance from the solidification front. Whereas under quiet casting conditions, overheating does not

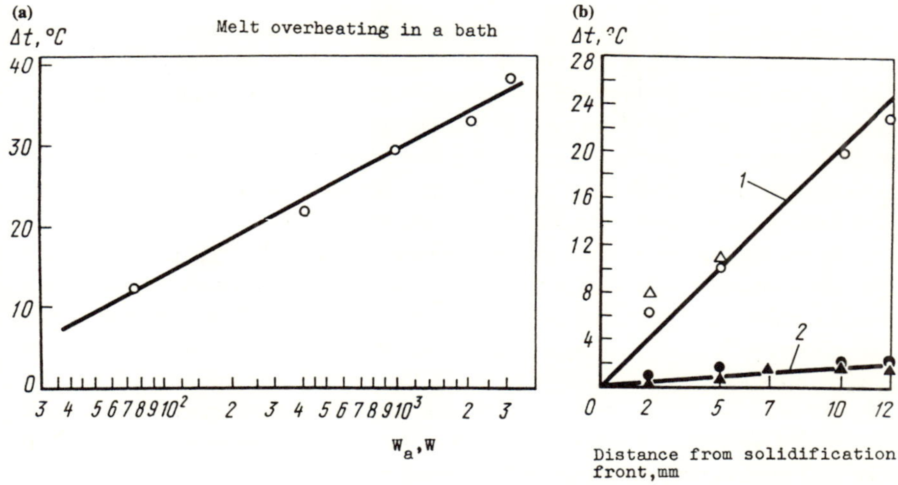

Figure 16. Effect of acoustic power (W_a) (**a**) and distance to the solidification front at W_a = 1.0 kW (**b**) on overheating of a liquid bath of an ingot from a 1960 grade alloy (270 mm in diameter). 1. UST; 2. Quiet casting.

exceed 2 °C at a distance of 15 mm from the solidification front. Accordingly, the supercooling zone adjacent to growing dendrites narrows dramatically with the ultrasonic action on a melt.

Similar investigations were performed using flat-shaped ingots from a MA2-1 grade magnesium alloy and a TsMP grade zinc alloy as well as large scale round ingots (650 to 1200 mm in diameter) from aluminum alloys [5, 12, 14, 65].

To summarize the results of the investigations on a temperature field in the liquid metal ingot, one can distinguish between the marked changes caused by the ultrasonic treatment from those corresponding to other dynamic actions on a solidifying melt such as stirring, vibration, etc. which do not cause cavitation.

Those traditional methods of dynamic action result, as a rule, in chilling of a melt in the liquid bath due to directed streams of relatively cold melt from the solidification surface to the liquid metal surface. With this, the supercooled zone extends from the solidification front into the liquid bath with a gradually decreasing degree of supercooling, which increases the probability of volume solidification. Correspondingly, together with grain refinement, the inhomogeneity of the internal constitution of a grain increases because branching of a dendrite occurs intensively with increasing distance from the solidification front.

But the ultrasonic overheating must result in the coarsening of an ingot structure but not in its refinement. Nevertheless, the structure gradually becomes finer with increasing ultrasonic intensity, the refinement degree increasing with development of cavitation phenomena in the liquid metal (Figure 17).

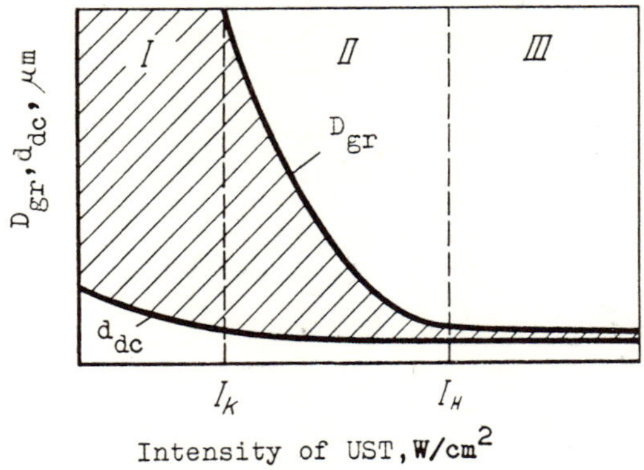

Figure 17. A diagram of grain refinement of ingots of 70 mm in diameter from alloys of the Al–Zn–Mg–Cu–Zr system (cooling rate with solidification of 60 K s^{-1}) in dependence on ultrasonic intensity and a cavitation mode. I. No cavitation; II. Developing cavitation; III. Intensive cavitation. D_{gr} is the grain size; d_{dc} is the dendritic cell size.

Moreover, under certain conditions of the ultrasonic treatment the special type of the as-cast structure is formed; that is referred to as the nondendritic structure with the grain size similar to that of a dendritic cell of the same ingot but cast without the ultrasonic treatment.

Recent investigations [5, 12, 14, 52, 62–65] show that a combined action of modifiers (producing new solidification sites) and intensive ultrasonic treatment in the developed cavitation mode are necessary to obtain the new type of an ingot structure. The input of ultrasonic energy to a liquid bath, ensures overheating of the melt and creates a temperature gradient near the solidification front, and the intensive cavitation treatment activates suspended particles of impurities and provides formation of a great quantity of solidification sites growing in the narrow zone of supercooling near the solidification front.

The structure formed under those conditions consists of polyhedral and virtually equiaxial grains and is referred to as nondendritic one. Ingots with the nondendritic structure can be obtained under industrial conditions for most of the high-alloyed aluminum alloys.

The term "nondendritic structure" reflects the origins of its formation when, under the specific conditions of ultrasonic treatment of a solidifying melt, a great quantity of solidification sites are generated. Their growth is hampered by temperature conditions and dendrite growth stops at an early stage of its formation.

4.2 Nuclei of Cavitation and Solidification Sites

As has been mentioned above, the occurrence of cavitation in a liquid metal provides strong catalytic action on the melt–nonmetallic solid impurities system, transforming those impurities to active solidification sites. The nonlinear character of cavitation bubble dynamics (according to the mechanism of a chain reaction) ensures sufficiently rapid (during several periods of a sound wave, which is about 100 to 200 μs at the frequency of 18 kHz) formation of the stable cavitation region near the surface of a radiator. A melt transferred through this region is enriched with active solidification sites.

The formation of solidification nuclei in solidifying melts occurs mainly on particles of insoluble impurities whose surfaces are fully or partly wetted by the solidifying substance. Although actual melts contain a host of submicroscopic particles of impurities, the quantity is insufficient to ensure the heterogeneous nucleation. Special additions (modifiers), promoting the formation of solidification sites, are usually added to the melt. Such modifiers for aluminum and its alloys are titanium, zirconium, boron, etc.; and for magnesium, iron, and zirconium. The modifying additions in aluminum alloys form aluminids which serve as nucleation centers. These are primarily solidifying intermetallics which form onto sublayers from fine dispersed particles of active impurities.

The investigations [5, 62–65] show that the presence of a modifier of nucleating action in an alloy is necessary to obtain the nondendritic structure. Introduction of small additions of transition metals (Ti, Zr, etc.) and activation of noncontrolling solid impurities ensure sharp increase in the quantity of solidification sites. So we may conclude that the ultrasonic treatment of a solidifying melt does not change the heterogeneous mode of solid solution crystallization on the surface of particles of intermetallics of transition metals but provides efficient multiplication of solidification sites due to the activation of impurity particles.

The role of nonsoluble impurities in a solidification process was considered in detail by Danilov et al. [60]. Kazachkovskii and Danilov [60] have shown that microscopic cracks and imperfections on the surface of solid particles are of great importance. Metallic crystals formed by any way inside those cracks can be preserved with heating above the melting temperature of free crystals. This interpretation offers an explanation for the role of overheating in decreasing the activity of the formation of solidification sites and is supported by the majority of scientists.

But the practice shows that only a small part of the submicroscopic particles (or so-called "plankton") are active. Most of the nonmetallic particles do not participate in the solidification process. Only the particles with deep cracks filled with a matrix alloy are sufficiently stable to serve as solidification sites [60]. The melting temperature of such an alloy appears to be significantly higher than that of the equilibrium liquidus. Filling of imperfections on the surface of nonmetallic inclusions is hampered by the presence of a gas phase within the imperfections, which does not allow the particles to serve as nucleation sites for the matrix and intermetallics.

The comparison of structures of continuously cast ingots with and without ultrasonic treatment in the developed cavitation mode clearly demonstrates that an

Table 13.

Ingot Diameter (mm)	Amount of Solidification Nuclei (cm^{-3}) Without UST	With UST
65	5×10^3	6×10^9
270	10^3	10^7
830	10^3	5×10^6

amount of active solidification nuclei after the cavitation treatment increases by several orders of magnitude as is shown in Table 13.

The development of cavitation in a solidifying melt can result in fracture of growing dendrites [6, 61] and multiplication of solidification nuclei in the form of dendrite fragments.

But such action of cavitation can be realized when the cavitation region locates close to the solidification front. In most cases of ultrasonic treatment involving the developed cavitation, the other mechanism of increasing nucleation is preferential.

As has been mentioned, the surfaces of noncontrolling particles of oxides, carbides, nitrides, etc. have imperfections of various kinds. The existence of the melt–gas–nonwettable particle system under conditions of an ultrasonic field results in formation of cavitation bubbles, developing by a chain reaction, onto the surface of the particles. A part of those primary bubbles and dynamically forming secondary bubbles can degenerate to gas bubbles, but one bubble always exists which can collapse and produce an impulse of high pressure. Chalmers [25] has shown that the size of a solidification nucleus decreases as the supercooling increases and the wetting angle is reduced. This dependence obtains a flat form of the nucleation sublayer. In the case of a curved surface of the sublayer (particularly when it is concave in the form of a crack), the size of the nucleus becomes even smaller and the nucleation is possible at lower supercooling. Moreover, if this crack is wetted by a melt and, hence, the wetting angle is close to zero, the solid phase inside the crack may be stable even at temperatures higher than that of the equilibrium melting.

It should be noted that the activation and wetting of noncontrolling nonmetallic inclusions in an acoustic cavitation field can occur due to the action of cumulative jets forming when cavitation bubbles collapse (see Section 2.4).

The effect of the surface relief is not restricted by the formation of the contact layer at active places. The existence of microscopic and submicroscopic cracks ensures a higher melting temperature for the crystalline phase inside these imperfections. Depending on the shapes and sizes of the cracks and on the value of surface tension, the crystal phase can exist at temperatures above the melting temperature of free crystals. Following supercooling during crystallization, solid particles of the phase serve as nucleation sites. Two conditions [60] limit the application of the crack theory of nucleation: (1) the activation of an impurity occurs only with formation of a solid phase on the impurity particles; (2) the activation of the impurity requires a certain time period dependent on the temperature of the process.

We should note that acoustic cavitation eliminates some restrictions stipulated by capillaries. The dynamics of a cavitation bubble—collapse and generation of a pressure impulse of $P_{max} \geq 10^2$ MPa or high-speed cumulative jets—provides accelerated filling of capillaries of submicroscopic sizes.

From the theory of irreversible thermodynamic processes, one can conclude that mass transfer in porous capillary walls occurs much more effectively with the pressure difference along the channel length rather than diffusion. In other words, the pressure difference in the capillary channel entrance can ensure abnormally rapid filling of the capillary with a liquid. Seemingly, this mechanism provides the filling of cracks of nonsoluble solid particles of "plankton" under action of ultrasonic cavitation. There is no necessity for long exposures for the activation of "plankton" particles because two or three periods of cavity pulsation (about 100–150 μs) are sufficient for the cavity collapse and filling of the capillary with a liquid metal under action of the impulse of 10^2 MPa.

The experimental data obtained on ingots from a 1960 grade alloy (0.2 mass% Zr) reveal the activation action of the ultrasonic treatment in the developed cavitation mode. The alloy melt was first overheated in a furnace, which resulted in deactivation of impurities and coarsening of the structure after casting under conventional conditions. The ultrasonic treatment of the solidifying metal allowed one to reactivate impurities in the whole temperature range studied. In practice it means that the ultrasonic treatment in the developed cavitation mode ensures reactivation of impurities and use of all of the modifier contained in the alloy for nucleation.

In conclusion we should mention that the mechanism of the cavitation action on the nucleation process is hypothetical (taking into consideration the views by Danilov and other scientists), but the fact of the impurities and modifiers activation with ultrasonic treatment in the developed cavitation mode is beyond question. The grain refinement with ultrasonic treatment is caused by the impurity particles activation, i.e. by transformation of the potential solidification sites into active nucleation centers—crystal nuclei.

4.3 Peculiarities of Nondendritic Solidification of Light Alloys[2]

The grain size of an Al-solid solution in continuously cast ingots is determined by a quantity of realized solidification sites which are represented by active solid particles of impurities or modifiers. In the case of uniaxial solidification, crystal nuclei forming are not initially bonded to each other. With their growth, these come into contact and, finally, form the solidification front eliminating the possibility of their movement within a melt.

Under conventional casting conditions (without ultrasonic cavitation), the supercooling of a melt arising at the solidification front can spread into the liquid metal with a simultaneously decrease of the degree of supercooling. This process is due to convective streams and, particularly, due to forced movement of the melt with electromagnetic or mechanical stirring.

Under those conditions, particles of inclusions (solidification sites) have sufficient time in the volume of the liquid bath and have an opportunity to grow under conditions of low supercooling.

The farther from the bath surface a crystal nucleus has formed, the longer is the period of its growth and the larger are the arms of the developing dendrite. Accordingly, the broadening of the supercooling region and corresponding grain refinement result not only in general coarsening of the dendritic cell, but in increased nonuniformity in dendritic cell size within a single dendrite and between different dendrites.

The ultrasonic treatment of a melt in the liquid bath of an ingot with developing intensive cavitation, as has been mentioned above, leads to an acoustic energy loss on the development of cavitation and to warming of the liquid bath. The actual value of the melt warming may be up to 20 °C for an aluminum alloy ingot of 270 mm in diameter and to 25 °C for a magnesium alloy ingot of 550 × 150 mm in section.

With the ultrasonic treatment of a melt in the ingot mold, the narrowing of the supercooling region near the solidification front with development of cavitation processes occurs due to an inner heat source without preliminary overheating of the melt in a holding furnace. Besides that, the melt in the liquid bath is continuously supplied by additional solidification nuclei as it passes through the zone of the cavitation treatment.

The typical feature of a nondendritic structure is the formation of polyhedral grain without branching with a size similar to that of a dendritic cell in an ingot solidified under the same conditions but without the ultrasonic action. This ultimately fine (under the given conditions) grain may be obtained only in the case when the quantity of actual solidification sites is so high that branching does not occur and the solidification process cancels out the stage of equiaxial dendrite nucleus formation.

Among earlier studies we should mention the work [66] where equiaxial grains were observed in laboratory-scale castings from a binary Mg—5.0 mass% Zn alloy and the work [67] where large nondendritic grains of 0.4 to 0.6 mm in size were found in a binary Al—4.0 mass% Cu alloy subjected to 20-Hz vibration at temperatures close to that of the liquidus. The authors [68] revealed that with rapid solidification of nickel alloys under similar conditions of solidification one can obtain small granules (20 to 50 μm) of two types—with a dendritic structure and with a nondendritic one which was called "microcrystalline." The formation of such a structure corresponds to significant supercooling of melt drops before the solidification starts.

To prove the differences in structure of dendritic and nondendritic types, studies were made involving modern and precise methods of structure evaluation. The aim of the studies was to confirm the facts that (1) the nondendritic structure consists of grains but not of dendrite arms, and that (2) a nondendritic grain has no inner dendritic branching [5].

The answer to the first problem is given by a metallographic study in a polarized light after color anodizing of samples and by electron fractography. One can easily see the significant difference in color of separate grains which reflects the difference in crystallographic orientation of nondendritic grains, a morphology of the grains is also indicative.

Methods of X-ray microdiffraction and of etch figures were used to quantitatively estimate the crystallographic orientation and values of misorientation angles of structure fragments (these were dendrite arms for small-angle boundaries and grains for large-angle boundaries).

The crystallographic orientations found were used to calculate the misorientation angles and to determine a statistical ratio of large-angle and small-angle boundaries. The maximum value of 10° for the misorientation angle was accepted for a small-angle boundary. Figure 18 represents the ratio of large- and small-angle boundaries for dendritic and nondendritic structures. Taking into account the inevitable presence of large-angle (intergrain) boundaries in a dendritic structure and incompleteness of a process of nondendritic solidification, one can conclude that the elements of the nondendritic structure have large-angle boundaries and are grains; whereas the elements of the dendritic structure have small-angle boundaries

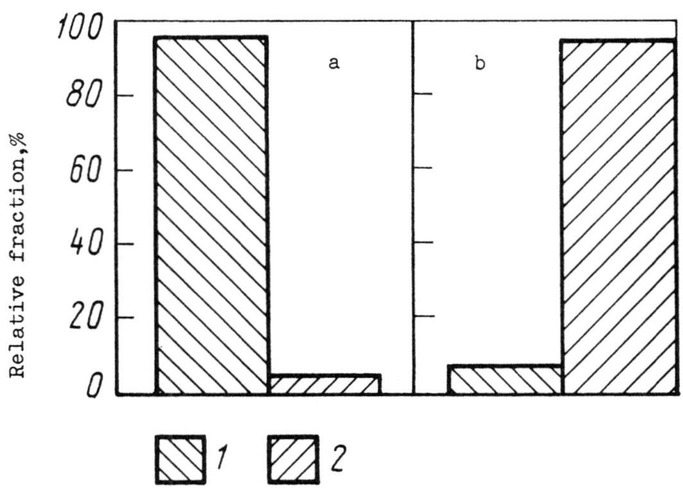

Figure 18. Proportion of large-angle (**1**) and small-angle (**2**) boundaries in nondendritic (**a**) and dendritic (**b**) structures of ingots of 74 mm in diameter from a 1960 grade alloy.

and are subgrains. Dendritic arms are separated by small-angle boundaries, and dendrites as a whole by large-angle boundaries.

To answer the second question (on inner branching), it is necessary and sufficient to determine whether there are interfaces within a nondendritic grain which strongly differ from the matrix by a content of alloying elements as is typical for dendritic cell boundaries. Methods of isocomposition (see Figure 22), X-ray diffraction, and microradiography were used to solve the problem.

One more consideration is for nonbranching of extremely fine nondendritic grains. As has been mentioned, sizes of nondendritic grains and of dendritic cells (in dendritic grains) are similar at the same crystallization rate. The branching of nondendritic grains violates one of the main regularities of solidification—the dependence of cross sizes of dendritic arms on the crystallization rate.

4.4 Effect of Cooling Rate of a Melt during Solidification on Formation of Nondendritic Structure in Light Alloys

Apparently, it is universally accepted that the degree of dendritic arm refinement increases with increasing cooling rate with solidification, and its value can be calculated.

For the first time this regularity was quantitatively estimated for aluminum alloys as the dependence of the dendritic cell size on the crystallization rate [59]. Later on, the analogous curve was obtained for steels by Gulyaev. The later work [69] is often referred to, here the dependence of dendritic arms spacing of the second order (dendritic parameter) on the cooling rate with solidification is given as a straight line in logarithmic axes according to the following formula,

$$d = av^{-n} \tag{14}$$

where d is the dendritic parameter; a and n are the constants; and v is the cooling rate in K s^{-1}.

Despite some terminology discrepancies—dendritic parameter, dendritic cell, etc.—the spacing between the axes of the second order dendritic arms is equivalent to the average cross-size of dendritic arms. Tens or, perhaps, hundreds of works confirming this dependence were published. It is so reliable that is often used for the back calculation of the cooling rate of an ingot section by the results of structural measurements. The dependence is not affected by casting modes (excluding those affecting the cooling or crystallization rates) and does not correspond to the grain size.

As for the dendrite grain size, it strongly depends on such factors as casting temperature, purity of a melt, preliminary overheating of the melt (deactivation of impurities) and other physical and technological parameters. Under such conditions the crystallization rate does not directly influence the grain size. This ambiguity reflects the fact that directly connected with the crystallization rate dendritic parameter virtually does not correspond to the grain size. But the experimental studies show that there is only one case when the grain size is quantitatively determined by the crystallization rate (and only by it). This is the case when the ultimately fine (under given conditions of casting) grain is formed due to various means of affecting a melt. Such a grain has no inner branching and is nondendritic.

It is rather difficult to obtain extremely fine grain structure with solidification. We manage to obtain such structure in light alloys by the combined action of modifying additions and the ultrasonic treatment on melts. But principally the nondendritic structure may be produced by other techniques increasing the quantity of active solidification nuclei in a melt.

One can consider the ultimate degree of grain refinement from the following point of view. It is well known that grain refinement by any method follows a damping curve. Obviously, the ultimate refinement is achieved when decreasing intensity of a parameter, which controls the process, results in no more refinement (see Figure 17).

With continuous or shape casting of aluminum and magnesium alloys, the ultimate degree of grain refinement was not achieved by any action applied separately: decreasing casting temperature, introduction of microcoolers, addition of modifiers, mechanical stirring of a melt, ultrasonic treatment, electromagnetic action, etc. The ultimate grain refinement was obtained only by combined action of the ultrasonic treatment of a solidifying melt and adding of modifiers such as zirconium, titanium, or other analogous additions.

It should be mentioned that under industrial conditions the size of a dendritic grain significantly (by a number of times) varies from melting to melting, from ingot to ingot. Whereas the size of the ultimately fine grain is stable and deviates from the average one by 20–30%.

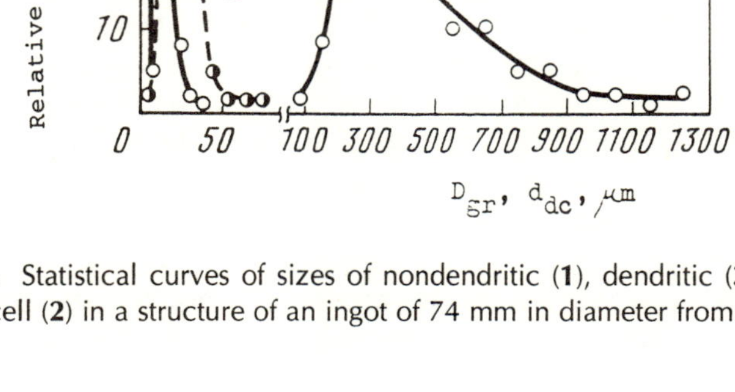

Figure 19. Statistical curves of sizes of nondendritic (**1**), dendritic (**3**) grains, and dendritic cell (**2**) in a structure of an ingot of 74 mm in diameter from a 1960 grade alloy.

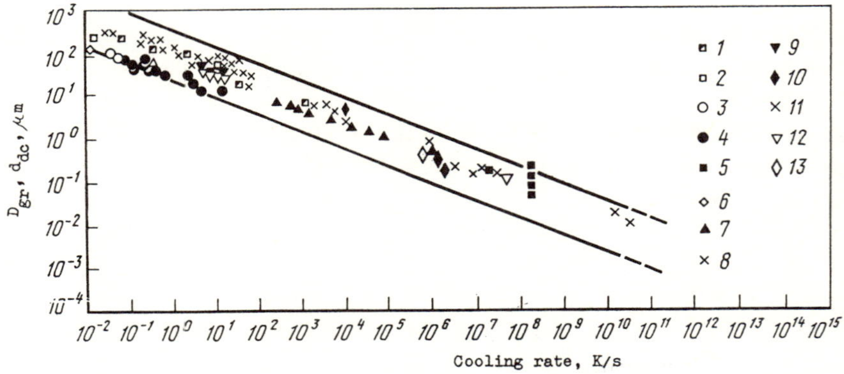

Figure 20. Dependence of the dendritic parameter and nondendritic grain size on a cooling rate for aluminum (1–8, 11), magnesium (9, 12), and nickel (10, 13) alloys. (1–10) Spacing between axes of 2nd order. (11–13) Nondendritic grain size.

The statistical distribution of sizes of dendritic and nondendritic (ultimately fine) grains in ingots (70 mm in diameter) from a 1960 grade alloy are given in Figure 19.

Investigations of microstructures of ingots, shape castings, and granules show that the ultimately fine grain is equiaxial and has no inner dendritic constitution. From correlation with microstructures of analogous castings and granules cast without the UST under similar conditions of heat removal from the outside surface, one can conclude that the average size of the ultimately refined grain (nondendritic grain) is close to that of dendritic arms in a coarse-grain metal structure. The statistical processing of our own results shows that the regularity of grain and dendritic cell dependence on cooling rate is reliable in the wide range of cooling rates for aluminum, magnesium, and nickel alloys as is shown in Figure 20. The smallest size of a dendritic parameter in the curve presented is of 0.1 μm [62–64].

Numerous investigations [5, 12, 14, 62–65] allow us to formulate the following points:

1. With any way of modifying action on a melt, a grain structure in the given casting (ingot, granule) can be refined to the certain extent.
2. This ultimately fine grain is nondendritic, i.e. has no inner branching.
3. The size of the nondendritic grain is unambiguously determined by the cooling rate during solidification being in reverse dependence on it.
4. The dependence mentioned is quantitatively similar with that of the dendritic arm size (in the case of dendritic grain formation).

Apparently, the nondendritic structure appears as a specific form of refinement of a growing solid phase when, due to high quantity of potential solidification sites, the nuclei formation is facilitated and occurs under the same supercooling conditions as the dendrite branching.

4.5 Effect of Refined (Nondendritic) Structure on Properties of As-Cast and Deformed Metal

The formation of a refined (nondendritic) structure ensures not only the refinement of solid solution grains but the obtaining of dispersed products of primary and secondary solidification, decreased hydrogen content, and improved density of an as-cast metal as well. Taking into account the homogeneous distribution of the nondendritic grain size in the volume of an ingot (shape casting), the positive effect of the nondendritic structure on properties of shape castings, ingots, and deformed semiproducts becomes evident.

The solidification of shape castings from aluminum alloys with casting into ceramic, gypsum-asbestos molds with low thermal conductivity or even into metallic molds (chills) occurs at a significantly lower rate than that with a continuous casting. Accordingly, structure and properties of shape castings differ infavorably from those of ingots. At the same time, numerous details such as turbine and fan plates can be produced only by means of fine shape casting allowing one to obtain simultaneously a massive hub and thick plates without mechanical processing. A lot of complex details can be obtained only by chill casting. Taking this into account, it is of great scientific and practical interest to develop a method of controlling structure and properties of a cast metal using ultrasound which will also improve the conditions for the filling of thin-wall plates with an edge of to 0.1 mm.

Preliminary experiments on the ultrasonic treatment of solidifying precise castings (into ceramic molds) have shown that significant warming of a melt in a mold occurs with development of cavitation and introduction of acoustic power above 100 W. As the volume of a casting is not large and the melt mass is hardly above 0.2 kg, the data obtained allow one to follow the improved filling of thick channels under ultrasonic action.

Figure 21 displays the data for temperature versus rate of solidification of a casting of a fan from an AL9 grade alloy; the experiment reveals significantly slowed down solidification in an ultrasonic field.

Investigation on the solidification process in all alloys studied shows that the action of ultrasound changes the temperature rate of solidification so significantly that the rate of overheating elimination in an ultrasonic field lowers by 5 to 6 times, the solidification rate slowing down by 6 to 8 times. It should be particularly stressed that the effect mentioned takes place under short-term ultrasonic treatment (1–2 min), which is significantly shorter than the solidification period for the given casting under silent conditions.

The data on the grain refinement of a shape casting of a turbine runner from an AL19 grade alloy under ultrasonic action are given in Table 14.

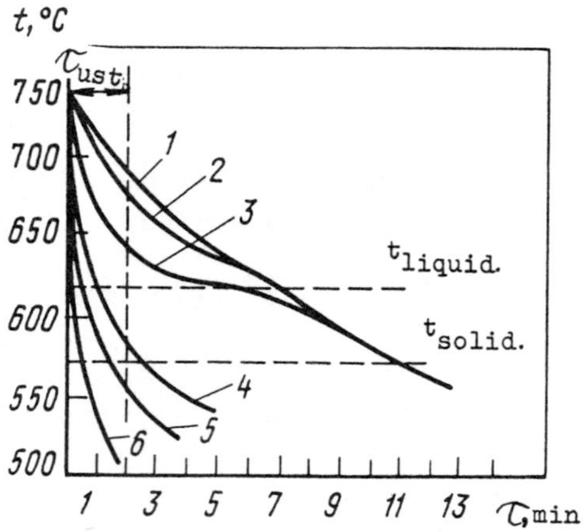

Figure 21. Cooling curves of different parts of a fan shape casting from an AL9 grade alloy. (1–3) With UST. (4–6) Without UST. (1, 4) The center of a casting. (2, 5) A base of a blade. (3, 6) A body of a blade.

The change in structure improves strength and ductility of shape castings from different casting aluminum alloys.

It is well known [59] that among the main elements of an ingot structure which are inherited by a deformed metal are grain and dendritic cell sizes and chemical segregation in the ingot volume.

The investigations have shown that the ultrasonic treatment of a melt and the formation of the nondendritic grain structure do not affect the dendritic segregation. A degree of microsegregation within a dendritic cell or a nondendritic grain is virtually the same; the nondendritic grain differs only by an increased volume proportion of sections with the lowest content of alloying elements (Figure 22), which can be explained by the character of growth of a spheric nondendritic grain as compared to a cylindric dendrite arm [5].

The influence of the ultrasonic treatment of a melt on the zone (macro) segregation is much stronger. As is known, the conditions necessary for zone segregation

Table 14.

	Grain Size[1] (μm)	
Section of Casting	*With UST*	*Without UST*
Center	75 (NDG)	800 (DG)
Hub	80 (NDG)	600 (DG)
Plate	140 (DG)	530 (DG)

[1]NDG = the nondendritic grain; DG = the dendritic grain.

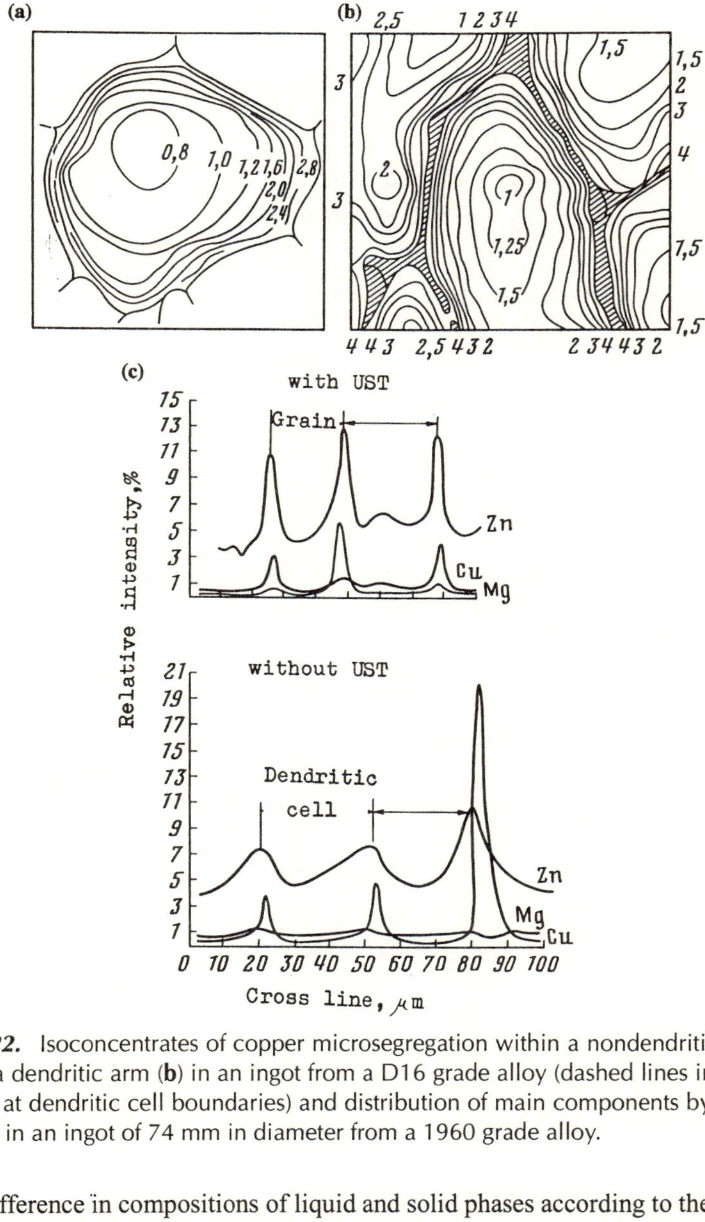

Figure 22. Isoconcentrates of copper microsegregation within a nondendritic grain (**a**) and a dendritic arm (**b**) in an ingot from a D16 grade alloy (dashed lines indicate eutectic at dendritic cell boundaries) and distribution of main components by a cast grain (**c**) in an ingot of 74 mm in diameter from a 1960 grade alloy.

is the difference in compositions of liquid and solid phases according to the phase diagram. The moving force of the zone segregation with a continuous casting of an ingot is the movement of the liquid matrix solution within a crystalline network under action of pressure difference induced by shrinkage. It looks like the cavitation treatment provides decreasing shrinkage due to increased quantity of solidification sites and narrowed region of their growth.

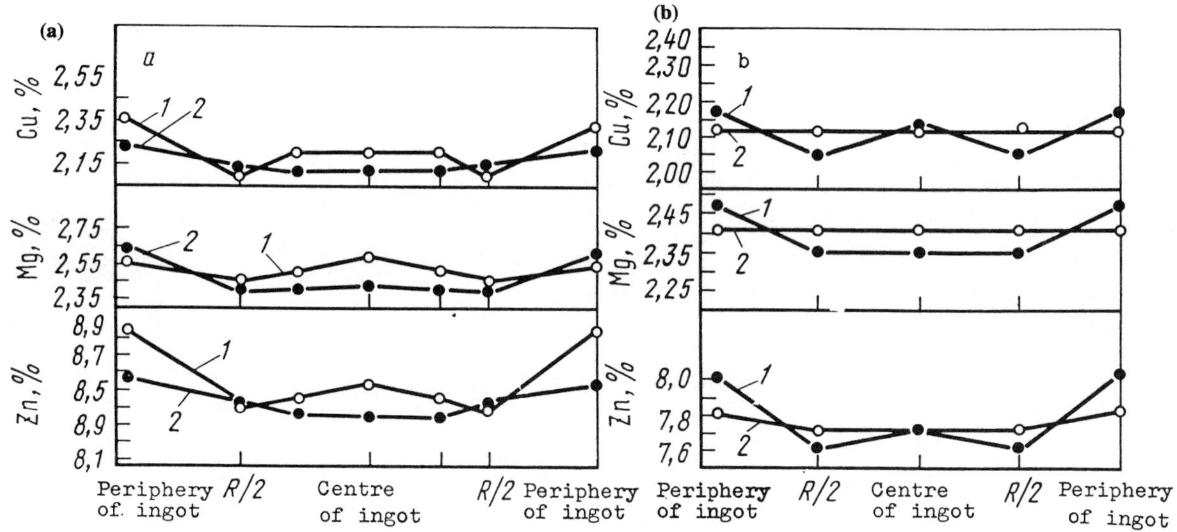

Figure 23. Effect of the ultrasonic treatment of a melt on zone segregation in ingots of 270 mm (**a**) and 112 mm (**b**) in diameter from a 1960 grade alloy. (**1**) Without UST; (**2**) With UST.

Actually, the linear shrinkage decreases and the surface quality of an ingot is improved with the ultrasonic treatment [62–65].

Figure 23 represents the distribution of main alloying elements in ingots of 112 and 270 mm in diameter from a 1960 grade alloy produced with and without the UST applied. It is clear that the ultrasonic treatment of the melt and transition to the nondendritic structure virtually equalize the chemical composition across the ingot.

Grain and dendritic cell refinement, particularly the formation of nondendritic structure with the average grain size equal to the dendritic cell size at the given solidification rate, opens the way to control structure and properties of deformed semiproducts of various kinds. First and foremost, the transition to the nondendritic structure sharply increases plasticity at room temperature (Figure 24) and in the temperature range used for hot working. The formation of the nondendritic structure results in lowered cracking during casting of large-scale ingots from aluminum alloys and promotes the process of hot working (deformation). Thus, the transition to the ultimately refined nondendritic structure with the high degree of structural and chemical homogeneity of an as-cast metal improves the reliability of producing large-scale ingots without cracking.

Nowadays, the application of the ultrasonic treatment ensures the reliable reproduction of the nondendritic structure in ingots (Figure 25) from the majority of aluminum structural alloys of the Al–Cu–Mg–Mn and Al–Zn–Mg–Cu systems (ingots of 65 to 1200 mm in diameter) and from magnesium alloys (ingots to 270 mm in diameter) containing modifying additions of Zr.

An effect of ingot structure on final properties of deformed semiproducts is widely described in the literature, the trends of investigations being aimed at lowering the influence of the coarse ingot structure by the development of optimum

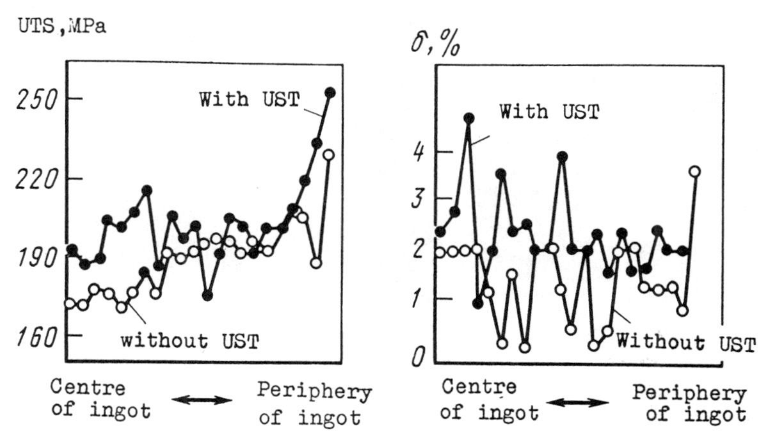

Figure 24. Diagram of mechanical properties change at 20 °C across annealed ingots from a D16 grade alloy containing 0.14% Zr (of the 2124 type).

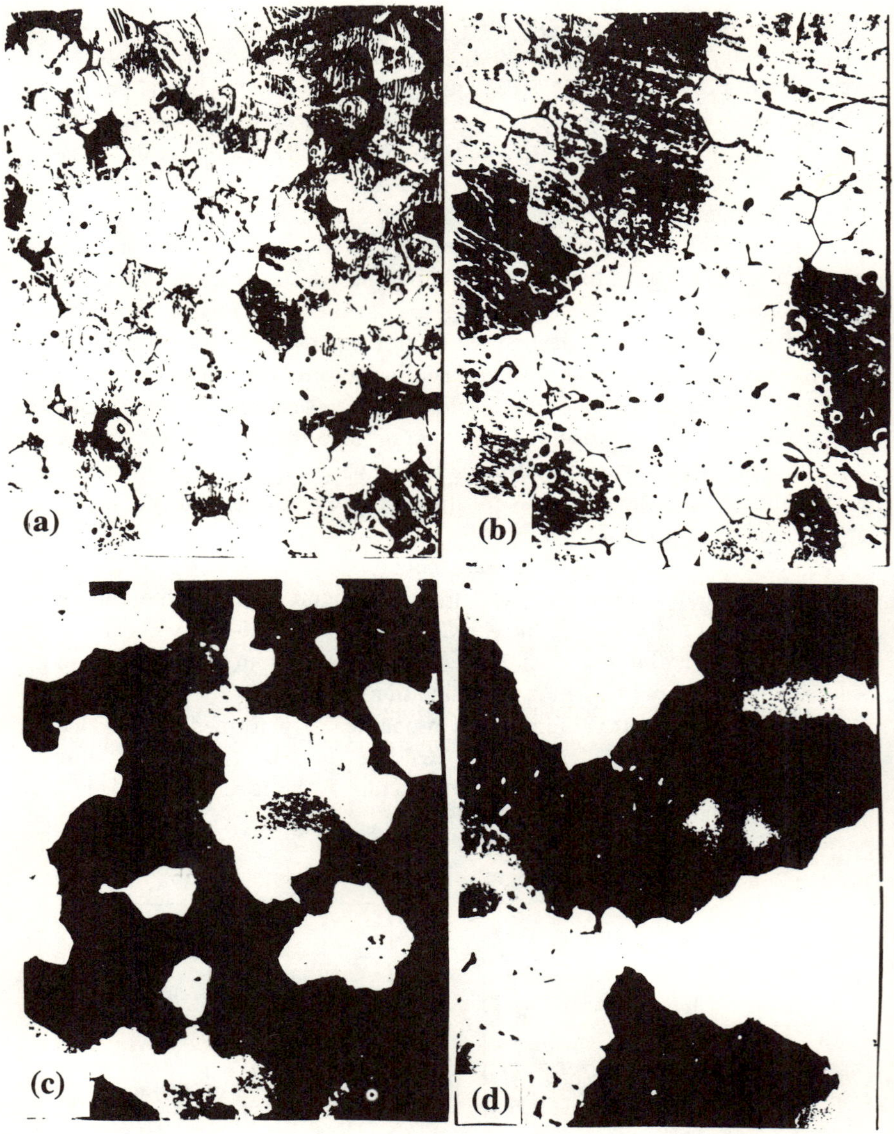

Figure 25. Microstructures of a central part of a flat-shaped ingot of 100 × 300 mm in section from a MA3-3 grade alloy (×100) and of large-scale round ingot of 1200 mm in diameter from an 1161 grade alloy (×50). (**a**, **c**) Casting with UST; (**b**, **d**) Casting without UST.

Table 15. Effect of Nondendritic Structure of an Ingot on Static and Dynamic Properties of Plates[a] from an 1161 Alloy

Property	*Nondendritic Structure*	*Dendritic Structure*
σ_b (MPa)	453	432
$\sigma_{0.2}$ (MPa)	317	293
δ (%)	16.2	14.7
LCF (kc)	239	132
K_{1C} (MPa $m^{0.5}$)	50.4	45.9

[a]60 mm in thickness.

deformation technique. The effect of the refined ingot structure on the properties of intermediate products is less studied. It is considered that at deformation degrees of 80–85% (extrusion coefficient of 6 to 8) one can obtain sufficiently high mechanical properties of a deformed metal despite the initial cast structure. But the widely performed investigations and more than 50-years technological experience allow one to consider that, in spite of the strong effect of deformation on refinement of structural elements, the initial structure of an ingot mainly determines the quality of deformed semiproducts.

To confirm this, let us consider the influence of the nondendritic structure of a flat-shaped ingot of 550 × 165 mm in section from an 1161 grade alloy on mechanical properties of plates produced from it [5, 65] (Table 15).

It is clear that structure-sensitive dynamic properties [low-cycle fatigue (LCF) and fracture toughness (K_{1C})] notably increase for the alloy with the nondendritic structure. One more proof of the advantage of the nondendritic structure are the results on mechanical properties measured in three directions of forgings from a 1933 grade alloy obtained from large-scale ingots of 845 mm in diameter (Table 16).

Table 16. Effect of Nondendritic Structure of Ingots[a] from a 1933 Alloy on Mechanical Properties of Large-Scale Forgings

	Nondendritic Structure[b]			*Dendritic Structure*[b]		
Property	*L*	*C*	*H*	*L*	*C*	*H*
σ_b (MPa)	476	476	463	449	488	462
$\sigma_{0.2}$ (MPa)	410	406	400	407	414	405
δ (%)	17.6	15.2	8.8	10.5	11.7	5.6
ψ (%)	52.2	36.0	30.0	41.0	25.0	10.1
K_{1C} (MPa $m^{0.5}$)	54.9	—	—	41.4	—	—

[a]Ø 845 mm.

[b]L - longitudinal direction; C - cross direction; and H - height direction.

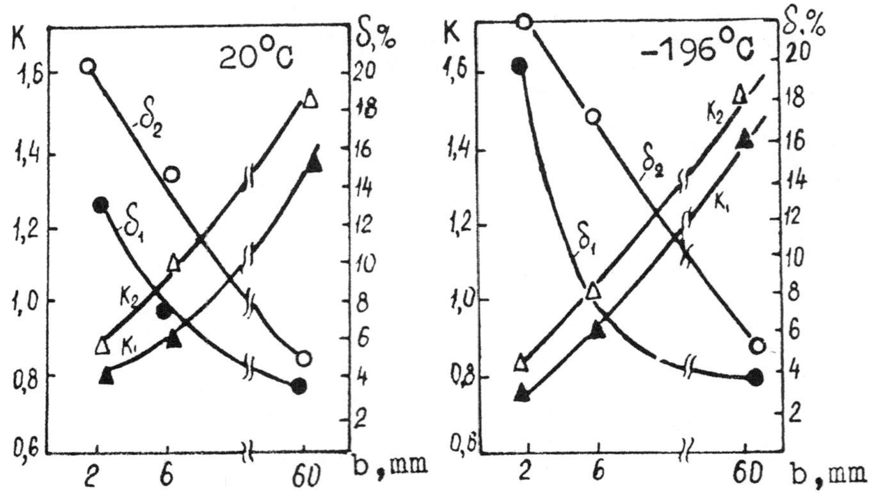

Figure 26. Effect of a nondendritic structure of an ingot and thickness of rolled semiproducts from a 1201 grade alloy on elongation (δ) and notch sensitivity (K) at 20 and –196 °C. (**1**) Semiproducts produced from ingots with a dendritic structure; (**2**) Those from ingots with a nondendritic structure.

In recent years interest in welded constructions from aluminum alloys working at cryogenic temperatures (below –196 °C) has growing [70]. It has been shown that the nondendritic structure of an ingot favorably affects mechanical properties of welded joints at 20 and –196 °C.

Tests of notched specimens (Figure 26) and determination of notch sensitivity coefficient (K) show that, despite thickness of rolled semiproducts, the K_1 coefficient (for a dendritic structure) is less than K_2 (for a nondendritic structure) by 20–25% for room and cryogenic temperatures. Thus, $K_2 \geq 1.5$ for plates and $K_2 = 1$ for thick sheets, which indicates their full insensitivity to a notch.

An effect of the nondendritic structure on properties of welded joints from a 1201 grade alloy was also studied. Welding wires produced from ingots with dendritic and nondendritic structures were used for welding of 3-mm sheets. It has been revealed that hot cracking is significantly reduced when using welding wires with a nondendritic structure. The welding rate can be increased from 7.0 to 10.0 mm min^{-1}. The bending angle and resilience absorbed energy increase from 68 to 84° and from 18 to 24 kJ m^{-2}, respectively.

5. CONCLUSION

The considered results of studies on the ultrasonic treatment of light alloy melts in the developed cavitation mode allow us to choose the most practicable branches of application of this nonpolluting advanced method of influence.

The ultrasonic methods of degassing and filtration should be used when there is a requirement for high purity in light alloys. An example of this is welding alloys of the Al–Mg system where there is a direct connection between the content of hydrogen and oxides and the quality of welded joints. The technology is of particular interest for the production of the base of hard disks (for computers) from alloys of this type.

The technology of ultrasonic degassing and filtration has a number of advantages for the production of light alloys containing volatile components (e.g. the Al–Zn–Mg–Cu system and some others) when effective vacuum degassing is hampered. An appropriate use of ultrasonic technology is in the shape casting of high-strength aluminum and magnesium alloys, used at strong static and dynamic loads. The combined ultrasonic treatment and cleaning from solid and gas impurities increase the fluidity of a metal and provide good filling of thin sections of castings.

The technology of the ultrasonic treatment of a solidifying melt with formation of the refined nondendritic grain in ingots and castings may be considered as one of the main trends in modern light alloy metallurgy. The use of acoustic cavitation as a mean of involving potential solidification nuclei in the solidification process (wetting of solid particles of impurities and modifiers and change of the temperature gradient near the solidification front in a liquid metal) results in formation of the nondendritic structure in most of light alloys. Thus, the ultimate degree of grain refinement may be obtained with sequential solidification of ingots and castings. The difference between the dendritic and nondendritic structures is in the following; instead of dendritic grains consisting of dendrite arms with small-angle boundaries and large-angle grain boundaries, nondendritic grains with large-angle boundaries without inner branching are formed.

The transition to the nondendritic solidification is very significant because of the strong influence on the deformation ability of a cast metal. The nondendritic structure, having virtually no effect on strength, sharply increases technological plasticity, which allows one to increase sizes of quality ingots from high-strength alloys without cracking and to improve deformation ability of the ingots with deformation. Finally, the nondendritic structure positively affects properties of hot worked semiproducts [71].

The regularity of the change of the solidification mode from dendritic to nondendritic has been experimentally confirmed in a broad range of solidification rates—from slow cooling of a large mass of a metal to rapid solidification of dispersed granules.

The ultrasonic methods of degassing, filtration, and grain refinement [72] can be successfully applied not only to aluminum and magnesium alloys but to other nonferrous alloys (on Sn, Zn, Cu etc. bases). For example, degassing and grain refinement of a melt of typographical zinc, using the ultrasonic treatment [73], provides not only increasing technological plasticity with casting and deformation but significant improvement of quality of printing sheets.

The wide industrial use of the ultrasonic treatment of melts requires further development of ultrasonic technique and equipment. Most of industrial sources of ultrasound work with automatic adjustment of frequency. Expansion of fields of application of the ultrasonic treatment requires the development of automatic control systems with automatic adjustment of amplitude, which may require generators and transducers of increased power.

APPENDIX

The correlation between Russian and U.S. standards of alloy grades mentioned in the text is given below.

Alloy System	*Russian Standard*	*U.S. Standard*
	Wrought Alloys	
Al	A7, A99	1060, 1199
Al–Mg–Mn	AMg2 (1520), AMg5 (1551), AMg6 (1560)	5052, 5056
Al–Cu–Mg	D16ch (1163), 1161	2124
Al–Cu–Mg–Mn–Si	AK6 (1360), AK8 (1380)	2014
Al–Zn–Mg–Cu	V95pch (1951), 1973, 1933, 1960	7475, 7012, 7049
Al–Cu–Mn	1201	2219
Mg–Al–Zn	MA2-1pch	AZ61A
Mg–Zr–Ce	MA3-3	—
	Cast Alloys	
Al–Si	AL4	A359
Al–Si–Mg	AL9	A356
Al–Si–Cu	AL3	A305
Al–Si–Cu–Fe–Ni	AL40	—
Al–Cu–Mn	AL19	A242

NOTES

1. The developed technique referred to as the UZFIRALS process is a part of the invention patented in USA (Pat. No. 4564059); Great Britain (Pat. No. 2100635); Switzerland (Pat. No. 651253); France (Pat. No. 2507933); Canada (Pat. No. 1194675); Australia (Pat. No. 540333); Germany (Pat. No. 3126590); Norway (Pat. No. 160832) and in other developed countries.

2. The technology of continuous casting with the ultrasonic treatment and formation of a nondendritic structure of an ingot is a part of the same patent as the UZFIRALS-process (see footnote 1).

REFERENCES

[1] Hiedemann, E.A. Metallurgical effects of ultrasonic waves. *J. Acoust. Soc. Am.*, 26 (1954) 831–842.

[2] Crowford, A.E. *Ultrasonic Engineering*. Butterworths Scientific, London, 1955.

[3] Eskin, G.I. *Ultrasound in Metallurgy*. Metallurgizdat, Moscow, 1960 (in Russian).

[4] Eskin, G.I. *Ultrasound Took a Step to Metallurgy*. First Edition, 1970; Second Edition (enlarged and revised), Metallurgia, Moscow, 1975 (in Russian).

[5] Eskin, G.I. *Ultrasonic Treatment of Molten Aluminium*. First Edition, 1965; Second Edition (enlarged and revised), Metallurgia, Moscow, 1988 (in Russian).

[6] Abramov, O.V. *Solidification of Metals in an Ultrasonic Field*. Metallurgia, Moscow, 1972 (in Russian).

[7] Campbell, J. Effects of vibration during solidification. *Int. Met. Rev.,* No. 2 (1981) 71–108.

[8] Abramov, O.V. and Teumin, I.I. Solidification of metals. In *Physics and Techniques of Powerful Ultrasound. Physical Basics of Ultrasonic Technique*. Nauka, Moscow, 1970 Vol. 3, pp. 427–514 (in Russian).

[9] Balandin, G.F. *Formation of Crystal Structure of an Ingot*. Mashinistroenie, Moscow, 1973 (in Russian).

[10] Richards, R.S. and Rostoker, W. The influence of vibration on the solidification of an aluminium alloy. *Trans. Am. Soc. Met.*, 68 (1956) 884–903.

[11] Seemann, H.J. and Menzel, H. *Z. Metallkd.*, No. 1 (1947) 39–43.

[12] Dobatkin, V.I. and Eskin, G.I. Ultrasonic treatment of melts of nonferrous metals and alloys. In *Action of Powerful Ultrasound on Metal Interfaces*. Nauka, Moscow, 1986 pp. 6–51 (in Russian).

[13] Agranat, B.A., Dubrovin, M.A., Khavskii, N.N., and Eskin, G.I. *Basics of Physics and Technique of Ultrasound*. Vysshaya Shkola, Moscow, 1987 (in Russian); Mir, Moscow, 1990 (in Spanish); Nisso Cusincy, Tokyo, 1993 (in Japanese).

[14] Dobatkin, V.I. and Eskin, G.I. Casting using ultrasonic treatment. In *Special Methods of Casting*. Mashinistoenie, Moscow, 1991 pp. 448–489 (in Russian).

[15] Buxmann, K. *Metallwissenschaft und Technik*, 25 (1971) 127–133.

[16] Angelov, G. Application of ultrasound in casting industry. In *Application of Ultrasound in Industry*. Sofia, Tekhnika, 1975, pp. 28–82 (in Bulgarian).

[17] Krátky, J. et al. *Trend VUMA*, 10 (1979) 28–39.

[18] Bondarek, Z. et al. *Przeglad Odlewnictwa*., No. 7 (1981) 229–232.

[19] Lane, D.H., Cunningham, J.W., and Tiller, W.A. *Metal Progress*, 76 (1959) 108–110.

[20] Mortensen, A., Cornie, J.A., and Flemings, M.C. *J. Met*., 40 (1988) 12–19.

[21] Suslick, K.S. *Science*, 247 (1990) 1939–1445.

[22] Rai, G., Lavernia, E., and Grant, N.I. *J. Met*., 37 (1985) 22–26.

[23] Flemings, M.C. *Solidification Processing*. McGraw-Hill Book, New York, 1974.

[24] Notlingk, B.E. and Neppiras, E.A. *Proc. Phys. Soc*., 63B (1950) 674–685.

[25] Chalmers, B. *Principles of Solidification*. John Wiley and Sons, London, 1964.

[26] Mason, T.J. and Lorimer, J.P. *Somochemistry: Theory, Application and Uses of Ultrasound in Chemistry*. Elis Horwood Ltd., Chichester, 1989.

[27] Lin Zhunjmao. *Wuli*, 8 (1979) 467–471.

[28] Frenkel', Ya.I. *The Kinetic Theory of Liquids*. Acad. Sci. USSR Publ. House, Moscow, 1959, Vol. 3 (in Russian).

[29] Harvey, E.W., Whitely, A.N., McElroy, W.D. et al. *Trans. Am. Chem. Soc*., 67 (1945) 156–162.

[30] Dobatkin, V.I., Gabidullin, R.M., Kolachev, B.A., and Makarov, G.S. *Gases and Oxides in Aluminium Wrought Alloys*. Metallurgia, Moscow, 1976 (in Russian).

[31] Sirotyuk, M.G. Experimental study of ultrasonic cavitation. In *Powerful Ultrasound Fields*. Nauka, Moscow, 1968, pp. 167–220 (in Russian).

[32] Eskin, G.I. and Shvetsov, P.N. *Tekhnol. Legk. Spl*., No. 1 (1971) 25–30.

[33] Abramov, O.V. and Astashkin, Yu.S. On peculiarities of ultrasound radiation in melts under cavitation conditions. In *Better Metallurgical Processes through Novel Physical Methods*. Metallurgia, Moscow, 1974, pp. 155–160 (in Russian).

[34] Margulis, M.A. *Sonochemical Reactions and Sonolumenescence*. Khimiya, Moscow, 1986 (in Russian).

[35] Kuttruff, H. *Acoustica*, 12 (1962) 230–235.

[36] Margulis, M.A., Grundel', L.M., Eskin, G.I., and Shevtsov, P.N. *Doklady Akad. Nauk SSSR*, 295 (1987) 1170–1173.
[37] Eskin, G.I., Ioffe, A.I., and Shevtsov, P.N. *Tekhnol. Legk. Spl.*, No. 1 (1974) 8–11.
[38] Knapp, R.T., Daily, J.W., and Hammitt, F.G. *Cavitation*. McGraw-Hill Book, New York, 1970.
[39] Lauterborm, W. and Hentschel, W. *Ultrasonics*, 23 (1985) 260–268; 24 (1986) 59–65.
[40] Grivnin, Yu.A., Zubrilov, S.P., and Larin, V.A. *Zh. Fiz. Khim.*, 54 (1980) 56–59.
[41] Boguslavskii, Yu. Ya. *Zh. Akust.*, 13 (1967) 23–26.
[42] Abramov, O.V. and Astashkin, Yu.S. On the estimation of pressures arising with cavitation development in light alloys melts. In *Better Metallurgical Processes through Novel Physical Methods*. Metallurgia, Moscow, 1974, pp. 161–167 (in Russian).
[43] Margulis, M.A. and Dmitrieva, A.F. *Zh. Fiz. Khim.*, 55 (1981) 159–163.
[44] Rozenberg, L.D. Cavitation region. In *Powerful Ultrasonic Fields*. Nauka, Moscow, 1968, pp. 221–265 (in Russian).
[45] Kikuchi, Y. *Ultrasonic Transducers*. Corona Publ., Tokyo, 1969.
[46] Bertnik, Yu.N., Trizna, Yu.P., Panov, L.I. et al. Studies on cavitation in a molten solder. In *Better Metallurgical Processes through Novel Physical Methods*. Metallurgia, Moscow, 1974, pp. 166–170 (in Russian).
[47] Makarov, G.S. *Refining of Aluminium Alloys with Gases*. Metallurgia, Moscow, 1983 (in Russian).
[48] Eskin, G.I., Shvetsov, P.N., and Ioffe, A.I. *Izv. Akad. Nauk SSSR. Metally*, No. 6 (1972) 69–74.
[49] Eskin, G.I., Slotin, V.I., and Katsman, S.T. *Precise Casting of Aircraft Details from Aluminium Alloys*. Mashinostroenie, Moscow, 1967 (in Russian).
[50] Novitskii, B.G. *Application of Ultrasonic Oscillations in Chemical-Technological Processes*. Khimiya, Moscow, 1983 (in Russian).
[51] Herrmann, E. and Hoffmann, D. *Handbook of Continuous Casting*. Aluminium-Verlag, Düsseldorf, 1980.
[52] Eskin, G.I. *Tsvetn. Met.*, No. 11 (1981) 35–40.
[53] Eskin, G.I. and Shvetsov, P.N. An analysis of efficiency of ultrasonic degassing of a melt with continuous casting of aluminium alloys ingots. In *Physical Metallurgy and Casting of Light Alloys*. Metallurgia, Moscow, 1977, pp. 17–31 (in Russian).
[54] Makarov, G.S., Pimenov, Yu.P., and Eskin, G.I. Regularities of filtration of aluminium alloys melts through fiber-glass multilayer filters. In *Physical Metallurgy and Technology of Light Alloys*. VILS, Moscow, 1990, pp. 135–143 (in Russian).
[55] Eskin, G.I. *Mechanisms of Fine Filtration of Molten Metals in Acoustic Cavitation Field*. Abstr. 2nd Meeting of European Soc. Sonochem. (ESS), Sept. 1991, Gargnano, Italy, 1991, pp. 62–63.
[56] Kitaigorodskii, Yu.P. and Drozhalova, V.I. Calculation of height and raise rate of liquid in capillaries under action of ultrasonic oscillations. In *Application of Ultrasound in Metallurgy*. Metallurgia, Moscow, 1977, pp. 12–16 (in Russian).
[57] Prokhorenko, P.P., Dezhkunov, N.V., and Konovalov, G.E. *Ultrasonic Capillary Effect*. Nauka i Tekhnika, Minsk, 1981 (in Russian).
[58] Eskin, G.I., Kudryashov, V.G., Shvetsov, P.N., and Kus'minskaya, Z.K. *Izv. Akad. Nauk SSSR. Metally*, No. 1 (1990) 53–56.
[59] Dobatkin, V.I. *Aluminium Alloys Ingots*. Metallurgizdat, Moscow, 1960 (in Russian).
[60] Danilov, V.I. *Constitution and Solidification of Liquids*. Izd. Akad. Nauk Ukr. SSR, Kiev, 1956 (in Russian).
[61] Abramov, O.V. The action of ultrasound on solidifying metals. In *Advances in Sonochemistry*. JAI Press Ltd., 1991, Vol. 2, pp. 135–186.
[62] Dobatkin, V.I. and Eskin, G.I. *Mechanisms of Nondendritic Solidification during Continuous Casting of Aluminium Alloys Ingots*. Proc. 2nd Int. Conf. Alum. Alloys, Oct. 1990, Beijing, Int. Acad. Publ., Beijing, 1990, pp. 278–282.
[63] Dobatkin, V.I. and Eskin, G.I. *Tsvetn. Met.*, No. 12 (1991) 64–67.
[64] Dobatkin, V.I. and Eskin, G.I. *Protsessy Lit'ya* (Kiev, *Ukr. Akad. Nauk*), No. 1 (1992) 31–37.

[65] Dobatkin, V.I., Eskin, G.I., Borovikova, S.I. et al. Regularities of structure formation in aluminium alloys ingots during continuous casting with the ultrasonic treatment of a solidifying melt. In *Processing of Light and Refractory Alloys*. Nauka, Moscow, 1976, pp. 151–161 (in Russian).

[66] Kattamis, T. and Wiliamson, R.B. *J. Inst. Met*., 96 (1968) 251–252.

[67] Kattamis, T., Holmberg, U.T., and Flemings, M.C. *J. Inst. Met*., 95 (1967) 343–347.

[68] Patterson, R.J., Cox, A.R., and van Reuth, E.C. *J. Met*., 32 (1980) 34–39.

[69] Matyia, H., Giessen, B.C., and Grant, N.I. *J. Inst. Met*., 96 (1968) 30–32.

[70] Eskin, G.I., Pimenov, Yu.P., Belova, M.V. et al. *Room Temperature and Cryogenic Properties of Rolled Semis and Welded Joints Produced from 1201 Alloy Cast Billet with Nondendritic Structure*. Abstr. 14th Int. Cryogen. Mater. Conf., June 1992, Kiev, Paton Electr. Weld. Inst., Kiev, 1992, p. 81.

[71] Eskin, G.I., *Ultrasonics Sonochem*., Vol. 1 (1994) 59–63.

[72] Eskin, G.I., *Ultrasonics Sonochem*., Vol. 2 (1995) 131–141.

[73] Eskin, G.I., Gur'ev, I.I., Soluyanov, Yu.F. et al. *Tsvetn. Met*., No. 1 (1981) 68–71.

SONOCHEMISTRY IN CHINA

Y. Zhao, C. Bao, J. Yin, and R. Feng

OUTLINE

Advances in Sonochemistry
Volume 4, pages 161–175

ISBN: 1-55938-793-9

1. INTRODUCTION

In China, as in other parts of the world, sonochemistry has been studied widely, especially since the 1980s, but the majority of these studies are not well known outside mainland China because most of the study results were published in Chinese science journals and, naturally, in the Chinese language. Therefore, we have undertaken the assembly of this review on the basis of searching information from the Chinese science literature since 1970. The review contains over 90 references on the varied studies on Chinese sonochemistry. The reports involve both fundamental studies and applications of power ultrasound in polymer science (degradation and copolymerization), organic synthesis, chemical analysis, the regeneration ion exchange resins, plating, and several others.

It is our wish to promote the cooperation and technology exchange between the West and East by providing this introduction to the study of sonochemistry in China for specialists in other parts of the world whose mother language is not Chinese. We believe that with the involvement of Chinese sonochemists, the study of sonochemistry would not only be more international but also lead to more rapid progress in the subject.

2. REVIEWS, EQUIPMENT, AND RESEARCH FUNDS

The study of the application of ultrasound to chemical processes began in the 1950s in China [1,2] but it did not receive much attention from Chinese chemists for many years. Part of this neglect may have been due to the fact that the term was often misunderstood due to its Chinese pronunciation being *Shenghuaxue*, which has, unfortunately, the same sound as biochemistry.

Since the early 1980s, based upon the rapid expansion of the number and scope of papers on sonochemistry generated across the world, several scientific reviews written in Chinese began to introduce periodically some of the achievements of Chinese chemists in this field. Among these there were six reviews on the application of ultrasound to organic synthesis [3–7], one on the organic chemistry of fluoro-compounds [8], three on advances in the design of sonoreactors [9–11], two on the basic mechanism and theoretical investigation of sonochemistry [12–13], seven on applications of ultrasound in chemistry and the chemical industry [14–20], one on copolymerization [21], and one on biochemistry [22]. The first Chinese book devoted to sonochemistry and its application was published in 1992 [23].

It was these reviews and the book that encouraged more and more Chinese chemists, especially those not versed in a foreign language, to first of all understand and subsequently to become interested in sonochemistry and to begin to use ultrasound in their research.

It is certainly true that the manufacturing technology for ultrasonic equipment has improved greatly in the last few years in step with the advancement of electronic and material science in China. Many ultrasonic instrument factories were built in large cities, such as the Wuxi Ultrasonic and Electronic Equipment Factory and the Shanghai Ultrasonic Instrument Factory which are currently the largest. Many models of ultrasonic cleaning baths and probes on both a laboratory and an industrial scale have become available.

Probably the most important factor in terms of university research is that sonochemistry has received a great deal of attention from the Chinese government and science departments. The Committee of the National Natural Science Foundation of China has now begun to provide research funds for studies in this field.

It is the combination of all of the above mentioned factors that has established a firm basis for the development of sonochemistry in China.

3. POLYMER SCIENCE

Polymer science is one of the earliest studied fields in sonochemistry in China with most of the studies carried out by a group in Chengdu. Supported by the National Natural Science Foundation of China, the group carried out a series of studies on the copolymerization and degradation of polymers under sonication [24–33]. The polymers which were produced by the group included some potential novel products for use in the development of oil fields [24,25].

3.1 Polyethylene Oxide and Polymethacrylate Copolymer Systems

The mechanism and kinetics of the degradation of polyethylene oxide (PEO) in water, benzene, and chloroform and its copolymerization with sodium methacrylate (NaMA) in water under sonication were studied using a 21-kHz ultrasonic probe [26]. The degradation rates were directly proportional to the vaporization enthalpy and viscosity of solvent. In a further study [27], the kinetics and mechanism of block copolymerization of PEO with NaMA in water under 21.5-kHz ultrasonic irradiation was studied. The ultrasonic copolymerization of PEO–NaMA in aqueous solution follows the kinetic relationship:

$$-d[M]/dt = k[R\cdot]^{1/6}[M]$$

As irradiation time or monomer concentration increased, the copolymer yield and the NaMA content in the copolymer increased. The structure of the copolymer was identified as having a block form by infrared, nuclear magnetic resonance, mass spectrometry, etc.

The heterogeneous copolymerization of PEO with hexyl methacrylate (HMA) was also studied under 21.5-kHz ultrasonic irradiation [28]. The ultrasonic irradiation degraded PEO into macroradicals which initiated the polymerization of HMA and formed PEO–HMA. The copolymer was again identified as a block copolymer by the methods quoted above.

3.2 Polyvinyl Alcohol and Polyacrylonitrile Copolymer Systems

The degradation of polyvinyl alcohol (PVA) and the copolymerization of PVA with acrylonitrile (AN) under 21.5-kHz ultrasonic irradiation was studied [29]. Both water-soluble and water-insoluble copolymers were obtained by changing the irradiation time or the amount of AN. The structure of the copolymer was again identified as having a block form by the usual techniques. A 25.4% yield of water-soluble copolymer (containing 14.0% AN) was obtained from 2% aqueous mixture of PVA and AN (1:1.6 by mass) under 21.5-kHz, 490-W ultrasonic irradiation at 19–21 °C for 28 min. A water-insoluble copolymer (containing 75.6% AN) was obtained using an aqueous mixture containing 2% PVA and AN (1:4 by mass) when the irradiation period was extended to 100 min.

Furthermore, the ultrasonic copolymerization was studied in terms of pH [30]. It was found that with a decrease in pH, both the copolymer yield and AN content in the copolymer were increased. In a later study, the mixture ultrasonically irradiated was a stable water-based dispersion and the particle size of the copolymer increased with decreasing pH. No acrylonitrile homopolymer was found in the reaction product.

3.3 Polyacrylamide and Acrylonitrile/Polyvinylacetate Copolymer Systems

The degradation of partially hydrolyzed polyacrylamide (PAM) and its copolymerization with AN under 18.2-kHz ultrasonic irradiation has been studied [31]. The copolymer obtained was water-soluble when the acrylonitrile content was less than 12% and the degree of hydrolysis of PAM was 16.1%. It was discovered that under sonication partially hydrolyzed PAM was degraded more easily than PAM itself.

In a later study, the ultrasonic degradation and copolymerization of partially hydrolyzed PAM and polyvinyl acetate (PVAc) in ethanol/water (55/45) was studied [32]. The degradation rate of PVAc follows Ovenall's equation approximately. Sonication of 0.50% aqueous mixture of PVAc and partially hydrolyzed PAM (1:1 by weight) at 25 °C using an 18.2-kHz probe for 10 min gave a yield of copolymer which amounted to 4.3%.

3.4 Other Polymer Systems

The degradation of methyl cellulose and its copolymerization with AN under sonication has been studied [33]. This degradation resulted from the cavitational effects of ultrasound but the degradation rate was not found to increase with irradiation time. A water-soluble copolymer was obtained by irradiating methyl cellulose and AN in a mixed H_2O/HCOOH solvent at 15 °C for 30 min using 21.5-kHz ultrasound. The copolymer was identified as both block and branch with joints mainly located at methoxyl groups of methyl cellulose by means of differential thermal analysis (DTA), IR, gel permeation chromatography (GPC) and mass spectrometry (MS).

The degradation and copolymerization of hydroxy cellulose (HEC) and PEO under ultrasonic irradiation has also been studied [34]. The degradation of HEC and PEO increased with the increase of ultrasonic intensity and decrease of solution concentration. The yield of copolymer reached maximum 55.07% by irradiating 0.5% HEC/PEO for 10 min at 25 °C with 18.2-kHz ultrasound. The copolymer was identified as a mainly block by DTA, IR, MS, X-ray diffraction, and polarizing microscopy.

The ultrasonic degradation of Chitosan (deacetochitin, found in crab shells) was studied by another research group [35]. Chitosan was degraded in acetic acid at 60 °C under ultrasonic irradiation. The viscosity of the Chitosan decreased by 80% after irradiation for 15 h, compared with 55% decrease by reflux under acid condition for 24 h. This was taken as evidence that the degradation under sonication is faster, purer, and more complete than under conventional conditions.

4. ORGANIC SYNTHESIS

4.1 Solid–Liquid System Involving Carbene

In this field, a group in Shanghai studied three-membered ring carbenes and dihalogeno-carbenes. Financial support was given by the National Natural Science Foundation of China in 1987. They studied the reaction of eight *gem*-dihalogeno-cyclopropanes with lithium, magnesium, or sodium under sonication using a 20-kHz ultrasonic bath [36,37]. Three-membered ring carbene intermediates were rapidly formed under sonication and further changed into propadienes derivatives through rearrangement [Eq. (1)] or into bicyclo-[4,1,0]-butanes through insertion into the adjacent C–H. The effect of sonication for such systems is very marked; for example, the reaction of 7,7-dibromobicyclo[4,1,0]heptane and magnesium in THF at ambient temperature under nitrogen occurs immediately and goes to completion in 15–20 min [Eq. (2)]. Under silent conditions there is no product even after 8 h under otherwise identical reaction conditions.

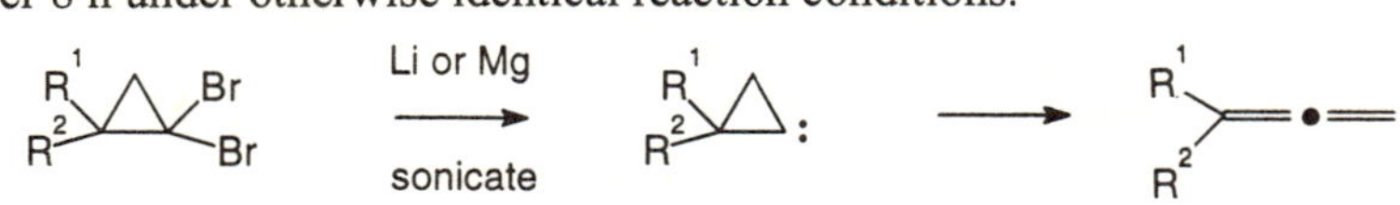

(1)

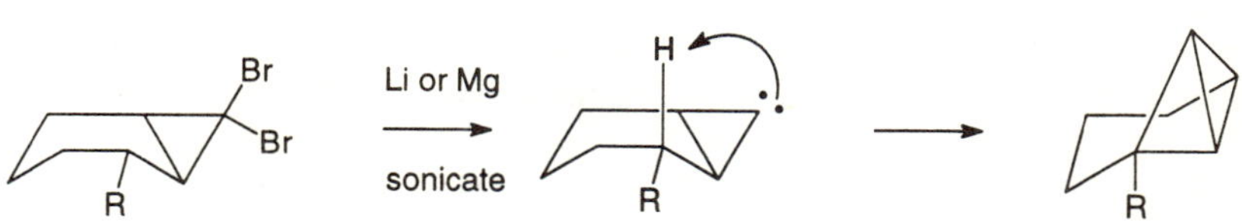

(2)

In order to verify the presence of carbene intermediate during the process, the reaction of 7,7- dibromobicyclo[4,1,0]heptane with lithium or magnesium was studied in the presence of alkene under sonication [38]. The additional products of the three-membered ring carbene with cyclohexene, vinyl ethyl ether, and dihydropyrane were identified in the products. The study provided a new, simplified method for the preparation of spiro-cyclopropane derivatives [Eq. (3)].

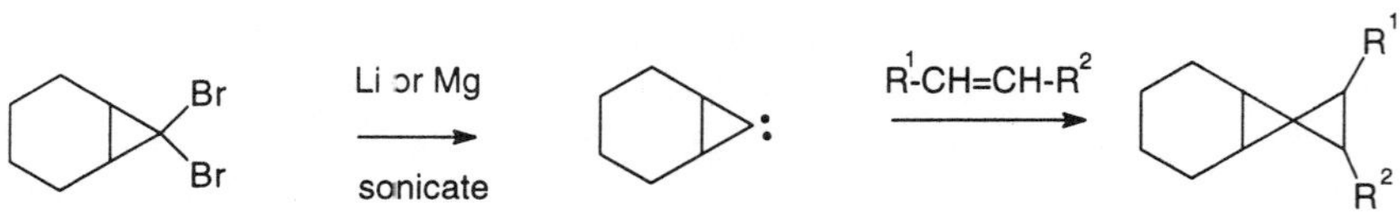

(3)

The mechanism of the reaction of *gem*-dihalogenocyclopropane and metal under sonication was studied in a further paper by this group [39]. It was thought that the 1-halogenocyclopropane free radical might be produced as an intermediate through the transference of a single electron and the loss of a halogen ion [Eq. (4)]. The carbene would result from the addition of an electron and the loss of the halogen ion. The carbene could then undergo further insertion, addition, or ring opening and rearrangement to yield corresponding derivatives.

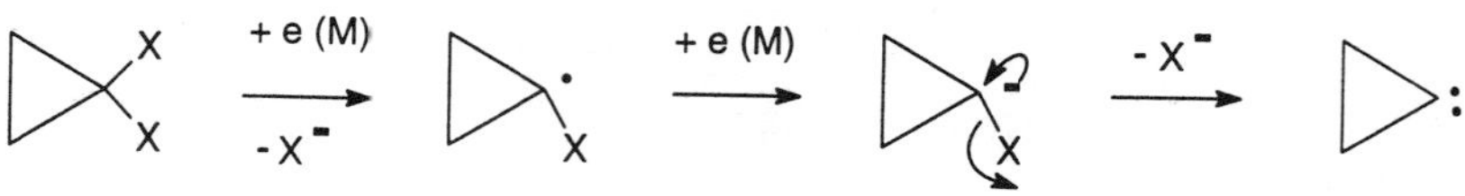

(4)

Further studies showed that using a combination of sonication and phase-transfer catalyst (PTC) the rate and yield of the reaction of alkene with dichlorocarbene which resulted from chloroform and sodium hydroxide pellets *in situ* could be improved efficiently [40]. Compared with the results reported by Regen [41] where sonication alone was used, they found that mechanical stirring was not necessary under high-power sonication. Other findings included the fact that the ratio of NaOH:alkene could be decreased from 10:1 to 3:1 and the reaction period could be shortened from 5 h to 10–15 min in the presence of 0.1–0.05% PTC.

A similar method was used to generate dibromocarbene using a combination of sonication and PTC for the reaction of alkene with dibromocarbene formed from bromoform and solid sodium hydroxide *in situ* [42]. Compared with the preparation of dichlorocarbene, however [39], the amount of PTC required had to be increased 10-fold. The combination of the effect of sonication and PTC was significant in that the presence of PTC allowed a shortening of the reaction period to 20–30 min and an increase in yield to 96% compared with 3 h and 40–50% using sonication alone.

An interesting result is given in a paper provided by this group on the reaction of benzaldehyde with dichlorocarbene generated *in situ* from chloroform and solid NaOH under sonication in the presence of PTC [43]. Normally α-hydroxybenzoic acid is the additional product in this reaction [44], but under sonication, benzoic acid and benzyl alcohol were obtained as the products of an ultrasonically induced Cannizzaro reaction. This work provides another example of the sonochemical switching of an established chemical reaction route.

4.2 Other Solid–Liquid Systems

The sonochemistry group in Shanghai have also reported that the fragmentation of aryl *tert*-phosphines in the presence of lithium can be promoted effectively using ultrasonic irradiation [45]. Triphenylphosphines could almost quantitatively be changed into lithium diphenylphosphides and phenyl lithiums in the presence of lithium metal under ultrasonic irradiation [Eq. (5)]. Sonochemical reaction was accomplished in 25 min at ambient temperature with no induction period. The reaction rate was increased by a factor of 6, compared with silent conditions. The method has been used for the preparation of some organophosphine ligands.

$$Ph_3P \xrightarrow[\text{sonicate 25 min}]{\text{Li / THF}} Ph_2PLi + LiPh \qquad (5)$$

The epoxidation of alkenes by means of either sodium perborate or sodium percarbonate and acetic anhydride can be enhanced by sonication [46]. When sodium perborate was used with sonication, the reaction time could be reduced significantly, but the yield was not significantly improved. With sodium percarbonate both a higher yield and a shorter reaction time were obtained. The mechanistic aspects of these reactions were discussed.

The cyclopropanation of dimethyl maleate with dribromomethane catalyzed by Co[0] or Ni[0] complexes generated *in situ* under sonication has been reported by a group in Changchun supported by the National Natural Science Foundation of China [47]. Ultrasound was found to accelerate the formation of the intermediate complexes, Co(*E*)-$CH_3O_2CCH{=}CHCO_2CH_3)_2(CH_3CN)_2$ and Ni(*Z*)-CH_3O_2-C CH=CH-$CO_2CH_3)_2(CH_3CN)_2$, and increase the reaction yield to 94% in 30 min, compared with 60–70% in several hours without sonication (Equation 6). It is interesting that if the reaction involving Ni is performed at lower temperature (0 °C), it is faster than when it is performed at the higher temperature of 50 °C.

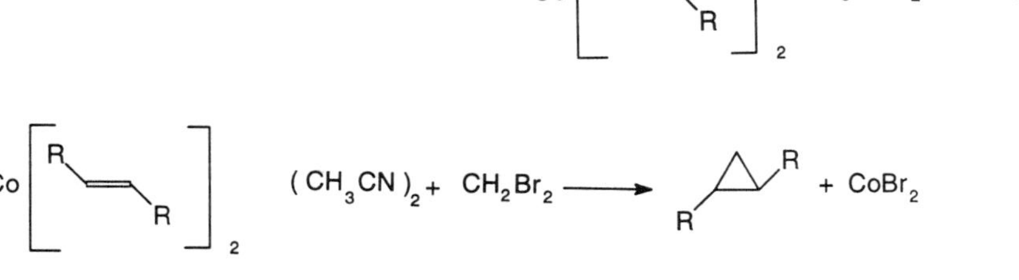

(6)

It was reported by a group from Hong Kong that the cyclopropanation of ethylenic fatty esters and triglycerides in the presence in zinc could be facilitated by sonication using a cleaning bath [48].

Azone, a penetration enhancer of a drug to the skin, was synthesized under 24-kHz ultrasonic irradiation at 58–62 °C in the presence of PTC by a group in Kunming [49]. The yield of azone was 71.8% under sonication for 6 h, compared with 71.5% within 10 h without ultrasound.

The effect of ultrasound on the hydrolysis of aluminum *iso*-propoxide has been studied [50]. The hydrolysis resulted in aluminum oxide particles whose size was 200–600 Å. The reaction was accelerated 10-fold by sonication and the hydrolysis rate could be increased further with increased ultrasonic intensity. This study provided a potential method for the preparation of ultrafine powders.

Some chemicals pretreated by ultrasound are known to be highly activated and more effective. A group in Taiwan performed interesting work studying the stereoselective reaction of ultrasonically dispersed potassium with organic sulfur compounds using a cleaning bath [51–54]. In a similar study, Zhao et al. reported the benefit of using presonication $CuBr_2/Al_2O_3$ in bromination of naphthalene [55].

4.3 Liquid–Liquid System

An example of using sonication and PTC to improve an aromatic nucleophilic substitution reaction has been provided by Wu et al. [56]. Nine diphenyl ether compounds were synthesized from chloronitrobenzene and alkyl-substituted phenols with higher yields and shorter reaction period (Equation 7).

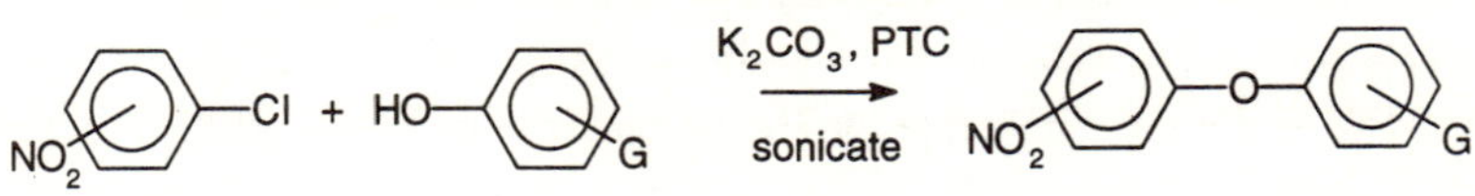

(7)

Another paper concerning the joint use of sonication and PTC to improve chemical reaction reported that eight alkoxy- or aryloxy-propanonitriles were synthesized by the addition of alcohols to acrylonitrile (Equation 8) [57]. In general the yield for such reactions was increased by 20–30% and the amount of PTC was decreased. It was also observed that waxy solid tetradecanol could be easily dispersed and reacted with propanonitrile at a temperature below its melting point using sonication.

$$ROH + CH_2{=}CHCN \underset{\text{sonicate}}{\overset{\text{PTC}}{\longrightarrow}} ROCH_2CH_2CN \qquad (8)$$

The oxidation of octanol with HNO_3 under sonication was studied by a group in Nanjing [58]. It was found that sonication for 5 min at 0 °C resulted in 84% capric acid but, in the absence of sonication, even after a reaction time of 2 h only 10% octanol nitrate and no capric acid was obtained. This is another example which shows that ultrasound can change the established chemical reaction route.

The group in Kunming have reported a synthetic approach to *d,l*-mandelic acid from benzaldehyde and aqueous solution of NaOH under the influence of ultrasound [59].

5. IN CHEMICAL ANALYSIS

An ultrasonic nebulizer has been designed and used for inductively coupled plasma atomic emission spectrometry [60] and microwave induced plasma–atomic emission spectrometry [61]. The apparatus is inexpensive and can be operated conveniently. Using this nebulizer, the detection limits of many elements, such as phosphorus, aluminum, and silver, were much reduced compared with the limits obtained using an aerodynamic nebulizer [62–64]. The ultrasonic nebulizer was found to be suitable for samples which have a high salt concentration.

A new separation and analytical procedure for the α and β phases in K136 alloy using ultrasonic sieving apparatus has also been reported [65].

6. IN BIOCHEMISTRY

Generally, low intensity ultrasound can increase the activity of enzymes and immobilized enzymes, or can improve the metabolism of cells by the improvement of mass transfer to substrates. One example is the enzymatic decomposition of mandelonitrile by mandelonitrile lyase in di-isopropyl ether which has been accelerated 10-fold by ultrasound [66]. The mandelonitrile lyase was not denatured by 30 kHz operating at 0.5 W cm^{-2} and pulsed at 15 s per min.

The absorption of trypsin onto a copolymer prepared from polystyrene and glycidyl methacrylate under ultrasonic irradiation was reported to be a simple and low-cost route to the immobilized enzyme [67].

7. REGENERATION WASTE ION EXCHANGE RESIN AND CHARCOAL

Two papers concerning pilot scale experiments have reported that waste ion exchange resin can be regenerated by ultrasonic cleaning [68,69]. Under the powerful cavitational effects induced by ultrasound, impurities which were absorbed on the ion exchange resin could be easily removed. This process extended the useful life of the ion exchange resin. The cleaning period of the resin was shortened and less chemicals were needed resulting in less pollution in the environment.

It has been reported that active charcoal from waste industrial catalyst $Zn(OAc)_2/C$ could be recovered by ultrasonic washing [70]. After ultrasonic washing in water for 15 min at ambient temperature followed by calcining at 650 °C, the zinc acetate could be removed effectively and the regenerated active charcoal was found to have a quality up to the standard of Forestry Department of

China, LY216-9. By comparing the DTA and TG curves of the active charcoal regenerated with hot water washing and ultrasonic washing, the enhanced effect of the ultrasonic washing was thought to be the result of some form of sonocapillary phenomena.

8. ISOLATION AND EXTRACTION OF MATERIALS

The isolation of the effective compositions from some traditional Chinese herbs under influence of ultrasound was reported [71]. The compounds Helicid (4-formylphenyl-allopyranoside), Berberin Hydrochloride, and Bergenin could each be isolated from different plant materials using ethanol at room temperature under the influence of sonication. Ultrasonic irradiation reduced the temperature and time required for the process, thereby increasing the efficiency of extraction. The products contained fewer impurities and the ultrasonic process appeared to involve simpler technology.

The extraction of tea leaf solids with water under the influence of ultrasound has been studied with respect to the effects of temperature, irradiation time, and power [72]. Sonication improved the extraction at 60 °C by nearly 20%, approaching the efficiency of that of thermal extraction at 100 °C.

9. IN CRYSTALLIZATION

Supported by the National Natural Science Foundation in China, a group in Guangzhou carried out a study of the crystallization of natural products in an ultrasonic field. In a pilot plant scale experiment using a 1000-W ultrasonic probe in a 25-kg solution of cane sugar, they found that ultrasonic irradiation could significantly affect the formation of crystal nuclei [73]. Compared with the nucleus formation induced by air sparging or the addition of sugar powders, the ultrasonic process could be carried out at lower saturation values so that the crystallization time and the energy needed were both reduced; in addition the crystal nuclei formed were more uniform, well-formed and refined.

10. METAL PLATING

The effect of ultrasonic irradiation on electroless copper coating of ceramic was analyzed statistically in terms of temperature, time, coating solution composition, acoustic power, and frequency [74]. Upon examination of SEM and metallography it was found that the use of ultrasonic irradiation during the plating process could produce a more uniform and better-bonded coating with a faster coating rate, compared to the control. The plating system was the electroless plating of copper on ceramic in a copper sulphate, potassium and sodium tartrate, and formaldehyde system. In one example, a copper coating with 40.5 kg cm^{-2} adhesion strength and 164-Knoop microhardness was obtained after sonication at 1020 kHz and 3.97-W acoustic power for 5 min, whereas a copper coating with 17.6 kg cm^{-2} adhesion strength and 110-Knoop microhardness was obtained in 30 min without sonication. The process could be used in large scale due to its convenience. A possible mechanism for the ultrasonically mediated coating rate increase was discussed in the paper.

11. FUNDAMENTAL STUDIES ON SONOCHEMISTRY

A series of fundamental studies on sonochemistry by Feng Ruo and his collaborators has been undertaken over a number of years. Their studies have focused on how the parameters of an ultrasonic field, such as sound intensity, frequency, shape of wave, etc., affect the cavitation yield which was detected by different methods.

The formation of free radial OH• and H• in a naturally air-saturated aqueous solution exposed to traveling ultrasonic wave of 820 kHz was investigated using a spin-trapping agent, 5,5-dimethyl-1-pyrroline-1-oxide (DMPO) and ESR techniques [75]. It was shown that the cavitation threshold occurred at 0.537–0.632 W cm^{-2}, and no further increase was observed above 3 W cm^{-2}. At a fixed sound intensity the yield of OH• increased linearly with the sonication time.

The chemical effects of 820-kHz diagnostic ultrasonic cavitation was studied and it was reported that the OH• radical produced under cavitation in water could react with non-fluorescent terephthalic acid (TA) to produce fluorescent hydroxy terephthalate (HTA) [76–78]. Thus cavitation induced by ultrasound could be detected through measuring the HTA fluorescence value produced. When the sound intensity was above the threshold value of ultrasonic cavitation, the yield of HTA increased rapidly with the sound intensity and the yield increased approximately linearly with exposure time. When the sound intensity reached a certain value, the yield of HTA reached saturation. The relationship between the yield of HTA and sonication period was described and explained by an equation given by the authors.

The sonochemical yield of 815-kHz ultrasound cavitation in a reverberating field was also studied by measuring HTA fluorescent value [79–80]. There are two characteristic effects of sonochemistry in the reverberating field. First, the cavitation threshold was about 0.3 W cm^{-2} in contrast to a value of 0.7 W cm^{-2} obtained in traveling field. Second, when the sound intensity was higher than the threshold, the sonochemical yield increased as the intensity increased and increased rapidly after the intensity was at 1.69–2.13 W cm^{-2}. Here there was an upturned point in the resultant curve (in the traveling field this would tend toward saturation). The theoretical analysis showed that the reason why the threshold decreased was that the sound energy density became high in the reverberating field, and the upturned point resulted from the disturbance of the radiation pressure on the liquid surface. Therefore, by experiment and theoretical analysis, this paper showed that to obtain higher sonochemical yields a reverberating field should be chosen.

A study was also made of the effects of pulsed ultrasound with a frequency of 832 kHz on a sample of terephthalic acid (TA) solution using a pulse intensity of

1.75 W cm^{-2}, an on:off ratio of 1:1, and a pulse width from 100 μs to 30 s [81–82]. The effect of the ultrasound pulse width on the cavitation in a small-size reverberating field was experimentally investigated and a corresponding theoretical analysis was made. It was found that the chemical yield of cavitation increased with an increase in pulse width and then approached a maximum value after about 10 ms. The chemical yield of cavitation was 43.9 when the pulse width was 180 s in a reverberating field, while it was 3.64 in a progressive wave field under identical experimental conditions. The dependence of the chemical yield of ultrasonic cavitation upon the ultrasound pulse width was deduced using some logical approximations in quantitative theoretical calculation. The paper offered a certain theoretical basis for further studies of cavitation effect in the reverberating field.

The detection of cavitation by an electrical parameter was used to study the chemical yield of cavitation in a reverberating field [83]. Nitric acid and nitrous acid are generated from nitrogen and oxygen dissolved in water under high pressure and high temperature, such as those formed during ultrasonic cavitation, so that the cavitation could be detected indirectly by detecting increases in electroconductivity of the aqueous solution. The relationship between the electrical parameter and sound cavitation was given by a simple theoretical analysis by the authors. These results have shown that laws of cavitation given by the two methods were the same.

The "pulse cavitation peak" phenomenon, where under an appropriate choice of the pulse width the cavitation yield reached an apparently maximum value, was first studied in a reverberating field [84,85]. Frequencies of 823-kHz and 1.7-MHz ultrasonic waves modulated by rectangular wave with an on:off ratio of 1:1 and pulse width from 100 μs to 270 s were used to irradiate both an aqueous terephthalate solution and deionized water, respectively. After sonication, the fluorescence and electroconductivity changes of the samples were measured. All the experimental data showed that a maximum value of cavitation yield occurred in the pulse width range between 10 and 100 ms. The theoretical result predicted by the authors based on the sonicated system resonance with the modulation envelope appeared to be in agreement with experimental data. In further investigations a similar phenomenon was observed for the 1.7-MHz ultrasound modulated by sine wave [86,87].

The frequency effect of ultrasonic waves with frequencies around 1 MHz (0.76, 1.0, and 1.7 MHz) was studied using the electrical detection method described above [88]. The experimental data and theoretical analysis of the results indicated that there was an optimum ultrasonic frequency corresponding to a maximum in sonochemical yield according to the bubble distribution in liquid. A Gaussian distribution of gas bubble radii was expected for a water sample exposed to a normal air atmosphere. In addition, experimental data also showed that any comparison of the frequency effect on the sonochemical efficiency should be under the conditions of not only the same sonic power but also the same sonic intensity.

A study on the cavitation effect of 28-kHz ultrasound on sonic intensity and sonication time was completed using the electrical method [89]. The experimental results appeared to be similar to those obtained at the higher frequencies (around 1 MHz), i.e. the cavitation yield increased nonlinearly with the sonic intensity but increased almost linearly with sonication time.

Using 1.2- and 1.7-MHz continuous wave ultrasound the sonoluminescence of alkali aqueous solution of luminol (a chemiluminescent agent) was investigated by means of a spectral analysis technique [90,91]. It was found that the sonoluminescence generated by the solution of luminol was mainly a visible light emission. On comparing the *sono*luminescent spectrum with the luminol-induced *photo*fluorescence spectrum of the solution, the authors found that there were some similarities between the *sono*excitation and *photo*excitation in respect to the emission wavelength.

Further studies on this phenomenon showed that the sonofluorescence spectra of three aqueous alkaline solutions of luminol were compared with their photofluorescence spectra [92]. The results showed that the fluorescence of the substance could be obtained by sonoexcitation, and both fluorescence emission wavelengths of sonoexcitation and photoexcitation were similar, in the range of 370–750 nm. Furthermore, the sonoluminescence of the alkali aqueous solution of luminol was sufficiently powerful to be used to produce ordinary photographic images on film and photopaper when exposed in a dark room. The photos and printing appeared to be as good as those simply exposed to daylight, except a longer exposure time was needed [93].

12. CONCLUSION

In China, sonochemistry has been studied widely and is thriving. It is hoped that a review such as this will stimulate international cooperation ventures and technology exchanges.

REFERENCES

Except where indicated all were published in the Chinese language.

[1] Li, P.S. *Chem. World*, 6 (1959) 301.
[2] Yuan, W.K. *Chem. World*, 5 (1959) 222.
[3] Xu, L.X. and Tao, F.G. *Org. Chem.*, 6 (1986) 415–421.
[4] Li, J.T. and Wang, Y.H. *Chemical Reagents*, 9 (1987) 98–102.
[5] Li, J.T. and Wang, Y.H. *J. Hebei Univ. (NS)*, 10 (1990) 85–90.
[6] Rong, J.H. *Chinese Chem. Bull.*, 2 (1991) 8–14.
[7] Xu, Q.H. and Pan, J. *Yunnan Chem. Eng.*, 2 (1993) 62–63.
[8] Zou, J.L. and Li, S. *Modern Chem. Eng.*, 7 (1987) 22–25.
[9] Cheng, C.M. *Chinese Chem. Bull.*, 6 (1992) 43–45.
[10] Zhao, Y.Y., Feng, R., Bao, C.G. et al. *Applied Acoustics*, 2 (1994) 44–48.
[11] Zhao, Y.Y., Feng, R., Xu, K.L. et al., *Chem. Eng.* 4 (1995) 6–11.
[12] Ding, D. *Tech. Acoustics*, 11 (1992) 41–44.
[13] Feng, R., Zhao, Y.Y., Cheng, Z.H. et al. *Tech. Acoustics*, 13 (1994) 56–61.
[14] Song, X.Q. *Univ. Chemistry*, 3 (1988) 6.

[15] Kuang, S.L. and Gong, C.S. *Chinese Chem. Bull.*, 6 (1994) 23–27.

[16] Kuang, S.L. and Gong, C.S. *Chinese J. Nature*, 13 (1990) 11–15.

[17] Lin, Z.M. *Applied Acoustics*, 12 (1993) 1–4.

[18] Zhao, Y.Y., Bao, C.G., Feng, R. et al. *Chinese Chem. Bull.*, 8 (1994) 26–29.

[19] Kuang, S.L. and Gong, C.S. *Modern Chemical Engineering*, 9 (1989) 23–27.

[20] Zhao, Y.Y., Bao, C.G., Xu, K.L. et al. *Yunnan Chemical Engineering*, 4 (1993) 49–52.

[21] Chen, K.Q. *Chinese Chem. Bull.*, 7 (1986) 27.

[22] Feng, R. and Zhao, Y.Y. *Progress Biochem. & Biophys.*, 2 (1994) 500–503.

[23] Feng, R. and Li, H.M. *Sonochemistry and its Applications*. Anhui Press of Science and Technology, 1992.

[24] Xu, X., Shen, Y., and Chen, K.Q. *J. Chengdu Sci. Tech. Univ.*, 3 (1982) 1.

[25] Li, W.R. and Xu, X. *Lett. of Micromol.*, 1 (1983) 31–38.

[26] Xu, J., Shen, Y., and Xu, X. *J. Chengdu Sci. Tech. Univ.*, 4 (1984) 1.

[27] Xu, J., Shen, Y., and Xu, X. *Chem. J. Chinese Univ.*, 7 (1986) 947.

[28] Zhang, J., Chen, K.Q., Liu, Q.R. et al. *Chinese J. Polymer*, 3 (1990) 271.

[29] Shen, Y., Chen, K.Q., Wang, Q. et al. *Chinese J. Chem. Eng.*, 4 (1988) 478.

[30] Liu, Q.R., Chen, K.Q., Xu, K.T. et al. *Chinese J. Chem.*, 2 (1990) 249.

[31] Xu, W. and Xu, X. *J. Chengdu Sci. Tech. Univ.*, 1 (1982) 69.

[32] Cao, Z.Q. and Xu, X. *J. Chem. Ind. Eng.*, 1 (1985) 56–63.

[33] Gong, X.Y., Chen, K.Q., and Xu, X. *J. Chem. Ind. Eng.*, 3 (1987) 318.

[34] Chen, K.Q., Shen, Y. et al. *J. Macromol. Sci. Chem.*, A22 (1985) 455–469 (Eng).

[35] Wang, W. and Qin, W. *Chinese Chem. Bull.*, 9 (1989) 44.

[36] Xu, L., Tao, F., and Yu, T. *Tetrahedron Lett.*, *26* (1985) 4231 (Eng).

[37] Xu, L.X., Tao, F.G., and Yu, T.Y. *Chinese J. Chem.*, 44 (1986) 1134–1138.

[38] Xu, L.X., Tao, F.G., and Yu, T.Y. *Chinese J. Nature*, 9 (1986) 315–316.

[39] Xu, L.X., Yu, T.Y., and Tao, F.G. *Chinese Sci. (B)*, 2 (1988) 113–121.

[40] Xu, L.X., Tao, F.G., and Yu, T.Y. *Chinese J. Chem.*, 46 (1988) 340–344.

[41] Regen, S.L. and Singh, A. *J. Org. Chem.*, 47 (1982) 1587 (Eng).

[42] Xu, L.X., Tao, F.G., and Yu, T.Y. *Chinese J. Chem.*, 46 (1988) 608–611.

[43] Tao, F.G., Huang, H., Sun, M. et al. *Chinese J. Applied Chem.*, 5 (1988) 91–93.

[44] Merz, A. *Synthesis*, (1974) 724 (Eng).

[45] Xu, L.X., Tao, F.G., and Yu, T.Y. *Chem. J. Chinese Univ.*, 9 (1988) 36–40.

[46] Tao, F.G., Xu, L.X., Lu, Y.Z. et al. *Chinese Org. Chem.*, 8 (1988) 441–442.

[47] Xu, X.L., Li, Z., and Na, Y.X. *Chinese J. Applied Chem.*, 4 (1987) 73–75.

[48] Lie Ken Jie, M.S.F. and Lam, W.L.K. *J.A.O.C.S.*, 65 (1988) 118–121 (Eng).

[49] Xu, K.L., Bao, C.G., Yang, M. et al. *J. Yunnan Univ. (NS)*, 15 (1993) 106–107.

[50] Du, S.K., Yu, D.C., Zhu, D. et al. *Eng. Chem. Metallurgy*, 12 (1991) 319–325.

[51] Chou, T.S. and Chen, M.M. *Heterocycles*, 26 (1987) 2829–2834 (Eng).

[52] Chou, T.S. and You, M.L. *J. Org. Chem.*, 5 (1987) 2224–2226 (Eng).

[53] Chou, T.S., Hung, S.H., Peng, M.L. et al. *J. Chinese Chem. Soc. Taipei*, 38 (1991) 283–284.

[54] Chou, T.S., Hung, S.H., Peng, M.L. et al. *Tetrahedron Lett.*, 32 (1991) 3551–3554 (Eng).

[55] Zhao, Y., Mason, T.J., and Lindley, J. *Chinese J. Org.*, 14 (1994) 435–437.

[56] Wu, L.C., Tan, P., and Zhang, Q.P. *Chem. World*, 4 (1992) 157–159.

[57] Sun, X.Y. and Peng, G.C. *Chinese J. Applied Chem.*, 7 (1990) 88–90.

[58] Feng, R., Liu, Z.B., Gao, J. et al. *Tech. Acoustics*, 12 (1993) 13.

[59] Xu, K.L., Bao, C.G., Lu, Q.H. et al. *Yunnan Chem. Eng.*, 1 (1992) 15–16.

[60] Wang, Y.C. and Yang, W.B. *Chinese Anal. Chem.*, 19 (1991) 102–104.

[61] Jin, Q.H., Zhang, H.Q., and Yu, S.R. *Spectrosc. Spectrum Anal.*, 9 (1989) 31.

[62] Zhang, H.Q., Ye, D.M., Yu, A.M. et al. *Chinese Anal. Chem.*, 20 (1992) 1065–1068.

[63] Yu, S.R., Zhang, H.Q., and Jin, Q.H. *Chinese Anal. Chem.*, 18 (1990) 1052–1055.

[64] Duan, Y.X., Zhang, H.Q., Lu, H. et al. *Chinese Anal. Chem.*, 20 (1992) 383–387.

[65] Ma, X. and Luo, S.J. *Chinese Anal. Chem.*, 18 (1990) 49–52.
[66] Zeng, M.S. and Uwe, F. *14th International Conference of Acoustics Proceedings*, 1992, C6-5 (Eng).
[67] Bo, J. and Chen, X. *14th International Conference of Acoustics Proceedings*, 1992, CP-31 (Eng).
[68] Xia, D.Y. *Chem. Fertiliser Ind.*, 4 (1987) 29–31.
[69] Huang, Y.N. *Ind. Water Treat.*, 6 (1986) 39–40.
[70] Bao, C.G., Zhao, Y.Y., and Yang, X.Y. *J. Nanjing Univ. (NS)* 2 (1995) 242–247.
[71] Zhao, Y.Y., Bao, C.G., and Mason, T.J. *Ultrasonics International 91 Conference Proceedings*, 87–90 (Eng).
[72] Mason, T.J. and Zhao, Y.Y. *Ultrasonics*, 32 (1994) 375–377 (Eng).
[73] Qu, T.Q., Li, Y.H., and Chen, S.G. *Tech. Acoustics*, 1 (1993) 15–20.
[74] Zhao, Y.Y., Bao, C.G., Feng, R. et al. *Digest of the International Workshop on Modern Acoustics Nanjing China*, 1994, p. 110 (Eng).
[75] Feng, R., Qian, Y., Xu, J.Y. et al. *Chinese J. Acoustics*, 10 (1991) 131–138.
[76] Feng, R., Xu, J.Y., and Shi, Q. *Prog. Natural Sci.*, 1 (1990) 376–390 (Eng).
[77] Shi, Q., Qian, Y., Feng, R. et al. *Acta Biophys. Sci.*, 6 (1990) 484–489.
[78] Feng, R., Xu, J.Y., Shi, Q., and Li, H.M. *Tech. Acoustics*, 12 (1993) 37–39.
[79] Wang, S.W., Feng, R., and Shi, Q. *Prog. Natural Sci.*, 2 (1992) 461–464 (Eng).
[80] Wang, S.W., Feng, R., Mo, X.P. et al. *Chinese J. Acoustics*, 12 (1993) 149–158.
[81] Wang, S.W., Feng, R., Xu, J.Y. et al. *Prog. Natural Sci.*, 2 (1992) 275–279 (Eng).
[82] Wang, S.W., Feng, R., Xu, J.Y. et al. *Ultrasonics*, 31 (1993) 39–44 (Eng).
[83] Mo, X.P., Feng, R., Zhou, H. et al. *Acoustics Lett.*, 15 (1992) 257–260 (Eng).
[84] Feng, R. and Li, H.M. *Sonochemistry and its Applications Anhui Press of Science and Technology*, 1992, pp. 144–147.
[85] Feng, R., Huang, J.L., Chen, Z.H. et al. *Digest of the International Workshop on Modern Acoustics Nanjing China*, 1994, 105 (Eng).
[86] Wang, S.W., Feng, R., Mo, X.P. et al. *Digest of the International Workshop on Modern Acoustics Nanjing China*, 1994, p. 108 (Eng).
[87] Feng, R., Wang, S.W., Zhu, C.P. et al. *Annual Science Report—Supplement of JNU (NS)*, Vol. 30, English Series 2, Nov. 1994, pp. 17–23 (Eng).
[88] Huang, J.L., Feng, R., Zhu, C.P. et al. *Digest of the International Workshop on Modern Acoustics Nanjing China*, 1994, p. 106 (Eng).
[89] Zhu, C.P., Feng, R., Huang, J.L. et al. *Digest of the International Workshop on Modern Acoustics Nanging China*, 1994, p. 107 (Eng).
[90] Li, H.M., Feng, R., and Chen, Z.H. *Chinese J. Acoustics*, 13 (1994) 148–152.
[91] Li, H.M., Li, Z.S., and Zhong, F. et al. *Applied Acoustics*, 12 (1993) 41–43.
[92] Li, H.M., Feng, R., Zhong, F. et al. *Digest of the International Workshop on Modern Acoustics Nanjing China*, 1994, 109 (Eng).
[93] Li, H.M., Zhong, F., Xi, A.D. et al. *Digest of the International Workshop on Modern Acoustics Nanjing China*, 1994, p. 111 (Eng).

THE USES OF ULTRASOUND IN FOOD PROCESSING

Timothy J. Mason and Larysa Paniwnyk

OUTLINE

Advances in Sonochemistry
Volume 4, pages 177–203

ISBN: 1-55938-793-9

1. INTRODUCTION

The use of ultrasound within the food industry has been a subject of research and development for many years and, as is the case in other areas, the sound ranges used can be divided basically into diagnostic and power ultrasound (Figure 1). The majority of studies are restricted to the range of 20 to 40 kHz, i.e. the usual ranges employed in conventional ultrasonic equipment.

Up to a few years ago the majority of applications and developments involved ultrasonic frequencies in the MHz range for noninvasive analysis. A number of uses have been developed for the measurement of such factors as the degree of emulsification or extent of particle distribution within dispersions. More recently the interest of food technologists has turned to the use of power ultrasound in processing. In this case the mechanical and chemical effects of cavitation are important and applications are very wide ranging (Table 1).

2. HIGH-FREQUENCY DIAGNOSTIC ULTRASOUND

Diagnostic ultrasound has advanced to a very exact science, particularly in the field of medical imaging; indeed it is this field which has led the drive for increased accuracy in measurements [1]. Other areas have benefited from such developments; for example diagnostic ultrasound can be used in polymer science. A review of this application has been presented by Pethrick in a chapter published previously in this series [2].

High-frequency, low-power ultrasound generally within the frequency range 0.5–20 MHz can be used to evaluate foodstuffs in terms of physical characteristics such as the degree of emulsification or the concentration of solids or gas. In

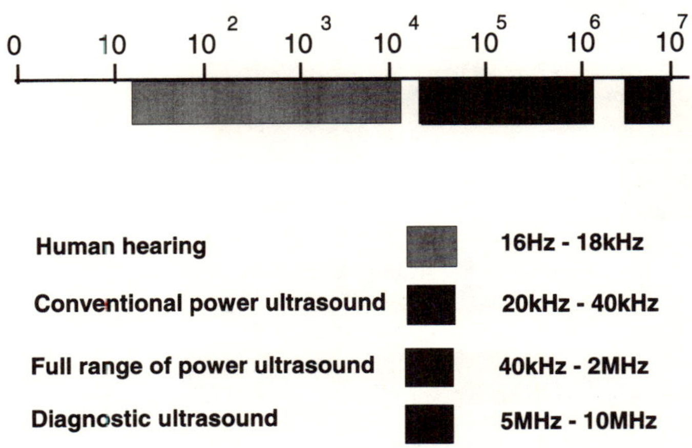

Figure 1. Ultrasonic sound frequency ranges.

Table 1. Some Uses of Power Ultrasound in Food Technology

Mechanical Action Used in Processing
enhanced mixing and blending
extraction of flavorings
meat tenderization
crystallization of fats and chocolate
improved food freezing
destruction of foams
precipation of airborne powders
filtration and drying
degassing
viscosity reduction
Chemical and Biochemical Action
oxidation processes
bactericidal action in liquids
sterilization of equipment
effluent treatment
selective effects on enzyme activity
enhanced growth of living cells

principle, both sound velocity and sound attenuation may be used for the examination of food products (Figure 2), the most simple application being in the determination of the quantity of material remaining in a vat or storage container (Figure

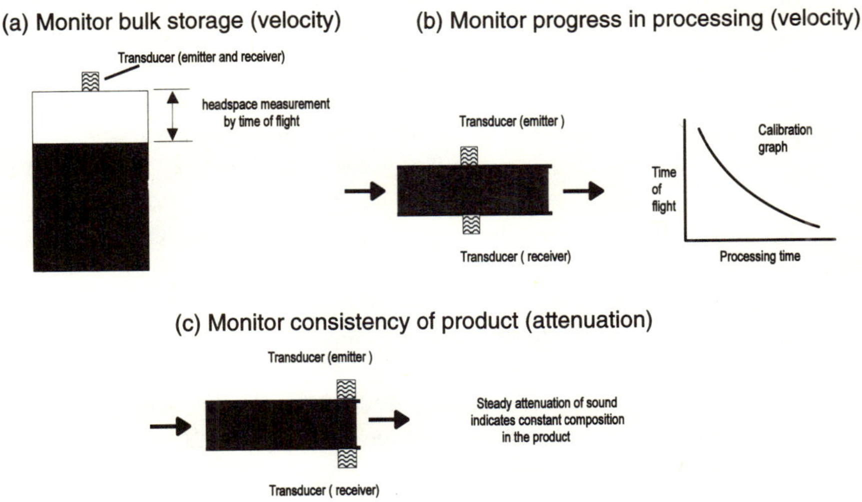

Figure 2. Uses of diagnostic ultrasound in food technology.

2a). In this case the "pulse–echo" technique will allow rapid (and remote) sensing of the headspace in a container since the time from emission of the sound pulse until its return as an echo from the surface can be electronically converted into distance using the velocity of sound in the air above the material (the velocity of sound in air is approximately 1500 ms^{-1} but it can be determined extremely accurately for the system to be used).

For homogeneous mixtures it is possible to use sound velocity in the conventional pulse–echo mode to determine the degree of emulsification or concentration of a sample. The "time of flight" of a sound signal through the sample can be monitored using a transmitter and receiver system which need not be in direct contact with the liquid (Figure 2b). The sensor can be another transducer on the opposite side of the reaction chamber (or pipe through which the material flows), although in many cases the emitting transducer also serves as the receiver. In order to interpolate the results from such measurements it is common practice to calibrate the instrument against a series of materials of previously determined physical characteristics.

Foods such as creams and sauces produce high scatter of ultrasonic waves because of the incorporated solid or gas and, in this situation, it is often easier to measure sound attenuation (Figure 2c). Thus by monitoring the attenuation of an ultrasound pulse it has proved possible to determine the degree of homogenization of fat within milk [3]. A standard for the determination of the degree of attenuation due to fat content was established by comparing the attenuation of whole milk with that of skimmed milk. Sound attenuation in milk and cream is mainly due to heat transfer in and out of fat globules and so, as the degree of fat homogenization increases (i.e. as fat droplet size decreases) sound attenuation also increases. Thermal losses account for most of the sound attenuation observed and other losses, e.g. those due to viscosity factors and signal scattering only account for a minimal amount of sound attenuation in this type of system.

The degree of emulsification in such materials can also be estimated by the measurement of ultrasound velocity in conjunction with attenuation [4]. It is possible to determine factors such as the degree of "creaming" (or "settling") of a sample, i.e. the movement of solid particles/fat droplets to the surface (or to the base) [5]. Such information gives details, for example, of the long-term stability of fruit juices and the stability of emulsions such as mayonnaise. The combination of velocity and attenuation measurements shows promise as a method for the analysis of edible fats and oils [6], and for the determination of the extent of crystallization and melting in dispersed emulsion droplets [7].

These methods of analysis are noninvasive and therefore nondestructive and non-hazardous. They can be connected to "on-line" production processes, thus offering the possibility of automation. They are also economic since the equipment is relatively inexpensive often involving only a single transducer acting as transmitter and receiver in a pulse–echo mode.

3. LOW-FREQUENCY, HIGH-POWER ULTRASOUND

Low-frequency, high-power ultrasound, with frequencies between 20 kHz and 1 MHz, can be employed directly in a wide variety of food processing applications (see Table 1).

3.1 Oxidation Processes

It is well known that sonication of aqueous solutions results in the production of hydrogen and hydroxyl radicals which undergo a number of subsequent reactions involving the formation of hydrogen peroxide (Figure 3). The presence of radicals within food samples may effect oxidation processes which could be either advantageous or detrimental to the final characteristic properties of the food. It is therefore essential to consider the type of food material which is to be treated with ultrasound and the dose (or exposure time) which is to be used.

A good example of potentially useful enhanced oxidation is in the aging of fermented products such as spirits where limited oxidation is potentially useful for flavor development and early maturation of the product. Sonication has been successfully employed to enhance the flavor of some wines and spirits. The effect of irradiation with 1 MHz ultrasound has been shown to alter the alcohol/ester balance in such products to give the appearance of an aged product [8]. This seems to be particularly successful for whisky where a patented method has been described which reduces the normally prolonged aging time to less than 1 year by applying ultrasound to the liquor in a standard barrel [9].

$$H_2O \rightarrow H^{\bullet} + OH^{\bullet}$$
$$OH^{\bullet} + OH^{\bullet} \rightarrow H_2O_2$$
$$OH^{\bullet} + OH^{\bullet} \rightarrow H_2O + O^{\bullet}$$
$$OH^{\bullet} + OH^{\bullet} \rightarrow H_2 + O_2$$
$$H^{\bullet} + O_2 \rightarrow HO_2^{\bullet}$$
$$HO_2^{\bullet} + H^{\bullet} \rightarrow H_2O_2$$
$$HO_2^{\bullet} + HO_2^{\bullet} \rightarrow H_2O_2 + O_2$$
$$OH^{\bullet} + H_2O \rightarrow H_2O_2 + O^{\bullet}$$
$$H^{\bullet} + H^{\bullet} \rightarrow H_2$$
$$H^{\bullet} + OH^{\bullet} \rightarrow H_2O$$

Figure 3. Decomposition of water with power ultrasound.

Although oxidation is essential for a number of processes involving flavor and/or color development, it is not always an advantage. Oxidation can have detrimental effects as it does with fats. In fact one of the most common paths to flavor deterioration in fats is the autooxidation of triacylglycerols [10]. Autooxidation is a free radical reaction in which the main initiators are hydroperoxides which decompose above 150 °C, although some decomposition does occur slowly below this temperature. Decomposition results in the formation of various volatile products which cause changes in the flavor of the fat. Any sonication process involving foods containing fats must therefore be carefully monitored.

3.2 Enzyme Reactions

One of the original uses of power ultrasound in biochemistry was to break down biological cell walls to liberate the contents (indeed many ultrasonic horn systems were first marketed as cell disruptors). Subsequently it has been shown that power ultrasound can also be used to produce a positive effect on enzyme activity. If the intensity is too high however the enzymes can be denatured.

Many examples are to be found in the chemical or biochemical literature; e.g. ultrasonically induced emulsification/mixing has been utilized in the two-phase enzymatic synthesis of dipeptides [11]. For the dipeptide synthesis shown in reaction (1) the source of ultrasound was an ultrasonic bath (38 kHz). The importance of sonication in such a system is that it promotes biphasic reaction in solvent mixtures such as petroleum ether/water which are not effective under conventional conditions (Table 2).

$$\text{BOC–Gly with Phe–N}_2\text{H}_2\text{Ph} \rightarrow \text{BOC–Gly–Phe–N}_2\text{H}_2\text{Ph} \quad (1)$$

Another fruitful area of research has been that of the sonochemical activation of immobilized enzymes where ultrasound appears to be particularly useful in increasing the transport of substrate to the enzyme. Using α-chymotrypsin (on agarose gel) and casein as substrate, a two-fold increase in activity was observed at 20 kHz [12]. Here the origin of the enhancement was thought to be associated with increased penetration of the casein into the support gel induced by cavitational effects close to the surface. However an increase in the activity of α-amylase (on porous polystyrene) was produced on irradiation with 7 MHz ultrasound [13]. This is a very significant result since at this high-frequency cavitation cannot occur and

Table 2. Sonochemically Assisted Dipeptide Synthesis[a]

Organic Phase	*Stirred*	*Sonicated*
diethyl ether	71	89
petroleum ether	12	62

[a]Solvent composition: water (75%); organic solvent (25%); 37 °C; 12 h.

so, in this case, the increased activity is thought to be associated with greater microstreaming of reagents to the surface.

It has been known for many years that ultrasound can be employed as a method of inhibiting enzyme activity. Nearly 60 years ago Chambers reported that pure pepsin was inactivated by sonication probably as a result of cavitation [14]. Enzyme inactivation through sonication is also considered to be responsible for the inhibition of sucrose inversion [15].

Peroxidase, which is found in most raw and unblanched fruit and vegetables, is particularly associated with the development of off-flavors and brown pigments. The effect of ultrasound (20 kHz, 371 W cm^{-2}) on peroxidase:sigma-P8000 dissolved in 0.1 M potassium phosphate at 20 °C buffer pH 7 has been reported (Figure 4) [16]. The activity of the peroxidase was estimated by the Bergmeyer method involving measurements of the decomposition of hydrogen peroxide with guaiacol as hydrogen donor through the rate of color development at 436 nm. The original activity of peroxidase was progressively reduced by 90% as ultrasound was applied over a 3-h period.

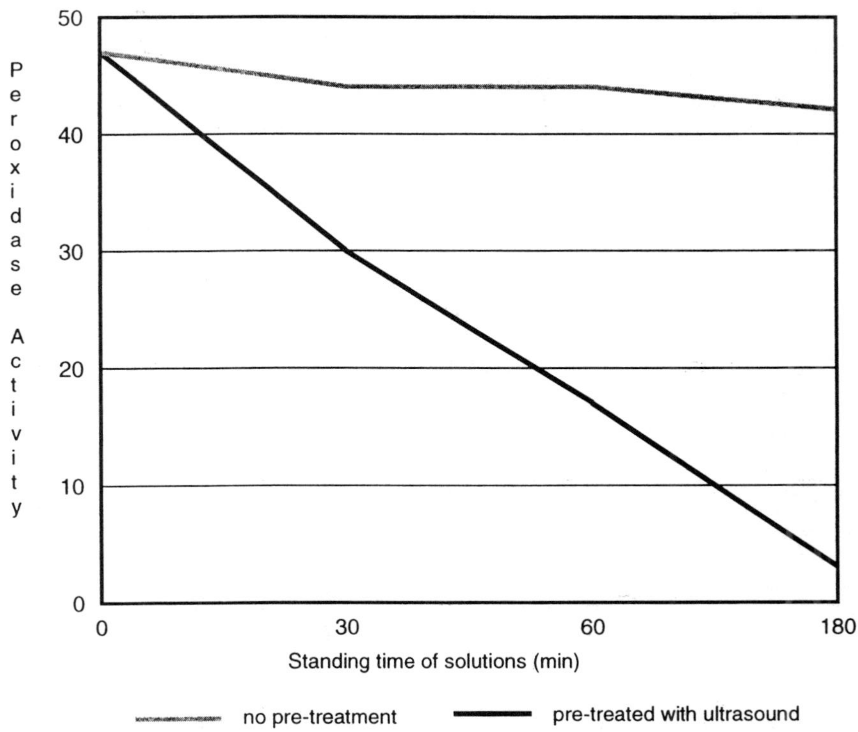

Figure 4. Influence of ultrasound on peroxidase activity (units/mg solid).

A convenient generalization concerning the effect of ultrasound on enzyme activity has been suggested. According to Naimark and Mosher [17], oxidases are usually inactivated by sonication while catalases are only affected at low concentrations. Reductases and amylases however appear to be highly resistant to sonication.

3.3 Stimulation of Living Cells

There are a number of examples of the use of ultrasound to increase the productivity of food products through the enhancement of efficiency of whole cells without disrupting the cell walls. A simple example of this is in the use of low-power ultrasonic activation of a liquid nutrient media to enhance the rate of growth of algal cells. Essentially this results in an increase in the production of protein (up to three-fold) and represents a real possibility for the production of food materials from unusual sources for human or animal consumption [18].

The production of yoghurt is an increasingly important process in the food industry. Recent investigations have shown that the use of ultrasound as a processing aid can lead to a total reduction of up to 40% in production time [19]. A very important finding in this work was that sonication reduced the normal dependence of the process on the origin of milk. Further it appeared to improve both the consistency and the texture of the product.

A remarkable influence of ultrasound on fish egg hatching has been reported [20]. The eggs were exposed to ultrasound of frequency 1 MHz for 35 min, 3 times a day and this led to a reduction in hatch time to 6 days compared to the normal 7 days. This, in itself, is of considerable industrial importance for fish-farming, but there are also two other benefits of ultrasonic treatment. It was also found that the ultrasound increased the fraction of the eggs which hatched and further that, once hatched, the fish demonstrated a higher survival rate.

Ultrasonically stimulated seed germination offers the possibility of increased productivity for large scale farm crops and in more general horticulture. Agricultural crop yields are dependent on the quality of the plant variety and on the percentage seed germination and growth. There are several reports in the literature which suggest that ultrasonic treatment of seeds before sowing is an effective method of improving crop yield [21]. In many of these studies the seeds were sonicated in a liquid medium [22].

The effect of the suspension media on the ultrasonically stimulated germination of seeds from a temperate *Cymbidium* species was studied. Pretreatment with 0.1 M KOH for 0.5 h or ultrasonic waves for 4 h resulted in the best germination, but the ultrasonic treatment induced quicker germination and faster rhizome growth [23]. A comparative study of the efficiency of ultrasound and some conventional modes of seed treatment revealed that ultrasound helped in resolving seed hardness [24]. A 10-min ultrasonic treatment at 0.7 Wcm^{-2} was most effective on *Trifolium repens* and *Lotus corniculatus*, resulting in 14 and 30% increased germination, respectively. Germination and the onset of flowering were promoted by sonication of the seeds of red clover (*Trifolium pratense*) resulting in an increase in both the number of established plants and of flowers per plant [25]. The greatest responses were obtained from a 10-min irradiation at 21.5 kHz and 0.7 W cm^{-2}. In experiments with wet seeds of cotton exposed to ultrasound at 0.5–2.5 W cm^{-2} for 1–10 min, an increase in ultrasound intensity decreased the exposure time required for stimulation of germination [26]. The highest germination resulted in seeds exposed to ultrasound intensity of 2.5 W cm^{-2} for 4 min.

More recent Russian studies have applied sonication under dry conditions which may be carried out up to several months before actual sowing [27]. The subsequent handling and packaging of the seeds is perfectly conventional with no special precautions required. The treatment unit itself uses ultrasonic equipment operating at 20 kHz with a vibrational amplitude of between 1–40 μm and is able to handle up to 600 kg h^{-1} with the possibility of installation in a continuous packing line. Ultrasonic treatment led to a threefold enhancement in sunflower seed germination in soil and a 10-day reduction in the ripening time of tomatoes.

The prospects for this technology are very good. Apart from offering improvements in germination, initial growth rate and reduced ripening times it is also a rival to chemical treatments and affords clean "organic" farming.

3.4 Sterilization

One of the major long-established industrial applications of power ultrasound is for cleaning and it has proved to be an extremely efficient technology. Ultrasound is particularly useful in surface decontamination where the inrush of fluid which accompanies cavitational collapse near a surface is nonsymmetric (Figure 5). The surface itself restricts the inrush of fluid from that side and so a jet of liquid is formed from the resultant major flow from the other side of the bubble. This powerful jet will dislodge dirt and bacteria from surfaces. The particular advantage of ultrasonic cleaning in this context is that it can reach crevices that are not easily reached by conventional cleaning methods. Indeed, a general patent has been applied for pertaining to the use of ultrasound as a method of pasteurization, sterilization, and decontamination of instruments and surfaces used within the medical, surgical, dental, and food processing industries [28]. The use of ultrasound allows the destruction of a variety of fungi, bacteria, and viruses in a much reduced processing time when compared to thermal treatment at similar temperatures. One recent example of surface cleaning is the use of a combination of ozone and ultrasound to kill *Salmonella* on deliberately contaminated egg shells [29].

High power ultrasound alone is known to damage or disrupt biological cell walls for the release of contents (*vide supra*). This will result in the destruction of the cells, i.e. sonication has the potential for the destruction of bacteria. Unfortunately very high intensities are required if ultrasound alone is to be used for complete sterilization. It has been shown however that low-power ultrasound is capable of

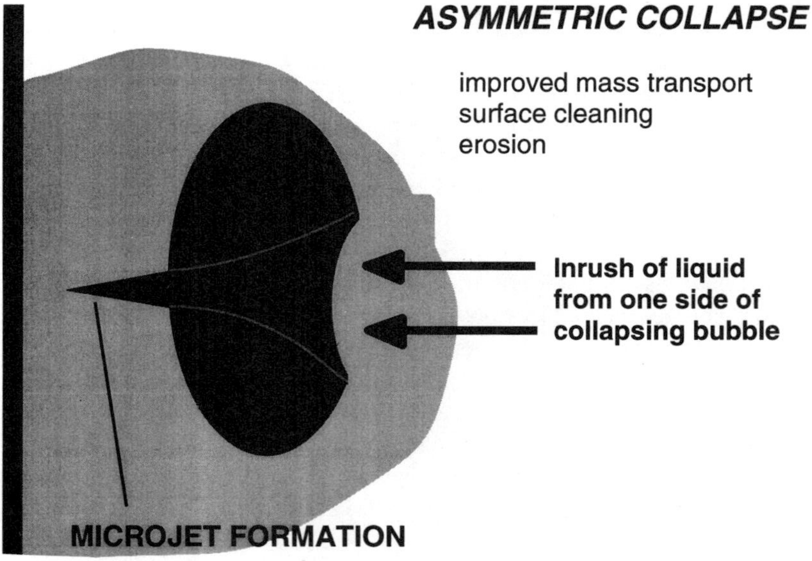

Figure 5. Collapse of a cavitation bubble near a solid surface.

enhancing the effects of chemical biocides; this is in part due to the breakup and dispersion of bacterial clumps which then renders the bacteria more susceptible to chemical attack. In the early 1960s researchers began to formulate possible mechanisms for physical disruption and the cause of cell death the proposed mechanisms related to cavitation phenomena which included shear disruption, localized heating, and free radical formation [30]. In 1975 it was shown that brief exposure to ultrasound caused a thinning of the cell wall [31]. This was attributed to the freeing of the cytoplasmic membrane from the cell wall. Such effects can be of great use in the sterilization and disinfection of food materials particularly where ultrasound is used in conjunction with a conventional sterilization technology.

In addition to the purely mechanical effects of cavitation there is another factor which makes sonication so efficient as a bactericide in aqueous conditions—the generation of radicals. Free radicals are generated during cavitation bubble collapse because the cavity, generated in any solvent, is unlikely to enclose a vacuum. During its formation in the rarefaction cycle of the wave the bubble will almost certainly draw in vapor from the liquid medium itself (or from any volatile reagents which are dissolved in it). On collapse, these vapors will be subjected to extremely large increases in temperature and pressure resulting in molecular fragmentation and the consequent generation of highly reactive radical species. Such radicals might then react either within the collapsing bubble or after migration into the bulk liquid. In the case of water, sonication gives rise to highly reactive OH· and H· radicals which can undergo a range of subsequent reactions [32]. An important product generated through the sonolysis of water is hydrogen peroxide (see Figure 3) which, together with the radical species provides a powerful bactericide and chemical oxidant.

Table 3. Effect of Ultrasonic Frequency on OH· Production at 30 °C

Frequency (kHz)	20	40	60
Fluorescence	30	40.2	29.3
Power (W)	50	26	11
Efficiency	0.6	1.6	2.7

The frequency of the ultrasonic irradiation has a significant effect on radical production. The frequencies generally employed for power ultrasound sonochemistry are in the narrow range between 20 and 100 kHz. There is, however, a considerable amount of information on sonochemistry performed at frequencies much higher in the ultrasonic range (around 1 MHz) [33]. In an important paper on the effect of ultrasonic frequency on radical production, Petrier has compared the effectiveness of 20- and 514-kHz irradiation for the generation of hydrogen peroxide in water at the same input power [34]. The rate of production of peroxide was about 12 times faster at the higher frequency. This result is ascribed to the fate of the OH· radical formed by the breakdown of water on cavitation bubble collapse. The OH· can react in the bubble or can migrate into the bulk and produce peroxide. At the higher frequency a shorter bubble lifetime allows more of the OH· to escape from the bubble into the bulk solution. This frequency effect is clear even over the much smaller range of 20 to 60 kHz in studies involving OH· detection by fluorescence [35]. When aqueous sodium terephthalate reacts with OH· it forms fluorescent hydroxyterephthalate, and its concentration can be estimated spectroscopically. The results in Table 3 show the fluorescence yield after 30 min irradiation for 10^{-3} M terephthalate at three different frequencies together with the energy input to the reaction estimated calorimetrically. The sonochemical efficiency for this reaction can be obtained by dividing OH· yield (directly proportional to fluorescence) by power input and the results clearly show that the efficiency of radical production increases with frequency.

Sterilization Using Sonochemically Assisted Bactericides

Ultrasound has been successfully employed in conjunction with bactericide for the destruction of microorganisms. In the treatment of raw water, ultrasound enhances chlorination as a method of biological sterilization [36]. The results (Figure 6) show survival rates of bacteria in raw water, using a fixed chlorine concentration of 1 ppm, with respect to two parameters: chlorine contact time and sonication time.

In a typical procedure, a sample of river water was taken and divided into three equal aliquots. One of the aliquots was then sonicated for 2 min. Another was sonicated for 1 min and the last, the control, was not sonicated. Following this procedure hypochlorite was added to all three aliquots to give a concentration of 1 ppm chlorine. These solutions are then left for 5 min to give a total chlorine contact

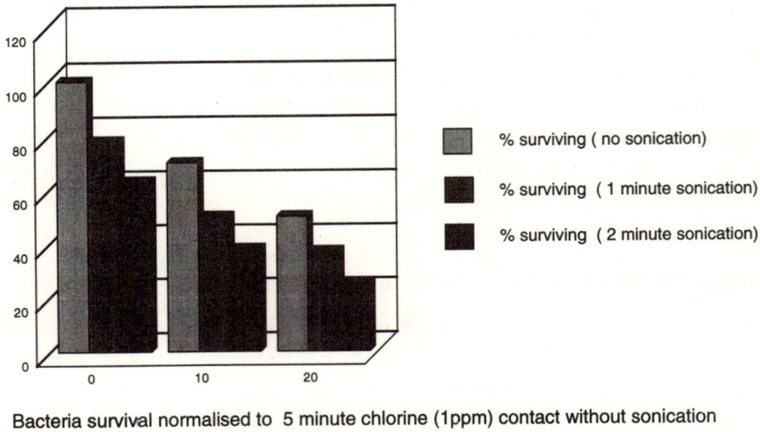

Figure 6. Combined effect of sonication and chlorination on bacteria survival.

time of 5 min. This was then repeated with differing chlorine contact times of 10 and 20 min. Samples (1 ml) were then taken from each aliquot, mixed with agar, and incubated at 37 °C for 24 h. The bacterial colonies arising were then counted.

The results of simply treating the raw water with chlorine for 5 min has been taken as the standard to which all other results have been normalized. On this basis it is clear from the results that a 2-min period of sonication using a Sonics and Materials VC-50 probe system (power level ca. 18 W cm^{-2}) results in about 40% improvement in bacterial kill compared with chlorination alone.

Sterilization Using Sonochemically Assisted Photolysis

The use of photochemical procedures for the removal of biological contamination is a technology which has reached pilot plant scale over the last few years in wastewater treatment. In combination with oxidants such as hydrogen peroxide, on a laboratory scale, it has been shown to be effective in the destruction of some pesticide residues [37]. Halocarbons in dilute aqueous solution (e.g. 1,1,1-trichloroethane) can be more efficiently destroyed through a combination of ultrasound and UV light rather than the single application of either technique [38].

One problem with photochemical treatment is that the method relies on good light transmission into the reaction medium. The slow coating of the transparent walls of the photoreactor vessel with biological or chemical deposits or coloration in the material to be treated causes loss in efficiency. A possible solution to the first of these problems is the use of ultrasound to protect the photoreactor walls against deposition of such materials [39]. Very general claims have been made in the patent literature relating to the possible combined use of photochemistry with sonochemistry some of which involve the purely mechanical effects of ultrasound in bacterial clump destruction before UV irradiation [40].

It is evident that food materials which are opaque are not likely to be effected, except superficially, by UV light. The combination of light and ultrasound is thus not likely to figure very highly in food sterilization as such. It is much more likely to be of use either in the pre-sterilization of aqueous additives or in the treatment of food factory effluent where rapid and efficient bacterial kill and waste product decomposition is of importance.

Thermal Sterilization Assisted by Sonication

The traditional method of sterilizing foodstuffs is thermal treatment and some efforts have been made to improve such treatment with the concurrent use of ultrasonic irradiation. The first report on the synergy between ultrasound and heat as a mechanism for killing bacteria, published in 1987 by Ordonez [41], was concerned with the vegetative bacterium *Staphylococcus aureus*. This report was followed by a further paper from the same group in 1989 [42] which provided results on the thermosonication (a term now given to the combined application of heat and ultrasound) of *Bacillus subtilis* spores. This treatment produced a substantial improvement in treatment time using a small scale reactor operating at 20 kHz and 150 W. Subsequent (but as yet unpublished) work has shown that the effect also extends to the inactivation of food enzymes. These findings are very significant to the food and other industries which operate processes to sterilize or pasteurize fluids with heat. It can be envisaged that process times and/or temperatures could be reduced to achieve the same lethality values. Environmental benefits from energy efficiency might also ensue.

When ultrasound was applied at moderate temperatures—below those used in pasteurization—to brain/heart infusion contaminated with *Salmonella typhimurium*, a reduction of contamination of 99% was observed [43]. Once again the enhancement to the efficiency of the process induced by sonication decreased as temperature increased. Reductions in contamination were also observed in the similar treatment of skimmed milk.

A new treatment process for food sterilization involving a combination of heat and sonication under pressure has been reported [44]. This methodology was found to decrease the heat resistance of *Staphylococcus aureus* by 63% and *Bacillus subtilis* by 43% as compared to heat treatment alone. This effect decreased with increase in temperature towards boiling point due to the fact that cavitational collapse becomes much less violent at these temperatures. Under pressure the boiling point is raised and lethality of the microbes above boiling point is maintained with values 5 to 30 times greater than those achieved with heat treatment alone. Spores appeared to be most resistant and yeasts the most susceptible to this type of treatment.

The increase in efficiency of chlorine in the destruction of *Salmonella* induced by ultrasound has been reported [45]. Attached *Salmonella* is often very difficult to remove from the skin of poultry. It was found that sonication produced a larger effect than shaking on the efficiency of decontamination with chlorine in this process. It seems that sonication serves to detach the bacteria from the surface making them more susceptible to disinfection by chlorine.

3.5 Ultrasonic Emulsification

One of the earliest uses of power ultrasound in processing was in emulsification. If a bubble collapses near the phase boundary of two immiscible liquids, the resultant shock wave can provide a very efficient mixing of the layers. Stable emulsions generated with ultrasound have been used in the textile, cosmetic, pharmaceutical, and food industries. Such emulsions are often more stable than those produced conventionally and often require little, if any, surfactant. Emulsions with smaller droplet sizes within a narrow size distribution are obtained when compared to other methods.

A type of mechanical transducer, which is used mainly for homogenization and emulsification, is the "liquid whistle" [46]. As its name implies this is a device for converting fluid motion into sound rather than the more conventional whistle which operates on gas motion. If a liquid is forced rapidly from a jet across a clamped thin metal blade the blade is caused to vibrate with a frequency dependent on the flow rate (Figure 7). The flow is adjusted to obtain ultrasonic frequencies and under these circumstances the liquid undergoes cavitation as it passes across the blade. A close analogy for this effect is the cavitation that is produced in water around a ship's propeller. When a mixture of immiscible liquids is forced across the blade of the liquid whistle the resulting cavitational mixing produces extremely efficient homogenization. Liquid whistle devices differ markedly from the more common

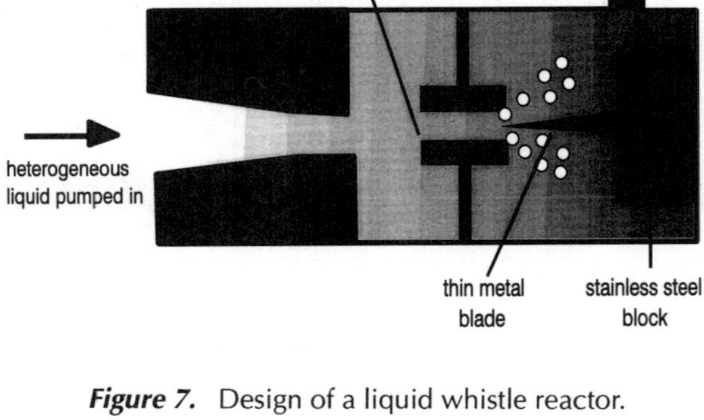

Figure 7. Design of a liquid whistle reactor.

sources of ultrasound (baths and probes) in that they derive their power from the medium (by mechanical flow across the blade) rather than by the transfer of energy from an external source to the medium.

The use of the liquid whistle type of ultrasonic generators for homogenization has increased dramatically since World War II. It was as far back as 1927 when Wood and Loomis reported that oil and water could be emulsified on sonication in the same beaker using quartz piezoelectric transducers [47]. In 1948 Janovski and Pohlmann highlighted the economic advantages to be gained from the use of a liquid whistle compared with magnetostrictive transducers of the type available at that time [48]. In 1960 a series of experiments was undertaken to compare four methods then in common usage for the emulsification of mineral oil, peanut oil, and safflower oil [49]. The results proved that a homogenizer, which operated via a liquid whistle, was superior to three other types of apparatus, namely a colloidal mill and two types of sonicator, one of which employed a quartz crystal and the other a barium titanate transducer. Subsequently the liquid whistle was adopted as an ideal method of homogenization.

The hydrolysis of wool wax and the subsequent refining of wool wax alcohols provides an excellent example of the use of a laboratory liquid whistle reactor. Generally it is with liquid–liquid reactions involving phase transfer catalysts that the full benefit of ultrasonic homogenizers becomes evident. Sonication produces homogenization (i.e. very fine emulsions) which greatly increases the reactive interfacial area and allows faster reaction at lower temperatures. Davidson has reported an example of this with the ultrasonically enhanced saponification of wool waxes by aqueous sodium hydroxide using a phase transfer catalyst [50]. The products obtained using this technique were in higher yield than obtained using a conventional hydrolysis and, perhaps of even more importance, the alcohols were cleaner and less colored. This indicates that the ultrasonic method leads to less decomposition.

An obvious benefit of the liquid whistle for emulsification is that it can be used for flow processing and can be installed "on-line." In this way volumes up to 12,000 liters per hour can be processed, as is the case in the manufacture of such items as fruit juices, tomato ketchup, and mayonnaise.

3.6 Extraction

The classical techniques for the solvent extraction of materials from vegetable sources are based upon the correct choice of solvent coupled with the use of heat and/or agitation. The extraction of organic compounds contained within the body of plants and seeds by a solvent is significantly improved by the use of power ultrasound. The mechanical effects of ultrasound provide a greater penetration of solvent into cellular materials and improves mass transfer. There is an additional benefit for the use of power ultrasound in extractive processes which derives from its use by microbiologists for the disruption and release of contents of biological cell walls.

The rate of ultrasonic extraction of sugar from sugar beets has been investigated [51]. It was found that the disruption of cells via cavitation effects results in the subsequent release of cell contents into the bulk medium. This combined with enhanced mass transfer, due to the effects of microstreaming, all resulted in a more efficient method for sugar extraction. Ultrasonically assisted extraction can also be applied to the production of medicinal compounds such as helicid, berberine hydrochloride, and bergenin from Chinese plants [52]. In some cases sonication increased the efficiency of extraction at lower temperatures producing a purer product in a shorter time. Thus helicid, which is normally extracted by refluxing in ethanol, can be obtained in a 50% higher yield in half the extraction time at room temperature using ultrasound. Once again efficient cell disruption and effective mass transfer are cited as the major factors leading to this enhancement.

Protein extraction from defatted soya beans was studied by Wang [53,54]. A continuous process was developed where sonication of the slurry by a 550-W probe operating at 20-kHz frequency resulted in an efficient extraction which exceeded any previously available technology. This was scaled up to a pilot plant for the extraction of soya bean protein [55].

The effects of ultrasonic treatment and ultrasonic irradiation on the extraction of cottonseed meal have been assessed [56]. Results indicated that a combination of enzyme treatment and sonication produced a significantly greater yield of extracted protein than traditional methodologies.

The extraction of tea solids from leaves is commercially important because it is the starting point for the production of instant tea. Instant tea is a powder derived from pure tea infusion from which water has been removed by spray drying. The use of ultrasound improves extraction at 60 °C by nearly 20% [57]. The effectiveness of ultrasonic extraction is greater than normal thermal extraction and the time taken is reduced, the majority of material is extracted in the first 10 min of sonication.

Power ultrasound has also been found to be effective in the extraction of protein from meat [58]. Ultrasound disrupts the meat myofibrils and this releases a sticky exudate which binds the meat together. The binding strength, water holding capacity, product color, and yields were examined after treatment either with salt tumbling, sonication, or both. Samples which received both salt treatment and sonication were superior in all qualities. Similar results were obtained from a study of the effect of sonication on cured rolled ham [59]. Ultrasonic treatment enhanced the extraction of myofibrillar proteins leading to an increase in the strength of the reformed meat.

An increase in both the yield and activity of the enzyme rennin from milk has been achieved using ultrasound [60]. Rennin is an important material used in the production of cheese.

3.7 Crystallization

Power ultrasound has proved to be extremely useful in crystallization processes. It serves a number of roles in the initiation of seeding and subsequent crystal formation and growth. It also has a property which is particularly beneficial in processing application, namely that the cleaning action of the cavitation effectively stops the encrustation of crystals on cooling elements and thereby ensures continuous efficient heat transfer. It is of considerable practical importance to be able to control the point at which crystallization occurs in any large-scale production process. Often crystallization can occur in an uncontrolled manner simply due to a slight drop in temperature or pressure. On the other hand, to control the initiation of crystallization is often difficult since problems may occur due to incorrect external factors such as temperature and pressure settings. Sonication is thought to enhance both the nucleation rate and rate of crystal growth in a saturated or supercooled medium by producing fresh and/or more nucleation sites in the medium. This may be due to cavitation bubbles acting as nuclei for crystal growth or by the disruption of seeds/nuclei already present within the medium, thus increasing the number of nuclei present in the solution/melt.

Sonication is believed to enhance both the nucleation rate and rate of crystal growth in a saturated or supercooled medium. There are thought to be two types of mechanism involved in sonically enhanced crystallization. The first is called *homogeneous crystallization* in which nucleation and crystallization in a pure supercooled melt cavitation produces fresh nucleation sites in the medium. Thus it has been observed that nucleation occurred after the collapse of the cavitation bubble [61]. It was suggested that the localized high pressures formed during cavitational collapse may well be strong enough to increase the melting point of the melt, thus initiating nucleation. Other research has suggested that sonication induces spontaneous crystallization via the formation of fresh nuclei [62] which arise spontaneously within the melt when it is close to the metastable point.

The second type called *heterogeneous crystallization* involves the mechanical effect of cavitation which breaks up crystals to generate new small "seeds" for further crystallization. Kapustin [63] observed small particles breaking off a thymol crystal when it was suspended into a saturated mixture of thymol and water and subjected to sonication; however crystallization did not occur in the absence of sonication. Many years ago Sokolov [64] also found that sonicating a seeded solution of sugar and $ZnSO_4$ resulted in an increase in rate of crystallization. This is supported by the observation that purification and removal of all possible nuclei from a solution of sucrose did not result in an acceleration of crystallization rate when the solution was subjected to sonication [65]. This suggests that, in the case of sucrose solutions, a heterogeneous mechanism is most likely.

Of course the possibility of obtaining a seed-free saturated solution of many of these materials under normal conditions is remote. It seems more likely that sonication can simultaneously enhance nuclei generation and crystal growth by taking advantage of the seeds already present in the solution.

There are a number of examples of the use of ultrasound in crystallization in the chemical field. Insonation of gels formed from sodium aluminate and sodium silicate leads to increases in the nucleation and crystallization rates for the formation of zeolites by up to six-fold and three-fold, respectively, at 85 °C [66]. The zeolite formed in the ultrasonic field show reduced particle size and a narrower size distribution compared with those produced by conventional methodology. The reduction in particle size was shown not to be caused by fracture of the zeolite particles by the effect of cavitation and is thought to be due to an increase in the number of crystallization nuclei and their dispersion by the acoustic field. The rate of nucleation was shown to increase, whereas the particle size decreased with increasing irradiation intensity. Results for a series of Zeolite NaA syntheses using different sources of ultrasound (low-power bath and high-power probe) show significant increases in the nucleation and overall reaction rates relative to control reactions with mechanical stirring (Table 4).

In these examples the mother liquor was unseeded before the crystallization process was initiated. The reactions performed in an ultrasonic field gave significant reductions in both particle size and in the breadth of particle-size distributions. Particle size was found to decrease with increasing acoustic power, but scanning electron microscopy (SEM) revealed that the crystal morphology was largely unaffected by the acoustic field although there was a reduction in crystal size with increasing acoustic power. In a series of separate experiments it was demonstrated that preformed zeolite crystals were not cleaved in an ultrasonic field. This confirmed that the observed reduction in particle size was due to an increase in the number of nuclei which are able to grow into separate crystals rather than subsequent fragmentation of larger crystals. The fact that the influence of the ultrasonic field is such that it leads to increased nucleation rates for the seeded system and that it gives rise to reductions in particle sizes similar to those in the unseeded reaction would seem to support a liquid phase nucleation mechanism [67]. This is one in which the ultrasound facilitates dissolution and dispersion of some of the seed which then provide germ nuclei rather than a solid phase nucleation mechanism which has been suggested by several groups of workers [68]. The increase in the nucleation rate in ultrasonic fields and the generation of more nuclei is most probably related to the very high rates of agitation in the ultrasonic fields.

Table 4. Effect of Ultrasound on the Crystallization of Zeolite NaA

Ultrasonic Source	*Nucleation Time (h)*	*Completion Time (h)*
control	5	10
bath (38 kHz)	3	7
Probe (20 kHz)	1	3.5

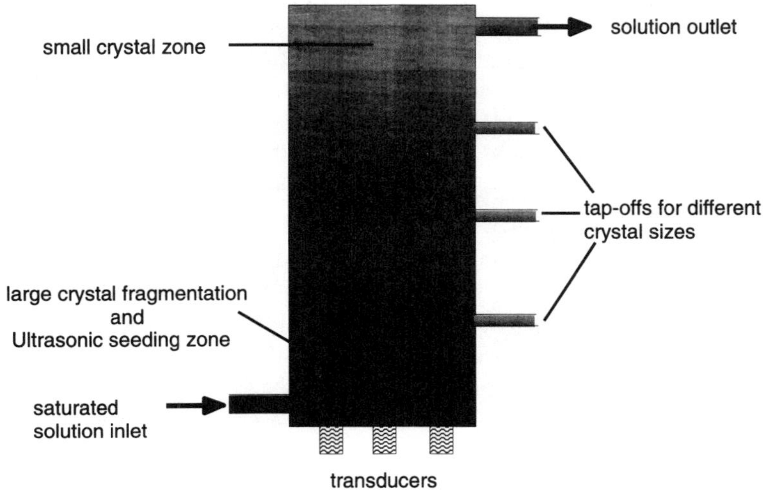

Figure 8. Large scale ultrasonic crystallization apparatus.

The scale-up of ultrasonically assisted crystallization has been cited in the literature. One U.S patent describes a unit which was operated successfully some years ago in the production of a crystalline drug (Figure 8) [69]. In this design the ultrasound, introduced through an array of horns at the base of the vessel, serves two purposes. First, the saturated liquid is ultrasonically seeded as it enters the bottom. The small seeds rise with the liquid flow and continue to grow and then become large enough to start to sink under the influence of gravity. Second, the large crystals are then fragmented as they approach the ultrasonic sources and rise again providing large numbers of fragments which themselves act as seeds.

It has been reported that ultrasound can be used to clarify wines through the precipitation of potassium bitartrate [70]. The treatment reduces the precipitation time from 4 to 10 days to 1.5–2 h.

One very important area related to crystallization in the food industry is the formation of ice crystals during the freezing of water. The quality of "fresh" thawed foods preserved through freezing can be somewhat disappointing in terms of texture. This is particularly true of soft fruits such as strawberries. The problem arises because the small ice crystals which are formed initially inside of the cellular material of the food continue to grow. As these crystals increase in size they break some of the cell walls leading to a partial destruction of the structure of the material. There is a considerable "dwell time" between the initiation of crystallization (usually at about –3 °C) and complete freezing at which point the temperature of the whole item can fall. Under the influence of ultrasound a much more rapid and even seeding occurs and this leads to a much shorter dwell time. In addition, since there are a greater number of seeds, the final size of the ice crystals is smaller and cell damage is reduced. In a somewhat related field the effect of ultrasound on the production of ice lollipops has been investigated [16]. Sonication generates ice crystals of a significantly reduced size with a more even distribution throughout the solid. This gives rise to a lollipop which is much harder than the conventional product which makes it less customer friendly in that it is far less easy to bite. A true advantage however seems to be that it adheres much more strongly to its supporting wooden stick.

3.8 Viscosity Reduction

Many food systems exhibit complex flow behavior and are thixotropic, that is their viscosity decreases as their molecules shear. This depends upon the spherical nature of the particles in the suspension. Initially they are randomly orientated but begin to line up in one direction as shear/stress is applied. Sonication results in reduced viscosity probably due to the particles ordering themselves uniformly in the path of the ultrasound.

This can be useful in processing where, for example, polysaccharides need to be "thinned" in order to be processed efficiently. In this context a reduction in viscosity due to ultrasound has been demonstrated in the cases of aqueous dextran, gelatin and agar [71].

3.9 Airborne Ultrasound

When a sonic standing wave is set up in air the particles suspended in the air will migrate to the nodes of the sound wave and this phenomenon has been used in a variety of applications. One remarkable consequence of this is the ability to construct a levitation reactor. When the source of ultrasound is above a planar surface, small particles will migrate to the nodes from the surface, i.e. levitate. Studies in such a system are not dissimilar from studies in the zero-gravity conditions of a space shuttle.

An application which is based on the effect of the migration of particles into the nodes of a standing acoustic field is the destruction of smokes. Smoke particles normally remain suspended in air for a considerable period because they are extremely light. In an acoustic field they will become concentrated in the nodal zones which will lead to an increased possibility of collision resulting in the formation of larger fragments. As these increase in size through further collisions they will become heavier than air and fall to the floor of the chamber. Full commercialization of such a process will require that the emitting face of the sound source is large and the design of such equipment is complex [72]. Developments of this process would certainly include the use of airborne sound for the precipitation of finely dispersed powders.

This same airborne ultrasound which can be used to destroy smokes is also capable of defoaming liquids by placing the acoustic source above the liquid surface upon which the foam is being generated. The methodology has proved so effective in defoaming that commercialization is already taking place in the bottling and canning process of various carbonated liquids (e.g. Coca Cola). At the time of this writing the mechanism of defoaming is not fully understood, although it is assumed to be connected in some way with the establishment of standing waves.

3.10 Degassing

The cavitational effects which are the basis of sonochemistry are also the reason for the extremely effective uses of ultrasound for the degassing of liquids. Any dissolved gases or gas bubbles in the medium act as nuclei for the formation of cavitation bubbles. Such bubbles are not easily collapsed in the compression cycle of the wave due to the fact that they contain gas and they will continue to grow on further rarefaction cycles, filling with more gas and eventually floating to the surface. Since the rarefaction cycles are taking place extremely rapidly (around 40,000 times per second using an ultrasonic bath) the bubbles grow so quickly that degassing appears to occur almost instantaneously.

The removal of unwanted air or gas is important in many food processes and it can be extremely difficult especially in very viscous liquids such as chocolate. Ultrasonically assisted degassing is particularly rapid in aqueous systems and can be used to remove any dissolved gas down to a very low level, thus is of great benefit in situations requiring the rapid and controlled removal of a gas or gases from a system. However ultrasound has also been employed to degas liquids as viscous as molten glass [73] and as dense as molten metals [74] during their solidification process.

A somewhat different development involves use of ultrasonic equipment for the removal of volatiles (such as methane and ethane) from brine which is a by-product of the offshore drilling for crude oil [75]. The particular advantage in the use of ultrasound in this situation is that the volatiles can be degassed while the brine remains under pressure, thus it avoids the traditional methodology of depressurizing to atmospheric and subsequently boiling off.

The electroplating of metals invariably involves gas generation at one or other of the electrodes and the physical presence of the gas acts as a barrier to efficient passage of current (discharge of ions). One of the ways in which ultrasound can improve electroplating—and indeed general electrochemical processes—is by the removal of this gas "barrier" [76].

3.11 Filtration

The requirement to remove suspensions of solids from liquids is common to many industries including chemical, engineering, and food processing. This separation can be either for the production of solids-free liquid or to isolate the solid from its mother liquors. Ultrasonic filtration of particulate matter from a liquid is now arousing some interest since the rate of flow through a filter can be increased substantially on application of ultrasound.

Conventional membranes of various sorts have been employed ranging from the simple filter pad through semipermeable osmotic type membranes to those which are used on a size-exclusion principle for the purification of polymeric materials. Unfortunately the conventional methodologies often lead to "clogged" filters and, as a consequence, there will always be the need to replace filters on a regular basis. Obviously a considerable economic advantage would accrue if nonclogging (and therefore continuously operating) filters could be developed. In some cases ultrasonic irradiation completely prevents the formation of a filter cake on the filter. In this respect, the application of ultrasound provides an ideal solution to the problem. There are two specific effects of ultrasonic irradiation which can be harnessed to improve the filtration technique: (i) sonication will cause agglomeration of fine particles (i.e. more rapid filtration), yet (ii) will supply sufficient vibrational energy to the system to keep the particles partly suspended and therefore leave more free "channels" for solvent elution. Using ultrasonic assistance, Fairbanks and Chen [77] obtained an 18-fold increase in filtration rate of motor oil through a sandstone filter when insonating the sample 15 cm above the filter. Semmelink [78] examined the effect ultrasound has on tap water flow through a metal wire cloth resulting in a rate increase of up to 300-fold in the presence of ultrasound.

The combined influence of these effects has been successfully employed to enhance vacuum filtration of industrial mixtures such as coal slurry, a particularly time-consuming and difficult process [75]. With the application of ultrasound to filtration, known as "acoustic filtration," the moisture content of slurry containing 50% water can be rapidly reduced to 25%, whereas conventional filtration achieves a limit of only 40%. Since coal slurry is combustible at 30% moisture content, the potential for this process is clearly enormous when applied to a continuous belt drying process. An improvement on the acoustic method has been developed in which an electrical potential is applied across the slurry mixture while acoustic filtration is performed [79]. The filter itself is made the cathode while the anode, on the top of the slurry, functions as a source of attraction for the predominantly negatively charged particulate material. The additional mobility introduced by the electric charge—"electro-acoustic filtration"—increases the efficiency of drying of 50% coal slurry by a further 10%. When applied to fruit extracts and drinks this technique has been used to increase the apple juice extracted from pulp. Where conventional belt vacuum filtration achieves a reduction in moisture content from an initial value of 85% to 50%, electro-acoustic technology achieved 38%, a result with obvious commercial significance.

The use of ultrasound has also enabled the development of a new, efficient method of separating particulate matter from a flowing material without the need for a filter membrane [80]. One such process consists of a device comprised of two transducers set at a horizontal distance apart across the liquid-containing particles. If one transducer is activated and sets up a standing wave, then any particulate contaminant in the fluid is seen to collect rapidly in regions corresponding to 1/2 wavelength distances on the axis of the ultrasonic beam. If both transducers are

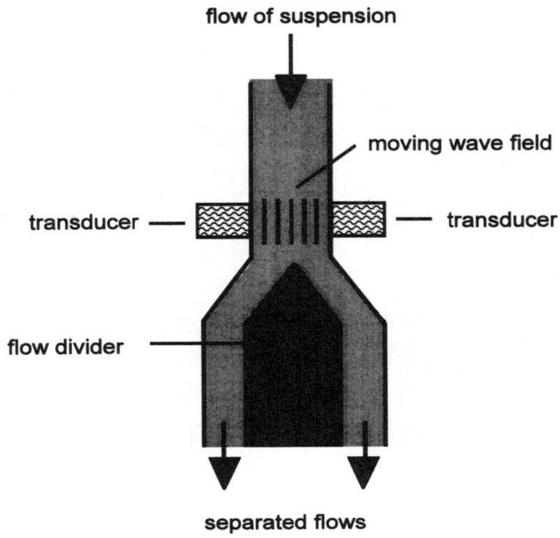

Figure 9. Separation of particles in an acoustic field.

energized, at the same frequency, the same effect occurs with positive interference resulting in the formation of a node of high-intensity sound and negative interference resulting in the formation of antinodes. On formation of the standing wave any particulate matter in the flow will gather at the nodal points. If there is a phase difference between the signals, a drifting wave may be established. This effect is the key to a separation process in which the standing waves traverse through the medium carrying the particles trapped at 1/2 wavelength distances. The particles thus become concentrated at one of the two transducers. It is but a short step from this simple idea of particle migration to the use of the principle for particle separation. All that is now required is that the transducers are placed across a flow system and the stream beyond the ultrasonic field is cut in two (Figure 9). A concentration of particulate matter will be produced in one of the two outflows. Recycling of these outflows may be required to achieve efficient separation.

Alternatively, if the standing waves are set up vertically using transducers on either side of a small vertical vessel (say a test tube), then any material which migrates and gathers at the nodes will be held static when the vessel is removed in a vertical direction from the ultrasonic field. Essentially the particles are physically being collected by the base of the vessel as it reaches them. This device is known as the "Wavecomb" [81].

3.12 Acoustic Drying

Acoustic drying is of potentially great commercial importance. Ultrasonically enhanced drying can be carried out at lower temperatures than conventional methodology which reduces the probability of oxidation or degradation in the

Table 5. Improvement in Throughput of a Rotary Drying Unit Using Ultrasound

Material for Drying	*Initial Water (%)*	*Target Final Moisture (%)*	*Increased Throughput Due to Ultrasound*
orange crystals	3.5	1.8	375
grated cheese	16.8	5.9	40
gelatin beads	12.9	3.7	83.3
rice grains	27.6	14.5	122.2

material. Additionally, unlike high-velocity gas drying the material is not blown about or damaged. In 1964 Soloff reported a study of acoustic drying of a number of materials using a rotary drier with a sonic source of 169 dB at 10.9 kHz [82]. The comparative results in terms of increased throughput due to the application of sonics is shown in Table 5.

Sugar crystals can be dried to a moisture content which is 50–75% less than can be obtained in the same time as conventional processing using ultrasound [83]. Similar technology reduced the drying time of fermentation sediment keeping the temperature below 40 °C [84]. The effect of sonication during the drying process of shelled corn and whole and crushed wheat increased the rate for all samples with the effect being more pronounced at lower temperatures [85]. A 130% drying rate increase was observed at 21 °C when sonic irradiation was applied. At 63 °C the rate was increased by 66%, and at 79.5 °C only a 6% rate increase was noted.

The drying of coal particles in air can be substantially accelerated using ultrasound [86]. Warm air at 65 °C was passed over the drying surface of the coal and ultrasound was applied directly to the surface of the coal particles. Not only was the rate of drying increased but also the moisture content of the final dried sample was reduced. This is thought to be initially due to a reduction of pressure above the sample to encourage water loss. The size of the particles greatly affected the characteristics of the liquid-filled channels present within the sample and may reduce the capillary action necessary for efficient drying.

Ensminger [87] examined the effects of sonic irradiation during the electrophoresis and electroosmosis drying of food materials. Electroosmosis involves the removal of water by encouraging water molecules to pass through porous membranes by the application of a direct current electrical potential. However, during such a process electrolysis of the water may occur which involves the formation of hydrogen at the cathode and oxygen at the anode. The presence of these gases reduces the electrode potential and restricts movement of water molecules. Application of ultrasound aids the removal of these gases from the electrodes and encourages higher potentials and faster drying rates.

Ultrasound can be employed to increase the rate of heat transfer between a solid heated surface and a liquid [87]. The ultrasound can either be introduced to the liquid itself or by vibrating the solid heated surface. It is thought that cavitation aids in the disruption of the boundary layer and as the temperature of the liquid increases the cavitation threshold is reduced. Heat transfer is increased by approximately 30 to 60%.

REFERENCES AND NOTES

[1] Wells, P.N.T. *Ultrasonics*, 31 (1993) 345.
[2] Pethrick, R.A. Ultrasonic studies of polymeric solids and solutions. In Mason, T.J. (ed.), *Advances in Sonochemistry*. JAI Press, London, 1991, Vol. 2, pp. 65–134.
[3] Miles, C.A., Shore, D., and Langley, K.R. *Ultrasonics*, 28 (1990) 394.
[4] Javanaud, C., Gladwell, N.R., Gouldby, S.J., Hibberd, D.J., Thomas, A., and Robins, M.M. *Ultrasonics*, 29 (1991) 331.
[5] Gartside, C.S. and Robiňs, M.M. Ultrasound in diagnosis, inspection and monitoring. In Mason, T.J. (ed.), *Sonochemistry, the Uses of Ultrasound in Chemistry*. Royal Society of Chemistry, 1990, pp. 27–46.
[6] McClements, D.J. and Povey, M.J.W. *Ultrasonics*, 30 (1992) 383.
[7] McClements, D.J., Povey, M.J.W., and Dickinson, E. *Ultrasonics*, 31 (1993) 433.
[8] Spect, W. *Nuovo. Chim. Ser.*, (1951).
[9] Chambers, L.A. and Smith, E.W. U.S. Patent No. 20 88 585 (1937).
[10] Gordon, M.H. *Chemistry In Britain* 27 (1991) 1020.
[11] Tadasa, K., Yamamoto, Y., Shimoda, I., and Kayahara, H. *J. Fac. Agric. Shinshu Univ.*, 26 (1990) 21.
[12] Ishimori, Y., Karube, I., and Suzuki, S. *J. Molec. Catal.* 12(2) (1981), 253.
[13] Rosenfeld, E. and Schmidt, P. *Archives of Acoustics*, 9 (1984) 105.
[14] Chambers, L.A. *J. Biol. Chem.*, 117 (1937) 639.
[15] Crawford, A.E. *Ultrasonic Engineering with Particular Reference to High Power Applications*. Butterworths Scientific Publications, 1955.
[16] Wiltshire, M. Presented at Sonochemistry Symposium, R.S.C. Annual Congress, Manchester, U.K., 1992.
[17] Naimark, G.M. and Mosher, W.A. *J. Acoust. Soc. Am.*, 25 (1953) 289.
[18] Bioprocessing Technology, *Technical Insights*, 11 (1989) (ISSN 0885-5625).
[19] Information provided by Undatim Ultrasonics, Louvain-le-Neuve, Belgium (1993).
[20] Xie, R., Zhang, Z., Zhang, D., Cheng, J., and Xia, X. *Proceedings of ICA14 Congress*, IP-3, 1992.
[21] Davidov, G.K. *Dokladi AN SSSR*, 29 (1940) 384.
[22] Istoma, O. and Ostrovsky, E. *Dokladi AN SSSR*, 65 (1976) 155.
[23] Choi, S.O. and Chung, J.D. *J. Korean Soc. Horticultural Sci.*, 32 (1991) 525.
[24] Nagy, J. and Tatar, M.K. *Novenytermeles*, 33 (1984) 329.
[25] Nagy, J. and Tatar, M.K. *Novenytermeles*, 33 (1984) 437.
[26] Khabibullaev, P.K., Ben'yaminov, G.I., and Yabukov, L.M. *Khlopkovodstvo*, (1975) 40.
[27] Abramov, O.V. Personal communication, 1994.
[28] Boucher, R.M.G. U.S. Patent 4,211,744 (1980).
[29] Akopyan, V.B. and Abramov, O.V. Report for IUS Ltd., Langford Ultrasonics, Birmingham, U.K. (1993).
[30] Hughes, D.E. and Nyborg. *Science*, 138 (1962) 108.
[31] Alliger, H. *American Laboratory*, 10 (1975) 75.
[32] Riesz, P. In Mason, T.J. (ed.), *Advances in Sonochemistry*, Second Edition. JAI Press, London, 1991, p. 23.
[33] Henglein, A. In Mason, T.J. (ed.), *Advances in Sonochemistry*, Third Edition. JAI Press, London, 1993, p. 17.
[34] Petrier, C., Jeunet, A., Luche, J-L., and Reverdy, G. *J. Amer. Chem. Soc.*, 114 (1992) 3148.

[35] Mason, T.J., Lorimer, J.P., Bates, D.M., and Zhao, Y. *Ultrasonics Sonochemistry*, 1 (1994) S91.
[36] Mason, T.J., Newman, A.P., and Phull, S.S. *World Water Environ. Engr.*, (1994),16.
[37] Rott, U. and Knehr, B. *Adv. Water and Effl. Treatment*, BHR (1993) 309.
[38] Toy, M.S., Carter, M.K., and Passell, T.O. *Environ. Tech.* 11 (1990) 837.
[39] Geplovasky, A., Donovalova, J., Toma, S., and Henciar, P. Czech Patent, CS 264,021 (1988).
[40] Savluk, O.S. *Khim. Tekhnol. Vody*, 4 (1982) 79, see also Gelzhaeuser, P. Ger. Offen. DE 3,739,979 (1989).
[41] Ordonez, J.A. *J. Dairy Res.*, 54 (1987) 61.
[42] Ordonez, J.A., Sanz, B., Burgos, J., and Garcia, M.L. *J. Applied Bacteriol.*, 67 (1989) 619.
[43] Wrigley, D.M. and Llorca, N.G. *J. Food Protection*, 56 (1992) 678.
[44] Sala, F.J., Burgos, J., Condon, S., Lopez, P., and Raso, J. Effect of heat and ultrasound on microorganisms and enzymes. In Gould, G.W. (ed.), *New Methods of Food Preparation*, Blackie, 1995.
[45] Lillard, H.S. *J. Food Protection*, 56 (1993) 716.
[46] Branson Ultrasonics, The Fairview Estate, Clayton Road, Hayes, Middlesex UB3 1AN, U.K.
[47] Wood, R.W. and Loomis, A.L. *Phil. Mag.*, 4 (1927) 417.
[48] Janovski, W. and Pohlmann, R.Z. *Angew. Phys.*, 1 (1948) 22.
[49] Singiser, R.E. and Beal, H.M. *J. Am. Pharm. Assoc. (Scient. Ed.)*, 49 (1960) 482.
[50] Davidson, R.S., Safdar, A., Spencer, J.D., and Robinson, B. *Ultrasonics*, 25 (1987) 35.
[51] Chendke, P.K. and Fogler, H.S. *Ultrasonics*, (1975) 31.
[52] Zhao, Y., Bao, C., and Mason, T.J. *Ultrasonics International '91 Conference Proceedings.* Butterworths, 1991, p. 87.
[53] Wang, L.C. *J. Food Sci.*, 40 (1975) 549.
[54] Wang, L.C. *J. Agric. Food Chem.* 29 (1981) 177.
[55] Moulton, K.J. and Wang, L.C. *J. Food Sci.*, 47 (1982) 1127.
[56] Childs, E.A. and Forte, J.F. *J. Food Sci.*, 41 (1976) 652.
[57] Mason, T.J. and Zhao, Y. *Ultrasonics*, 32 (1994) 375.
[58] Vimini, R.J., Kemp, J.D., and Fox, J.D. *J. Food Sci.*, 48 (1983) 1572.
[59] Reynolds, J.B., Anderson, D.B., Schmidt, G.R., Theno, D.M., and Siegel, D.G. *J. Food Sci.*, 43 (1978) 866.
[60] Zayas, J.F. *J. Dairy Sci.*, 69 (1986) 1767.
[61] Hunt, J.D. and Jackson, K.A. *J. App. Phys.*, 37 (1966) 254.
[62] Chormonov, T. *Chem. Abstr.*, 60 (1964) 12727b.
[63] Kapustin, A.P. *The Effects of Ultrasound on the Kinetics of Crystallization*. Consultants Bureau, New York, 1963, p. 41.
[64] Sokolov, S. Ya. *Tech. Phys. USSR*, 3 (1936) 176.
[65] Van Hook, A,. Radle, W.F., Bujake, J.E., and Casazza, J.J. *J. Am. Soc. Sugar Beet Technologists*, 9 (1957) 590.
[66] Lindley, J., Lorimer, J.P., Maan, R., Mason, T.J., and Roberts, C.W. *Ultrasonics International '89, Conference Proceedings*. Butterworths, 1989, p. 1264.
[67] Zhdanov, S.P. *Adv. Chem. Ser.*, 101 (1971) 20.
[68] Stojkovic, S.R. and Adnadjevic, B. *Zeolites*, 8(6) (1988) 523.
[69] Midler, M. U.S. Patent 3,510,266 (1970).
[70] Spirov, N., Goravon, N., Mitev, D., and Lesichkov, V. *Khranit Prom.*, 22 (1973) 30; *Chem. Abstr.* 79 (1973) 51798e.
[71] Lockwood, A.R., James, A.E., and Pautard, F.B. *Research London*, 4 (1951) 46.
[72] Gallego-Juarez, J.A. Presented at a power ultrasound symposium, Leatherhead Food Research Institute, U.K. (1992).
[73] Spinosa, E.D. and Ensminger, D.E. *Glass Industry*, (1987) 10.
[74] Degassing, filtration and grain refinement processes of light alloys in a field of acoustic cavitation, Eskin G.I. in this volume (1996).

[75] Senapati, N. Ultrasound in chemical processing. In Mason, T.J. (ed.), *Advances in Sonochemistry*. JAI Press, London, 1991, Vol. 2.
[76] Mason, T.J., Lorimer, J.P., and Walton, D.J. *Ultrasonics*, 28 (1990) 333.
[77] Fairbanks, H.V. and Chen, W.I. *Chem. Eng. Symp. Series*, 67 (1971) 108.
[78] Semmelink, A. *Ultrasonics International '73 Conference Proceedings*. IPC Science and Technology Press, Guildford, 1973, p. 7.
[79] Muralidhara, H., Parekh, B., and Senapati, N. U.S. Patent 4,561,953 (1985).
[80] Schramm, C.J. The manipulation of particles in an acoustic field. In Mason, T.J. (ed.), *Advances in Sonochemistry*, JAI Press, London, 1991, Vol. 2, p. 293.
[81] British Technology Group, 101 Newington Causeway, London SE1 6BU, U.K.
[82] Soloff, R.S. *J. Acoustic Soc. Am.*, 36 (1964) 961.
[83] Boucher, R.M.G. U.S. Patent 3 175 255 (1970).
[84] Boucher, R.M.G. *Ultrasonic News*, 3 (1959), 8–9, 14–16.
[85] Huxsoll, C.C. and Hall, C.W. *Trans. ASAE.*, 13 (1970) 21.
[86] Fairbanks, H.V. *Ultrasonics*, (1974) 260.
[87] Ensminger, D. *Drying Tech.*, 6 (1988) 473.

SONOELECTROCHEMISTRY

David J. Walton and Sukhvinder S. Phull

OUTLINE

Advances in Sonochemistry
Volume 4, pages 205–284

ISBN: 1-55938-793-9

1. INTRODUCTION

1.1 Background

The many benefits of ultrasound in chemical systems are well-known, and are reported elsewhere in this volume, but perhaps the most striking influence of ultrasound concerns heterogeneous reaction systems, particularly those with a solid–liquid interface where particle size modification, the cleaning of surfaces, or the formation of fresh surfaces are among the beneficial processes.

Another well-established heterogeneous interface is that between the electrode surface and the electrolyte in electrochemistry, where there are regimes of various degrees of order, characterized by differing mass transport phenomena and involving different kinetic and thermodynamic requirements. Adsorption and surface phenomena are important and in general it has been recognized for some time that vibration of an electrochemical system can produce a variety of effects.

Early studies in the field concerned electroanalytical and voltammetric systems, later exploited to improve the electrodeposition of metals in the electroplating industry; but until relatively recently very little had been reported about electroorganic systems. This situation is now being redressed, and there is now an upsurge of interest in the application of ultrasound to electrochemical systems of all sorts.

This has come about because of modern developments in both ultrasonic technology and in electrochemistry. Thus ultrasonic sources are now readily available in a variety of powers, frequencies, and geometries, capable of various pulse-styles, duty cycles, and other sophistications. Baths, horns, probes, and other reactors are now all available to the experimentalist.

Electrochemical methodology has now become extremely sophisticated, with a vast range of current/potential/frequency/time parameters available for manipulation, and with the development of microelectrodes, flowcells, and systems of controlled hydrodynamics which enables a sound theoretical basis for the interpretation of electroanalytical results, exploiting modern technology for data digitization and computational power. Electrochemistry can now be performed in media and with electrode systems and cell designs that could not have been employed earlier.

Ultrasound and electrochemistry provide a powerful combination, and this is a particularly appropriate time to review developments. This chapter is divided into sections concerning three main areas of electrochemistry, namely electroanalysis, inorganic electrochemistry, and organic electrochemistry. It concludes with a number of other applications grouped together. This review is mainly restricted to the application of ultrasound during the actual electrochemical events, which is the essence of sonoelectrochemistry. However, there is also a body of work in which ultrasound is used at some other point in the overall electrochemical procedure—for example in the cleaning of electrodes prior to making up the cell. This is particularly well-established in battery technology, but will not be discussed except on appropriate occasion. Sonochemical principles should be familiar to the reader, but before detailing specific sonoelectrochemical results it is perhaps worth summarizing fundamental principles of electrochemistry.

1.2 Electrochemical Principles

Electrochemistry concerns reactions in which a key process is the transfer of electrons at an electrode. There are three general types of system.

1. *Electroanalysis*, in which power from an external source drives measurable changes in system parameters such as current and potential as a function of time or some other variable; the data being processed to achieve a deduction without the isolation of reaction products.
2. *Electrosynthesis*, in which an external power source drives a preparative-scale electrolysis, the aim of which is to isolate and characterize a product, which may be a species in solution, a precipitate, a coating on the electrode, or some other phase. It should be noted that the conditions of electrosynthesis employ relatively high substrate concentration and large-area electrodes; and are often quite different from those of electroanalysis, with consequence upon the drawing of mechanistic comparisons between the techniques.
3. *Electrochemical power generation*, in which the electrochemical reaction itself drives the movement of electrons, and no external power source is involved. Typical systems include batteries and corrosion, and studies are often of an electroanalytical nature, with some examination of products.

There follows a short resume of the various aspects of electrochemistry, and the interested reader wishing to obtain further information is directed towards the plentiful reviews in the field. The following list is by no means exhaustive and other appropriate references are given throughout this review. Thus for general electrochemical laboratory practice [1], for electroanalytical methods [2] (the latter is part of a series [3]), for theoretical concepts [4], and for a recent view [5,6]. There are a number of useful introductory books, e.g. [7,8], and many series of reviews of "advances in the field," e.g. [9].

A useful introduction to the concepts of organic electrochemistry remains [10], while the most thorough introductory book and a classic in the field is [11]. A modern summary is given in [12], and there are more extensive series of volumes, e.g. [13].

There are reviews of other aspects, such as the industrial uses of electrochemistry [14], batteries [15], and corrosion [16] among others.

1.3 The Electrochemical Cell

In essence, all that is required for electrochemistry is two electrodes, which for an externally driven system are the anode which accepts electrons from species that approach it, and is the oxidizing electrode, and a cathode which donates electrons, and is the reducing electrode. The simplest system employs a constant flow of current through the cell. This galvanostatic procedure is experimentally straightforward: (1) the flow of charge can be directly integrated with time to monitor consumption of electricity, (2) the power source is just a constant-current generator, and (3) the whole arrangement can easily be scaled up for industrial processes. However, the true potentials at each electrode are not known, and the measured cell voltage across the cathode and anode includes the potentials of processes happening at both electrodes plus cell resistance and other losses. As a reaction proceeds and a substrate becomes consumed, the requirement for constant charge per unit time will initiate reaction of the next highest energy process. Thus for more controlled studies a third electrode, the reference electrode, is employed, and the potential of the working electrode is monitored against a known standard thermodynamically reversible couple, suitably chosen. This requires a potentiostat, a more sophisticated device that is essential for electroanalytical studies where accurate knowledge of potential, and its manipulation to induce changes in other parameters is the purpose of the experiment. Potentiostatic control is also used in electrosynthesis if a second reaction occurs at a higher potential making it is essential to keep the applied potential below this value. In a potentiostatic synthesis the cell current will die away as electroactive substrate is consumed, and the voltage to the counter electrode will drift to maintain the required potential between working and reference electrodes.

Since oxidized species are generated at the anode and at the same time reduced species evolve from the cathode these are likely to react together if allowed to meet. This is not usually a problem in electroanalytical studies where the total charge passed during a measurement is low, but must be taken into account for preparative electrolysis where stoichiometric amounts of charge are passed over significant time periods. It may be that the counter electrode reaction is gas evolution, in which case the products are removed from further reaction in the cell, and it may even be that the overall cell process requires the meeting of the reduced and oxidized species to form the final product. Such "coupled" systems are desirable since each charge to cross the cell is effectively used twice, both at the cathode and the anode, with enhanced efficiency; but careful selection of reactions is needed and these systems are relatively rare in practice. More commonly processes occurring at one of the electrodes (called the working electrode) are of interest and a separator is placed across the cell to compartmentalize the anolyte and catholyte; but this must still allow the transport of ions across the barrier or else current cannot flow. Any separator will of course affect overall cell resistance and undivided cells are to be preferred where possible on energy grounds. However, a separator can provide a useful sophistication if for example a permselective membrane is employed that only allows one type of ions to pass, such as for example Nafion® membranes in a

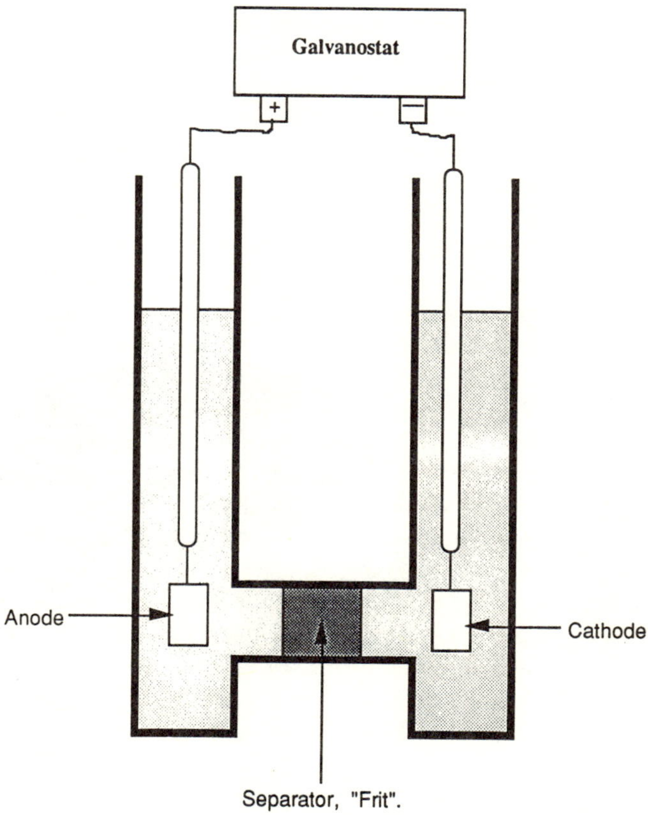

Figure 1. Typical H-cell for electrolysis using two-electrode setup (reference electrode can be inserted into working compartment for potentiostatic control).

chloralkali cell. Whether or not the cell is divided, from the point of view of overall efficiency it does not matter whether ultrasound improves processes at the working electrode or at the counter electrode since either or both will improve the cell performance.

Figure 1 shows a diagram of a typical laboratory scale preparative cell with separator widely used for academic studies (the H-cell). However, this is quite unrepresentative of the type of cell that would actually be used in a industrial scaled-up process. An example of a commercial cell is given in Figure 2. This shows an exploded diagram of the ICI FM21 cell, able to operate on a kilogram scale. ICI also produces a smaller laboratory scale test cell (FM01) of the same design for developmental studies prior to full scale-up. There are a number of other commercial designs available, and the reader is directed to ref. 12 for further details. The ICI type industrial cell has a very large electrode surface area, a very narrow interelectrode gap, and a wafer-thin separator, all to improve efficiency. In addition, commercial systems often employ flowing electrolytes and other means of forced hydrodynamics. Further sophistication is sometimes provided by use of multiphase electrolytes in which, for example, the anode and cathode are placed in different immiscible solvents. There are a number of engineering aspects to be addressed in the development of commercial electrochemical systems, and the reader should be aware that there may be substantial steps to be overcome in taking what appears a promising laboratory process toward scale-up.

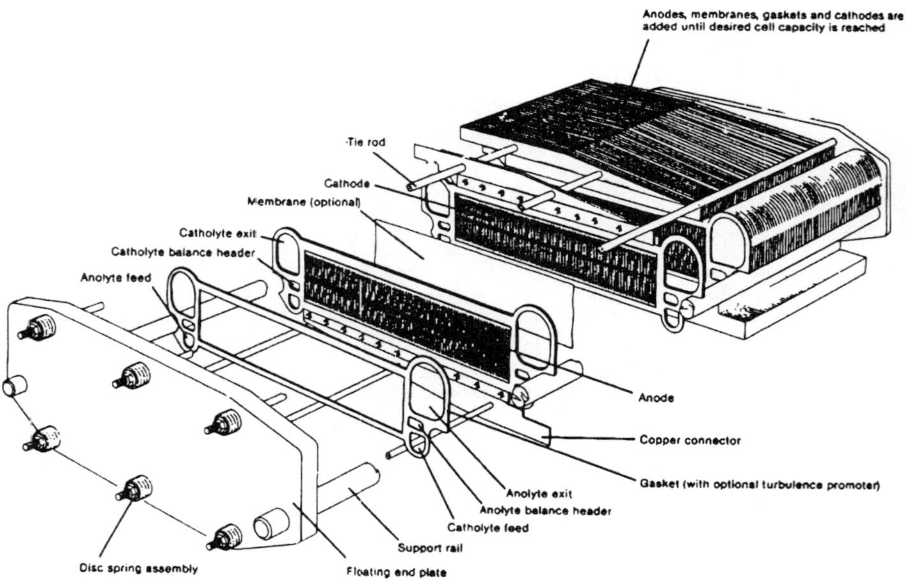

Figure 2. ICI FM20 industrial electrochemical cell (taken from ref. 8 with permission).

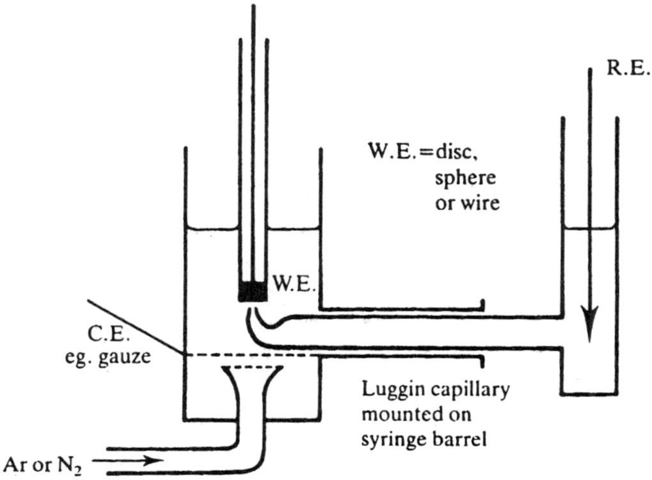

Figure 3. Schematic of simple cell setup for cyclic voltammetry (taken from ref. 8 with permission). W.E. = working electrode, C.E. = counter electrode, and R.E. = reference electrode.

There are also a number of criteria to be addressed in setting up an electroanalytical experiment. Figure 3 shows a typical laboratory cell arrangement for voltammetry. The working electrode could be a hanging wire of millimeter dimensions (a millielectrode), a flat disk of similar dimensions, or a disk of micron dimensions (microelectrode). The disk may be rotated in the horizontal plane to give the "Rotating Disk Electrode" [17]. All these electrode systems give different hydrodynamics, of importance to the interpretation of data. There are other modern developments in electroanalysis—for example using channel electrodes in which the electrolyte flows down a narrow tube cell with the electrode painted on to the

Table 1. Advantages of Electrochemistry in Synthesis

- Externally variable reactivity by control of electrode potential or rate of current flow.
- Selective transformations in mild conditions where, for example, the temperature of the bulk cell can be kept low to avoid side reactions, decomposition of electrogenerated species, or similar processes.
- Powerful and relatively unusual species (e.g., ion-radicals and similar) are generated *in situ*. Can separate the rate of generation of species from conditions in the bulk medium.
- Immediate control of reaction since the current ceases immediately when switched off. This is not the same for reflux where there is substantial thermal inertia.
- Avoids stoichiometric use of reagents; for example, expensive metal ions in reactive valent states which often have associated toxicity problems.
- Environmentally friendly—electrochemistry is a clean technology.

Table 2. Potential Influences of Ultrasound upon Electrochemical Systems

- Enhancement of mass transport to/from an electrode and in bulk solution, which affects concentration-dependent phenomena.
- Diminution of the diffusion layer, which affects electron transfer.
- Alteration of the adsorption and surface properties of the electrode.
- Reduction of electrode fouling and removal of electrode coatings; cleaning of the electrode surface.
- Induction of sonochemically induced follow-up reactions, which are not found in the silent system: the behavior of, for example, ion radicals under ultrasound is not well-recognized.
- Sonochemical generation of electroactive species that are nonelectroactive in silent conditions.

surface [18], or wall jet electrodes in which electrolyte is sprayed on to an electrode [19], while for *in vivo* electrochemical sensors the whole cell could occupy the size of a pinhead.

In general there are a wide number of different cell geometries, electrode configurations, and other process considerations in any electrochemical system, together with different means of imposing power using galvanostatic, potentiostatic, constant ramp, pulsed, cycled, or other procedures. There is also the nature of the reaction system, the solvent system, the electrolyte salt, and the addition of quenchers or other reagents to trap intermediates or control reaction pathways. Ultrasound could interact in a variety of ways to alter cell performance, as discussed in the following sections of the review—for example, in electrosynthesis there are several existing and well-recognized advantages of electrochemical methodology listed in Table 1, and the possible benefits that may accrue from ultrasonic irradiation given in Table 2. Specific process considerations for scale-up are briefly addressed in Table 3.

Examples of all such possible influences of ultrasound upon electrochemical systems are found in the work described in this review. This is why sonoelectrochemistry is such a promising field of research.

Table 3. Electrochemical Process Considerations (Important for Scale-Up)

Enhanced Efficiency: Since 1 Faraday = 96,500 coulombs, for one mole of substrate in a one-electron per molecule reaction, one amp must be passed for 96,500 seconds, and that is if all electrons go toward the desired reaction: Efficiency improvements are particularly desirable.

Energy Losses: Lessened cell voltages are desirable, since energy is often lost as heat [energy (watts) = current × voltage].

Cell Economics: Use of cheaper or recoverable solvents, e.g. aqueous emulsions for water-immiscible substrates to obviate expensive organic solvents; also use of lower concentrations of cheaper or easier to recover electrolyte salts, since enhanced mobility requires less ions to maintain current.

Ease of Work-Up: Benefits of more straightforward work-up (e.g. lower amounts of electrolyte salt facilitates product separation).

2. SONOELECTROANALYSIS

2.1 Early Studies

The effect of ultrasound upon electroanalytical systems has been studied for a considerable time. Thus a review by Yeager and Hovorka in 1953 had 105 references in it [20], and the reader is directed towards this paper for a grounding in physicochemical aspects as perceived at the time. The review addressed the effects of ultrasonic waves on three main areas stated as: electrode processes, electrokinetic phenomena, and structural studies on electrolyte solutions. Prior to a summary of more recent developments, this review is now discussed in some detail and all references are found therein.

Early studies on electrode processes concerned several aspects including alternating components in electrode potentials from gas-evolution reactions, particularly the polarized hydrogen electrode where the effect was attributed to modulation of the *IR* drop in the solution in the vicinity of the electrode through periodic variation in the size of gas bubbles formed at the electrode surface. The dc changes in electrode potentials for a number of systems were discussed, including the ability of ultrasound to decrease the overvoltage for hydrogen evolution at a number of metal electrodes, attributed to the reduction of concentration gradients adjacent to the electrode and the stripping off of the irreversibly adsorbed materials which reduce the active surface area of the electrode. Ultrasound also modifies the potentials at which various metals are electrodeposited, and structural changes in the electrodeposits were also observed. Metal corrosion is also affected by ultrasound; these results are further discussed in Section 3 below.

The effect of ultrasound upon electrokinetic phenomena was quoted for a number of aspects. Thus membrane transducers operate by development of a streaming potential as a dilute electrolyte solution flows through a porus diaphragm. If dc flow is replaced by the ac flow of an ultrasonic field then an ac potential is produced. This effect can be enhanced by polarization of the membrane. Alternatively ac potentials may be generated between two metal probes sealed in a glass capillary when the wall is vibrated longitudinally. This also gives streaming potential data. There may also be mercury present in the capillary contacted externally, e.g. by platinum electrodes. Here ac potentials are connected with charging capacitance at mercury solution interfaces due to the periodic variations in surface area under vibration. The device was tested as a transducer for a hydrophone, microphone, and a gramophone pickup.

A further electrokinetic system also tested as a hydrophone employed a fiber-coated wire in an electrolyte solution. Ultrasound induced potential differences between the wire and the bulk solution depending on the nature of the fiber coating. All these electrokinetic systems give data that are in principle related to zeta potentials.

Ultrasound also produces vibration potentials when propagated through colloidal solutions due to greater displacement of the ions in the diffuse layer than of the central colloid particle, for reasons of differential mass and frictional coefficients. This effectively produces dipoles of periodic movement, and hence ac potentials which can be detected by inert metal electrodes, the effect being most pronounced when these are placed in the direction of sound propagation and separated by an odd-multiple of the half-wavelength. This latter requirement demonstrates the importance of geometry, ultrasonic frequency, and power in sonoelectrochemical effects. These parameters should be borne in mind in the setting up of any sonoelectroanalytical experiment and in the analysis of data obtained.

Colloid vibration potentials offer a means of measuring the zeta-potential, and hence charge, on colloid particles. Values of $\sim 10^{-4}$ $Vcm^{-1}s$ at frequencies of a few hundred kilohertz seem to be typical of this effect, and a range of colloids were examined, including silver, silver iodide, and arsenic trisulfide.

In electrolyte solutions a similar effect is observed when anions and cations, which have different masses, solvation shells and frictional coefficients, are exposed to ultrasound. As in the case of colloids, periodic relative displacements will produce ionic vibration potentials, which may be measured as potential differences between two points separated by a finite distance, or by monitoring potential variation at one point relative to the average potential of the solution. The effect is smaller than for colloids, typically below 10^{-5} $Vcm^{-1}s$ at 200 kHz, but great experimental care must be taken to minimize direct electromagnetic coupling between the ultrasonic generator and the detection apparatus since this influences empirical values obtained. This problem was minimized during measurements of ultrasonically induced periodic pressure and temperature variations in the conductivity of electrolyte solutions by passing an ac current at frequency ν_1 through a cell exposed to ultrasound at frequency ν_2 and using an amplifier tuned at the combination frequency $(\nu_1 + \nu_2)$ to measure the resultant ac potential.

It has also long been noted that ultrasonic treatment of electrolyte solutions can produce increases in the measured bulk time-averaged conductivity that can persist for some considerable time. This was seen in both aqueous and nonaqueous systems, with some dependence upon experimental procedure and parameters such as ultrasonic frequency, and a diversity of results were reported. A possible explanation is that sonochemical reactions may create ionic species which alter solution conductivity. The ultrasonic treatment of electrolytic solutions prior to electrolysis remains a useful strategy, particularly in the preparation of battery electrolytes.

Measurement of ultrasonic velocity through an electrolyte solution permits evaluation of the isothermal compressibility, which can be related to ionic solvation phenomena, though only by making a number of necessary assumptions in the treatment. Yeager's review [20] refers to a number of other reviews in this field and also to the absorption of ultrasound by electrolyte solutions, another field widely studied at the time. Again the aim was to elucidate ion solvation and solution interactions from relaxation effects, and again necessary assumptions were required in the treatment. Ultrasound throws up interesting phenomena in electrolytes, e.g. the anomalous absorption of aqueous magnesium sulphate solutions, and this methodology provides another means of probing solutions. These early studies were carried out before modern methods were developed for voltammetry using rotating disk electrodes, microelectrodes, and other systems of controlled hydrodynamics, but the fundamental principles of physical chemistry remain of interest to electrochemists.

This early review [20] summarized international work in the field, including results from the substantial effort at Case Western Reserve University by Yeager, Hovorka, Zana, and many others associated with the Ultrasonics Research Laboratory there. This effort was part of an extensive and lengthy study by Yeager's group funded by the U.S. Office of Naval Research, which commenced in 1948 and continued up to 1981. During this time Yeager produced 58 publications in the open literature and 48 technical reports to the Office of Naval Research. Details of these are given in the unclassified Final Report in 1982 [21]. The reader is directed towards this and another review [22] to find appropriate references for details of apparatus and methodology as well as a summary of the results of the program. These include the development of a pulse-modulated ultrasonic technique, which allowed the first absolute values of partial molal volumes, and also gave information on ion–solvent and other solution interactions. These studies were extended during the course of the program to colloids, micelles, polyelectrolytes, lattices, dispersions, suspensions, emulsions, polymers, and biological systems; also developed were the acousto–electrokinetic effect and the interface electromodulation acoustic effect to further probe electrolyte systems. In all, this program represented a substantial contribution to the understanding of physical chemical phenomena under ultrasound, though without addressing in detail modern voltammetric aspects of electroanalysis. These have been the subject of more recent sonoelectroanalytical experimentation as detailed below.

2.2 Hydrodynamics and Mass Transport Phenomena

In one of the earliest papers on sonoelectrochemistry, Moriguchi [23] found that ultrasonic irradiation lowered the water decomposition voltage at platinum. Ultrasound affects gas-evolving reactions by sweeping the electrode clean of bubbles as they form. This affects hydrogen evolution, and is particularly useful to enhance chlorine evolution in chloralkali cells [24] since adhesion of "sticky" chlorine bubbles is a well-known problem in this technology. Both hydrogen evolution and chlorine evolution are thermodynamically reversible processes, whereas oxygen evolution is not, and ultrasonic perturbation provides a means of distinguishing between these mechanisms since unlike the others, oxygen evolution from dilute acid on platinized platinum is not enhanced by ultrasound [25]. This result has

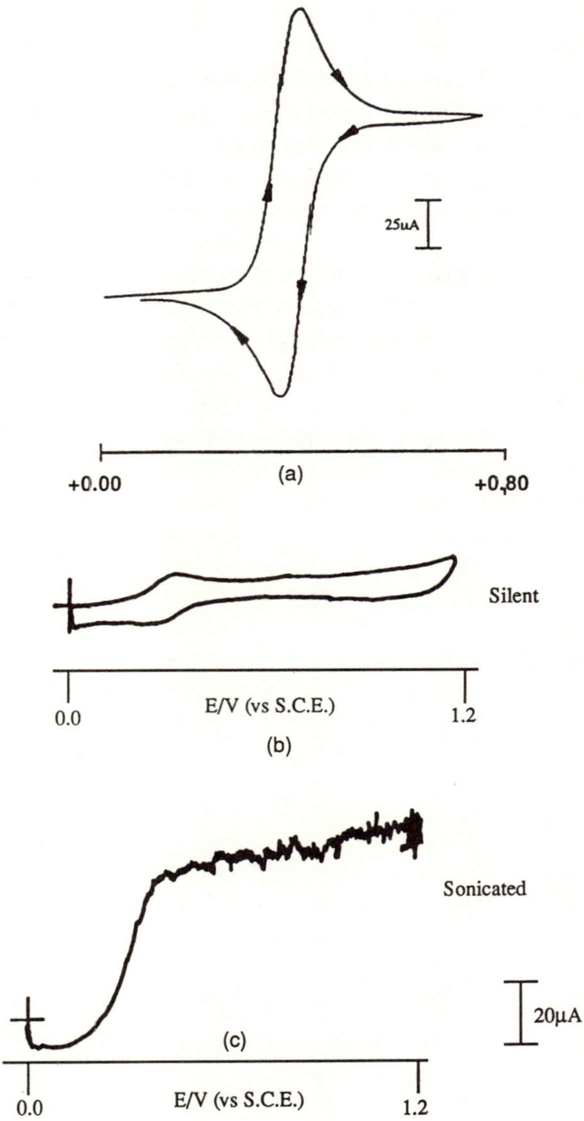

Figure 4. Voltammetry at platinum wire of ferrocyanide in aqueous solution (all traces taken from ref. 31). Curve (**a**) Typical "reversible" silent voltammogram showing classical shape. Curve (**b**) Trace from similar silent system to (**a**) but over extended potential range and on contracted current scale. Curve (**c**) Effect of 38-kHz ultrasound (bath) on same system as trace (**b**). Scan rate: 25 mV sec^{-1}.

implications for the sonication of systems in conditions where chlorine evolution and oxygen evolution compete.

However, the main effect of ultrasound is to increase the rate of mass transfer to the electrode raising the limiting current density and reducing the diffusion layer thickness. This is now an established phenomenon [26], and is for example well documented in the field of metal electrodeposition [27] (further detailed in Section 3 of this review). A discussion of the diffusion-layer thickness in the commercial "hydroson" electroplating tank gives a useful description of the principles [28]. There has also been substantial Rumanian work in the field, not so straightforwardly available to workers elsewhere. This includes among other reports a study into the diffusion kinetics under ultrasound of a number of redox systems [29], and also a relatively recent review of electrochemical and sonochemical perturbation upon diffusion-controlled phenomena [30].

The effect of ultrasound upon diffusion processes can readily be demonstrated by examination of the cyclic voltammetry of a typical reversible solution-phase redox couple such as ferrocyanide/ferricyanide at an inert electrode under both silent conditions and under sonication [31]. This is shown in Figure 4, where curve (a) demonstrates the characteristic shape of the oxidation and reduction waves at a platinum wire electrode of 1 mm diameter and a few mm in length. The data is representative of a typical reversible one-electron transfer voltammogram at an electrode of these dimensions. As the potential is scanned in the oxidation direction, as indicated in the Figure 4, there is a current observed once an appropriate potential is reached for electroactivity of the substrate. In principle this current should continue to increase since increasing the electrode potential promotes the electron transfer, but acting against this is the ease of mass transport to the electrode and in a still solution there comes a point at which demand for substrate diffusion becomes such that the curve turns over, and the process becomes under diffusion control, giving a characteristic peak shape. The same occurs on the return scan, provided no chemistry or other process interferes with the product of oxidation within the experimental timescale; the redox voltammogram has both forward and reverse peaks, the respective currents and potential values of which are connected by well-established identities leading to diagnostic criteria for electrochemical reversibility. Another important parameter here is the scan rate, and for a reversible system in solution the peak current should increase with its square root.

Curve (b) of Figure 4 shows the same silent system as curve (a) but now upon a contracted current scale, while curve (c) shows the effect of ultrasonic irradiation upon curve (b), scanned at the same rate and in the oxidation direction only. Note that curves (b) and (c) are on the same current scale, both taken from ref. 31. Ultrasound has produced a 10-fold increase in maximum current. The plateau shape shows a limiting current at the extreme of oxidation potential reflecting hydrodynamic control independent of the voltammetric sweep rate. (This shape is also seen in other voltammetric procedures, e.g. when using rotating disk electrodes or microelectrodes.) In Figure 4 curve (c) this limiting current is found to be inde-

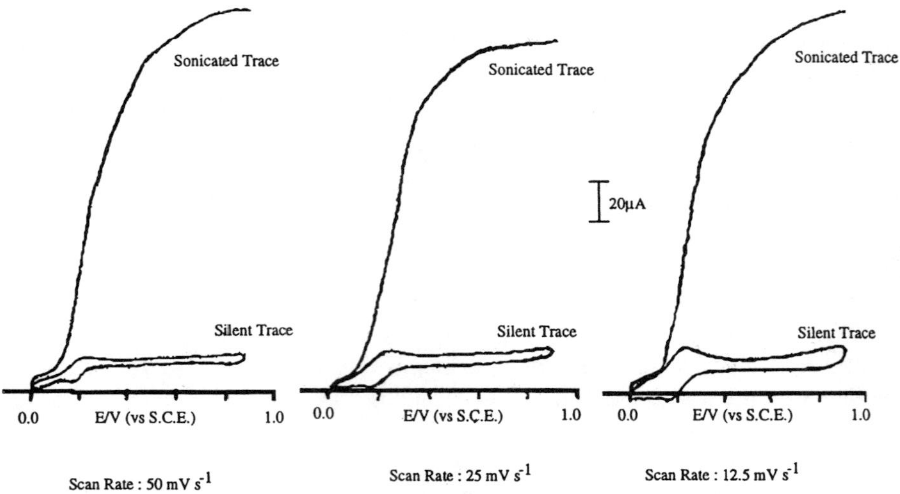

Figure 5. Scan rate independence of limiting current in voltammetry under ultrasound. System: Ferrocyanide in aqueous solution at platinum wire electrode (taken from ref. 31).

pendent of scan rate, at constant power and frequency of the ultrasonic irradiation. This is more clearly shown in Figure 5 where increasing scan rate enhances the currents from the silent traces while the limiting current under sonication is not increased.

In electroanalysis, electrodes of millimeter dimensions are termed "millielectrodes," while the more recently developed very small area electrodes of micron dimensions are termed "microelectrodes"; there are differences in properties beyond simply the change of dimension. Thus in millielectrode-scale experiments the enhancement of the diffusion-limited current plateau has been observed by a number of other workers—for example, in the reduction of methylviologen in aqueous acetonitrile [32], in the oxidation of bis(cyclopentadienyl) molybdenum dichloride in acetonitrile [33], as well as in several other studies on the aqueous ferrocyanide/ferricyanide couple using wire or disc millielectrodes to study diffusional phenomena [34–36]. Typical values of the diffusion layer thickness of approximately 5 μm are found under ultrasound [35] in contrast to the normal value of approximately 500 μm in silent conditions.

An important aspect in the experimental procedure for sonovoltammetry concerns the geometric setup of the cell, the electrodes and the ultrasonic source. This certainly alters observed phenomena, and care should be taken when drawing comparisons between different systems. The importance of the geometry of the experimental setup was demonstrated by examining diffusion-coefficient data obtained from ferrocyanide/ferricyanide redox voltammetry using a gold-sphere millielectrode moved around various positions under the tip of a sonic probe in a relatively large volume beaker cell (to avoid interference from sound waves reflected from the glass walls). The probe tip was also moved with respect to the distance to the base of the cell [37].

Both these geometric parameters altered diffusion data measured as Sherwood dimensionless number or as diffusion coefficients; maxima and minima in these parameters mirrored nodes and antinodes from the ultrasound. This involved relative motions between the various components of several centimeters since the wavelength of sound at 20 kHz is of this order, depending on the medium. These workers were using electrochemistry as a probe to monitor ultrasonic power, and a fuller account of this work is given in another chapter of this volume, but the effects of geometry upon behavior of the electrochemical probe are noteworthy.

Despite this, however, the general effect of ultrasound upon limiting currents in redox voltammetry seems to be similar over a range of systems. Thus the data in ref. 31 refers to irradiation from an ultrasonic bath with the wire electrode being suspended inside a voltammetric cell clamped several centimeters away from the external source of sound, whereas in ref. 34 a disk electrode is placed just 1.5 mm face-on to the vibrating tip of a sonic horn in the test solution. In the latter case the distance is well below a quarter-wavelength, this comparison brings into question the role of cavitation and microstreaming upon the Nernstian diffusion layer.

Another example of similar effects in quite different geometries concerns a comparison of data from ultrasonic hydrodynamic voltammetry (UHMV). This technique involves modulating the rate of convection to an electrode surface [38], and here this is achieved by imposing on/off cycles of ~1 Hz on top of ultrasonic irradiation at typically 20 kHz. This technique was employed to enhance the sensitivity of electrochemical detection of micromolar ferrocyanide using a glassy-carbon disk millielectrode 1 cm away and face-on to a 20-kHz ultrasonic horn [39]. However, the shape of the trace obtained was very similar to that in Figure 6, where ferrocyanide redox is shown at a platinum wire millielectrode in a cell irradiated by an ultrasonic cleaning bath at 25 kHz. Benefits of the modulated system include lessened temperature rise during the timescale of the voltammetric scan and greater ease of achieving a flat limiting-current plateau. This is sometimes a problem in sonovoltammetry, depending upon the experimental setup, and often a gradual and noisy increase of current with potential is observed without reaching a well-defined plateau. This shows poor coupling of ultrasound into the system, possibly due to reflections causing interference, or some other phenomena due to the experimental setup. As a rough rule of thumb, if running a voltammogram under mechanical agitation (such as nitrogen bubbling) produces a similar trace to one under sonication, then the ultrasound is not coupling properly into the system. Another feature of UHMV is that there is a point on the steeply changing part of the wave where the modulation produces no effect and where the direction of the sawtooth-shaped current pulses alters in sign (see Figure 6). This phenomenon may be related to the potential of zero charge for the system, and cannot be observed in the absence of modulation, such as from the continuously sonicated trace in Figure 4 curve (c).

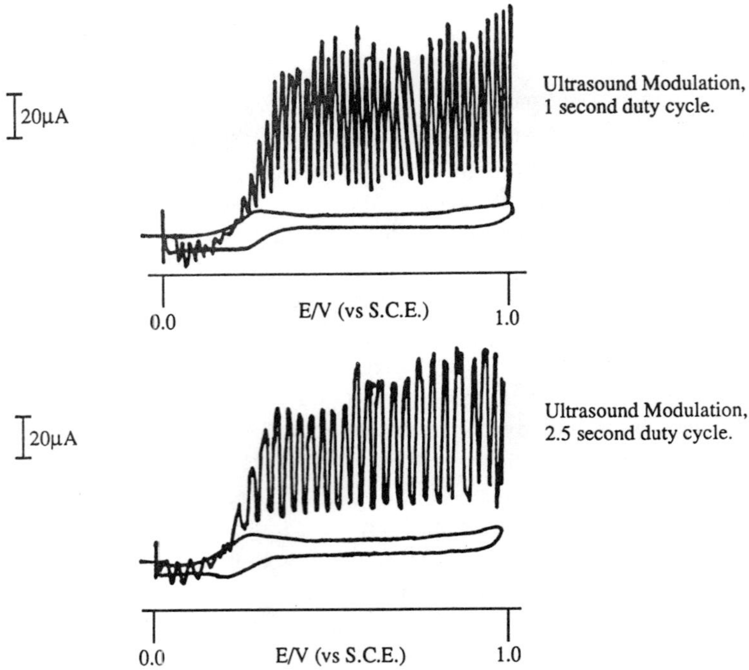

Figure 6. Ultrasonic hydrodynamic modulation voltammetry. System: Ferrocyanide in aqueous solution at platinum electrode (taken from ref. 31). Scan rate: 25 mV sec^{-1}. Silent traces are shown for comparison. Ulltrasound from 38-kHz bath.

Similar limiting-current enhancements are produced upon ultrasonic irradiation of voltammetry at microelectrodes. The significance of such small-area electrodes is that a plateau-shaped wave is obtained in silent conditions [40] because diffusion towards the electrode may be considered to be hemispherical. The trace is shown in Figure 7 curve (a), which shows the redox voltammetry of ferrocene in acetonitrile at an 27 μm-diameter platinum electrode in silent conditions for the oxidation wave only. The difference between this and the peak-shaped oxidation wave of $Fe(CN)_6^{4-}$ in Figure 4 curve (a) is because on a microscopic scale the millielectrode in the latter case may be considered flat and thus planar diffusion takes place. When ultrasound is applied to the microelectrode Figure 7 curve (b) there is little change of shape, but the limiting current increases, reflecting diminution of the diffusion layer and enhanced mass transport.

Similar results at microelectrodes have been reported for oxidations of bis(cyclopentadienyl) molybdenum dichloride [33] and chromium hexacarbonyl [41], both at platinum in acetonitrile. In the former instance, ultrasound also promoted the overall conversion of the substrate to the two-electron product and with greater efficiency at a potential where a one-electron process normally occurs, by inducing a follow-up disproportionation reaction. This provided a means to distinguish

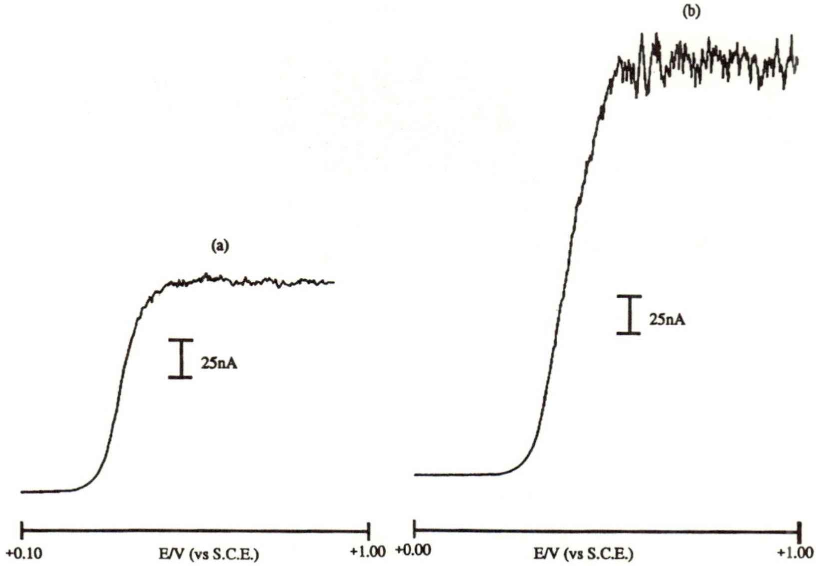

Figure 7. Voltammetry at platinum microelectrode. System: Ferrocene in acetonitrile. Ultrasound: 20 kHz (taken from ref. 31).

between two possible mechanisms that are difficult to resolve in silent conditions—the ece mechanism and DISP1 (disproportionation first order) [33]. In the latter instance, ultrasound also prevented any electrode fouling that hampers successive scans with this substrate in silent voltammetry [41]. Compton [42] also states, as does Yegnaraman [43], that ultrasound millielectrodes appear to behave as microelectrodes with uniform diffusion layers of the order of a few μm thickness. This behavioral similarity may be seen by comparing the sonicated traces in Figures 4 and 7.

Another system of defined hydrodynamics is the rotating disk electrode [17], and here the limiting-current depends upon the rate of rotation of the disk as shown in Figure 8. Since ultrasound also produces a step-shaped trace to a limiting current it is possible by comparing silent and sonicated traces at a disk electrode to calculate a "theoretical rotation speed" at which the disk would have to rotate to achieve the same transport limit as is found under ultrasound.

A number of assumptions must be made in such calculations, and Hagan and Coury [34] studying the ferrocyanide/ferricyanide redox couple in aqueous KCl at a ~1 mm radius platinum disk arrived at a figure of 160,000 rpm. This corresponds to a very fast rotation that would be situated in the turbulent flow regime at a conventional rotating disk, and suggests that ultrasound can achieve limiting-current conditions beyond those attainable in practice by rotation. These workers

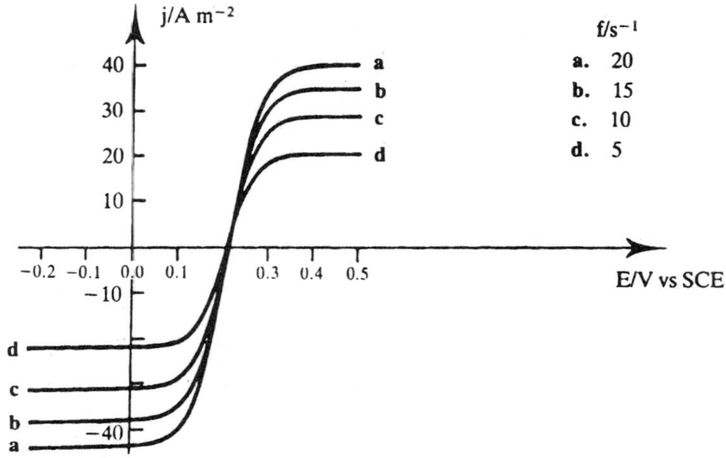

Figure 8. Voltammetry at rotating disc electrode. Limiting-current dependence upon rotation speed for typical reversible couple (taken from ref. 8 with permission).

also found that the greatest effect of ultrasound occurred when the tip of the 20-kHz probe was only 1 mm from the face of the electrode disk. This limiting-current enhancement falls away in a smooth curve as this separation distance is increased to 15 mm, and irrespective of the distance to the base of the cell (up to 15 mm), which is a somewhat different dependence in geometric parameters to that reported in ref. 37. However, all aspects of geometry must be taken into consideration, and one factor that is not always reported in detail is the overall size of the electrode mounting—that is to say the total solid area facing the sonic source rather than just the smaller active area of the electrode material. It may be some function of these relative areas, together with the area of the sonic source, the separation distance, and overall cell dimensions at a particular frequency, that are the determining factors, and an underpinning fundamental account of the physical aspects of these electroanalytical systems would now be of benefit.

Other workers have estimated the effective rotation speed to produce the same limiting-current as is observed under ultrasound, with values of similar orders of magnitudes. Thus Compton et al. [42] using a platinum electrode of millimeter dimensions to examine ferrocene redox in acetonitrile under 20-kHz insonation calculated some 12,000 rpm. Degrand et al. [32] calculated an apparent value of some 20,000 rpm for methylviologen reduction in wet acetonitrile at a platinum disk electrode of 0.2 mm^2 area, also at 20-kHz insonation. This latter group, by increasing the scan-rate to 10 V s^{-1} and by using a platinum microdisk of similar area to the calculated cross-section of cavitation bubbles in their system, were able to observe μs-long current pulses that they attributed to individual cavitation bubbles impinging on the electrode surface. These authors conclude that enhanced limiting-currents are due to a limited number of localized pulses rather than a general increase in current density across the whole electrode surface. These are interesting observations, but the role of cavitation remains under discussion, and other workers who observed similar current spikes [34] attributed this to vibrational modes from bubbles rather than individual bubble oscillations. Ideally to determine the effects of cavitation some independent assessment of the phenomenon should be attempted. It has been shown, for example, that at constant power (as measured by the calorimetric method [48]) for the aqueous ferrocyanide redox system at a platinum wire electrode insonated via ultrasonic baths, the observed enhanced limiting-current hardly changes in value when the irradiation frequency is increased from 38 to 800 kHz [31]. The nature of cavitation alters across this 40-fold range and a proper account of these sonoelectroanalytical phenomena should involve different ultrasonic powers, frequencies, and duty cycles, as well as describing different system geometries and allowing for other significant factors including solution volatility, viscosity, and the presence or absence of nucleation sites for the formation of cavitation bubbles. An early attempt was made to quantify the enhanced diffusion leading to increased limiting-current [47], involving three dimensionless terms, amplitude number, diffusion length number, and cavitation damping number, but the apparent independence upon frequency noted more recently [31] suggest that this treatment requires refinement.

2.3 Ultrasonic Effect on Electron Transfer at the Electrode

It is fair to say that the effect of ultrasound upon the fundamental electron transfer processes at an electrode have been less widely studied than the effects upon mass transport phenomena. Electrode kinetics is defined by the Butler–Volmer equation, which by a series of practical assumptions reduces to the Tafel equation [44],

$$\text{Butler–Volmer equation: } I = I_o\left[\exp\left(\frac{\alpha_A nF}{RT}\eta\right) - \exp\left(\frac{\alpha_C nF}{RT}\eta\right)\right]$$

$$\text{Tafel Equation: } \log I = \log I_o + \frac{\alpha_A nF}{2.3RT}\eta$$

where I = current density at overpotential η, I_o = exchange current densities, η = overpotential, α = transfer coefficient, n = number of electrons, R = gas constant, T = absolute temperature. This is usually depicted by an experimental Tafel plot, which involves the log of the modulus of current (in anodic and cathodic directions) as a function of potential, as shown in Figure 9.

The extreme left and right hand ends of the V-shape are horizontal as the system becomes diffusion-controlled, so that rate of electron transfer at the electrode no longer affects the current. However, as the system approaches E_E, the equilibrium potential for that particular electrochemical system, there is a sloping region where a linear approximation may be made before attaining values close to E_E, where

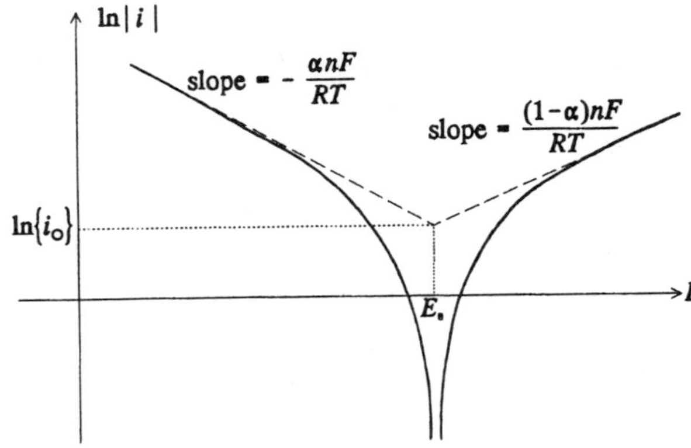

Figure 9. Typical Tafel plot (taken from ref. 8 with permission).

electron-transfer is the rate-determining factor. The significant parameters that may be obtained from such a plot are E_E itself, the slopes of the linear portions (respectively proportional to the anodic and cathodic transfer coefficients α_a and α_c, with mechanistic inference), and also the value of the exchange current I_o at the equilibrium potential E_E.

There is a current here because although an external measurement reads zero this reflects dynamic equilibrium in which the forward and reverse components are in balance. The choice of the electrode material is important. Mercury or lead are good electrode materials for reductions because the competing hydrogen-evolution reaction has a high overpotential, reflecting a low-exchange current density for this unwanted reaction. In the chloralkali cell the preferred reaction, chlorine evolution, competes against thermodynamically preferred oxygen evolution. An electrocatalytic electrode is one in which specific interactions (e.g. adsorption) influence electron transfer to promote a particular reaction [45]; ruthenium dioxide (often supported on titanium), which has a high exchange-current density for the desired chlorine evolution over oxygen evolution, is a preferred electrode material in commercial chloralkali systems [44].

In principle, ultrasound might alter any or all of these parameters, but there remains controversy in this area at present. Tafel plots are not usually employed for electroorganic systems where cyclic voltammetry is the preferred analytical technique, but are well-established in metal electrodeposition or metal dissolution studies or for simple inorganic redox systems such as hydrogen evolution. Recent studies on silver electrodeposition [46] suggest that sonication may alter transfer coefficients (Tafel slopes) and exchange current densities. This has practical implications in electrocatalysis; the sonoelectrochemical enhancement of hydrogen evolution and the comparative enhancement of chlorine evolution over oxygen evolution [24,25] have already been discussed.

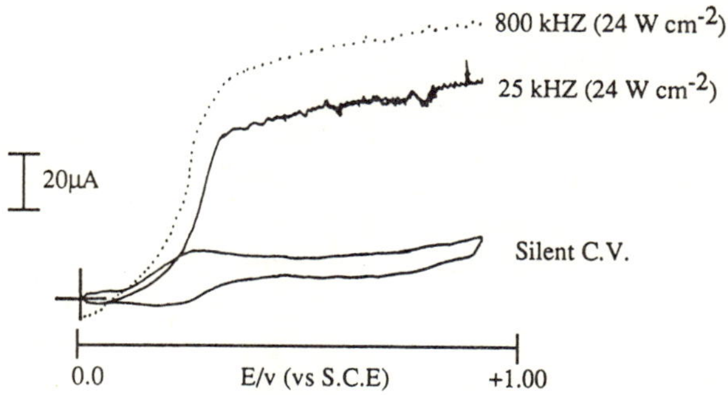

Figure 10. Effect of ultrasonic frequency upon voltammetry at platinum wire electrodes. System: Ferrocyanide in aqueous solution. Insonation was from different bath systems with different cell configurations but at constant power as determined by the calorimetric method. Scan rate: 25 mV sec^{-1} (taken from ref. 31).

Some controversy surrounds the effect of ultrasound upon thermodynamic half-wave potentials which are related to the equilibrium potentials in the Tafel plot. An apparent shift of the potential for the ferrocyanide redox couple in aqueous acid at a platinum wire was observed with frequency change in ultrasonic irradiation from 38 to 800 kHz (Figure 10), but the necessarily different source and cell configurations may influence this observation. This frequency change at constant power (as measured by the calorimetric method [48]) did not significantly influence the limiting-currents. However, change of power at constant frequency does affect the limiting-current. This has been recognized by many workers and is shown in Figure 11 from which it can also be seen that change of power does not appear to significantly influence the redox potential.

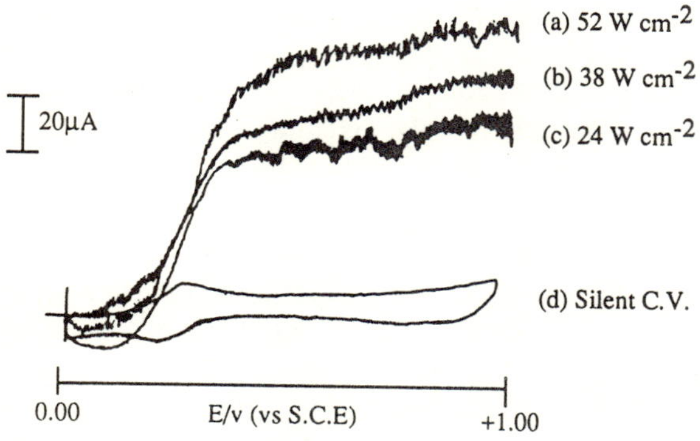

Figure 11. Effect of ultrasound power upon ferrocyanide voltammetry at platinum wire electrode. Insonation from 20-kHz horn probe system for all traces. The power was measured calorimetrically. Scan rate : 25 mV sec^{-1} (taken from ref. 31).

However, Compton et al. [35] have demonstrated an anodic shift in the reduction wave for the aqueous ferrocyanide redox system at a platinum disk electrode under 20-kHz irradiation. These workers confirmed the expected temperature dependence of this redox couple to be in the opposite direction to the observed sonoelectrochemical change, and therefore suggested that chemical irreversibility occurs under ultrasound with perhaps sonically generated hydroxyl radicals or similar species resulting in the formation of, for example, $Fe(OH)(CN)_5$. The distinction between thermodynamically reversible and irreversible systems is important in electrochemistry, and is not always easy to determine for step-shaped voltammetric traces that give a limiting-current plateau, especially if under ultrasound there are superimposed current spikes or alternating current signals from capacitative-switching synchronous with the applied insonation. The possibility that ultrasound switches reversible silent systems to irreversible ones (for example EC systems) should be borne in mind.

Half-wave potentials may also be shifted by temperature effects (hence the control studies in ref. 35 above). The "hot spot" theory of cavitation predicts very large microscopic short-lived temperature gradients [48,49], which would be well beyond attainment by external temperature manipulation of the cell. This theory also predicts very large microscopic pressure gradients, and while these might be expected to influence gas-forming or gas-consuming reactions, it may be that these gradients are high enough to also affect solvation phenomena, particularly if these are also affected by the electric fields in the region close to the electrode. There will be differential solvation or other ionic interactions since species involved in a redox electron transfer process have different charges; and since pressure terms are implicit in thermodynamic parameters, it may be that microscopic pressure phenomena under ultrasound are sufficient to alter factors that are normally assumed to be constant. The "hot-spot" theory predicts microscopically extreme conditions and it may be that in these circumstances other traditional assumptions are invalidated, e.g. the common use of concentrations to replace the more exact activities in thermodynamic equations.

These are fundamental considerations and are of interest not just to electrochemists and sonochemists, but care must be taken in correctly attributing an apparent shift in an experimentally observed potential under ultrasound. As already mentioned, system parameters and other factors may influence an observation beyond the effect under investigation. Thus there have been reports on the use of the titanium tip of the sonic horn itself, suitably electrically insulated, as the electrode material [50]. Dubbed the "sonotrode", this is a clever idea to combine the two active components of a sonoelectrochemical system; the authors noted the expected enhancements in limiting currents and an alteration in the morphology of copper electrodeposited from aqueous solution on to the titanium tip, which was the reaction under test. However, although titanium is widely used in sonochemistry because of its low-loss characteristics under vibration, it is not a common electrode material for electroanalysis because of its inferior electron transfer characteristics

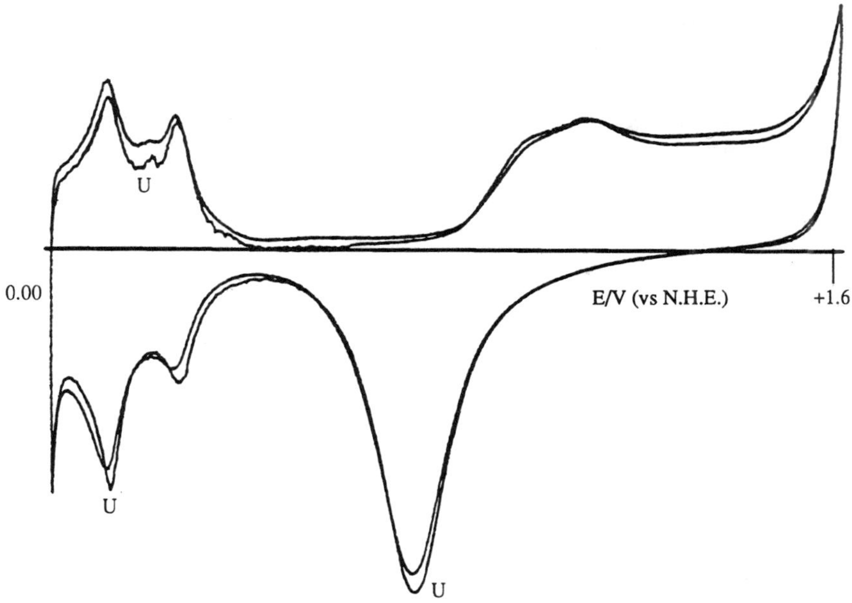

Figure 12. Cyclic voltammetry of platinized platinum in aqueous acid. Scan rate: 50 mV sec^{-1}; U refers to sonicated trace (taken from ref. 31).

and its reactivity (e.g. the ready formation of an oxide coating). This complicates the claim of these authors that overpotentials in copper electrodeposition are altered in this system. It may be the case, but should be viewed in respect to the above-mentioned difficulties in assessing sonoelectroanalytical phenomena. Of course, this does not preclude the use of titanium and other such materials in electrosynthetic systems where the isolated products are self-evident, and where reactive metal electrodes are proving to be a promising area of electrosynthetic research.

In general, the behavior of electrode surfaces under ultrasound is of great interest, and not just for reactive metals because processes at traditional "noble" metal electrodes are more complex than might be expected and are associated with electrocatalysis and other phenomena [45]. There is some effect in the behavior of surface species at nickel wire electrodes when sonicated from a 20-kHz horn [41], but interestingly the voltammetry of a platinized platinum wire electrode in dilute aqueous acid is little affected by ultrasound from a 38-kHz bath in the conditions employed [31]. The complexity of processes at this apparently noble metal in the region between hydrogen evolution and oxygen evolution is shown in Figure 12, with the sonicated trace superimposed. There remains discussion about the interpretation of this voltammetry, and it may be that the small changes under ultrasound will prove meaningful. Insonation of the platinum electrode in this work was carried out for relatively short periods of time. However, when platinum was exposed to ultrasound from a 20-kHz horn for numbers of minutes, ac impedance analysis showed evolution of the fractal dimension, with implications towards activation effects [41] reflecting progressive surface erosion and roughening. Thus processes at the electrode surface provide further factors to be taken into consideration in sonoelectroanalysis.

Rigorous electroanalytical studies on sonoelectrochemical systems are not necessarily straightforward to interpret, but offer great potential to analysts and physical chemists.

3. ULTRASOUND IN INORGANIC ELECTROCHEMISTRY

3.1 Introduction

This section deals with the use of ultrasound in the general area of "Inorganic Electrochemistry", and the areas that are reviewed here include:

- Cleaning processes,
- Electrodeposition and electroplating,
- Electrochemical corrosion/dissolution, erosion and passivation,
- Electropolishing, electromachining and electrochemical etching,
- Semiconductor systems including solar cells.
- Batteries, and
- Other inorganic systems.

Since the mid-1970s there has been a considerable amount of material published on the influence of ultrasound upon the electrochemistry of metal systems. Most of this work was carried out in former Eastern block countries and concentrated on such electrochemical processes as corrosion, electrodeposition, and electrochemical dissolution. Recently there has been an upsurge in the interest shown in sonoelectrochemical processes using both non-metal and metal systems worldwide. There have been a considerable number of publications in the employment of ultrasound in areas as diverse as semiconductor production to sono-electrochemical machining and metal finishing. A review by R. Walker [27] into the use of ultrasound in metal deposition systems, provides an introduction into the fundamental effects of ultrasound in plating and metal finishing.

Ultrasound has also been successfully employed for the production of novel inorganic compounds such as Si semiconductors in which case it is found that ultrasound increases yields of products and in some cases influences the reaction mechanism itself. For the traditional electrochemical process of metal electroplating the presence of ultrasound increases the thickness of the material deposited, as well as increasing the efficiency of the reaction. Ultrasound has also been successfully employed in the machining of complex electronic circuitry using electrochemical etching techniques. A majority of papers published in the literature still use ultrasound only as a cleaning tool, although this use has led to investigations into the effects of ultrasound upon electrode surface adsorption processes and even electron transfer reactions.

Research has been increasingly carried out in the fields of electroplating, electro-machining of metals, and other metal finishing techniques in order to improve both the cleanliness and rate of deposition as well as the quality of coating produced. Obviously the cost of any electrochemical process is of importance in industry. Therefore attention is directed towards increasing the rate of the electrochemical processes in order to raise throughput, and reduce the concentrations or volumes of the starting materials and solvents used, thus lowering the overall cost of the process. The latter is of great significance at present since recent legislation has eradicated or is about to eradicate many of the toxic substances used in electrochemical processes, thus alternative systems have been investigated and are becoming integral parts of production lines.

Ultrasound has been shown to be very beneficial in many industrial electrochemical processes and has been employed in overcoming some of these problems.

3.2 Cleaning Processes

Prior to any electrodeposition reaction it is very important that the plating surface is clean and free from grease so that there is a strong adhesion between the substance being plated and the plating surface. The presence of contamination on the surface prior to coating or electrodeposition is one of the most common causes of subsequent failure. These cleaning studies are not strictly sonoelectrochemistry, but are a significant aspect of metal electrodeposition.

The traditional methods of removing grease, oils, and other contaminants use an organic solvent to dissolve the contaminant. This is usually carried out by immersion into a hot or a cold liquid tank or by exposure to a vapor, thus producing a clean surface. However these processes are expensive and alterative methods have been employed using aqueous solutions of alkalies, which react with grease to form water-soluble salts (saponification).

Once the metal has undergone the degreasing procedure the metal oxide and/or scale is dissolved. For steel, cold hydrochloric or hot sulfuric acid is commonly used, although mixtures of acids may be necessary for certain corrosion-resistant alloys. Copper and its alloys are cleaned using nitric acid and aluminum is cleaned using caustic soda. The metal is then washed with water (deionized, distilled or tap, depending upon the degree of cleanliness required) to remove any residues remaining from the cleaning processes. The "wet" article is then immediately immersed in the required solution for electrodeposition or conversion treatment in order to minimize any possible oxide growth, or thoroughly dried for painting, hot dipping, or other dry/hot processes, thus producing good adhesion to the coating.

Ultrasound has been successfully used to assist most of these cleaning processes. Fuchs [51] has stated that when ultrasonic cleaning process is properly applied, it can save time, money, and increase cleanliness to a level that cannot be achieved by any other means.

Walker has shown that the rate of removal of wax from a glass surface, using an aqueous alkaline solution and ultrasonic agitation was up to 1500 times faster than with a similar solution under mechanical stirring [52].

Russian scientists have also looked at the degreasing of plating surfaces before electroplating using ultrasound and several commercial alkaline cleaners [53]. From their results it can be concluded that strong alkaline cleaners such as Synalod 50 and Radolod may be replaced by mild cleaners such as Synalod 60 if ultrasound is applied in conjunction with the cleaner.

The development of a cylindrical ultrasonic transducer or vibrator has revolutionized ultrasonic wire cleaning. The wire or narrow metal strip continuously passes through the ultrasonic source while being submerged in the cleaning solution with increased cleaning rate, improved cost effectiveness, and reduced pollution hazards [54,55,56].

Burstein [57] has given several examples of aqueous ultrasonic cleaning systems that can remove:

- atmospheric soils from brass parts, using mildly acidic tap water that contained detergent,
- particulates, forming and cutting lubricants from nickel parts using a strong, chelated alkaline detergent and tap water, and
- buffing and polishing compounds from stainless steel using a soap-free alkaline cleaner and tap water.

Ultrasonic systems using less harmful and corrosive materials have been successfully used as an alternative to vapor cleaning. For example, stainless steel, which used to be cleaned with distillate-spray wash and vapor rinse using 1,1,1-trichloroethane, has been replaced by immersion in ultrasonic bath containing trichlorotrifluoroethane and methanol. Hence the beneficial substitution of one organic solvent by a more "environmentally friendly" cleaner is possible.

Ultrasound can also be employed, after electrodeposition, to help remove electrolyte from crevices and recesses in the coating. Burstein [57] emphasizes the need to ultrasonically agitate both the rinse solution and the clean, plated particles immediately after plating and before the drying process.

3.3 Electrodeposition and Electroplating

Ultrasound has been used extensively in the electroplating industry for many years, and the literature is full of articles regarding the advantages of ultrasound upon electrodeposition and other electrochemical processes. These benefits include:

- increased hardness,
- coating thickness control,

- improved porosity,
- increased efficiencies of reactions and rates of deposition,
- use of less toxic electroplating procedures,
- better adhesion,
- altered morphology, and
- minimization of levelers, brightness, and other additives.

The electrodeposition process is complicated, but can be simply thought of as the transfer of ions to and/or from the metal surface [58]. It is well known that when a metal is immersed in an aqueous solution a diffusion layer (Nernst diffusion layer) forms at the metal/solution interface. If an electrochemical reaction is to occur at the metal surface it is therefore necessary for ions to be transported across this diffusion layer. Any process which can affect this layer will therefore influence the electrochemical process. Ultrasound is known to reduce the thickness of this diffusion layer [26] but is unlikely to completely remove it as was suggested by early Russian workers. Ultrasound can also effect electrochemical reactions since it produces surface cavitation and acoustic streaming both of which assist diffusion to and from the metal surface, this movement often being the rate-controlling step in electrochemical processes such as deposition.

The collapse of cavitation bubbles produced by ultrasound causes pressures of up to several thousand atmospheres to be generated [48,49] and the shock waves that are produced at the surface of an immersed metal give an increase in the dislocation density, which results in hardening of the surface layers.

A metal ion in aqueous solution consists of a positively charged cation surrounded by several water molecules. This ion is attracted from the homogeneous bulk solution towards the negatively charged cathode by electrostatic forces during the plating process, and enters the diffusion layer surrounding the electrode. It loses its solvation molecules and the ion approaches the metal surface where it is adsorbed as a bare ion. It then diffuses across the surface to a suitable site where it combines with electrons and is discharged. Hence, ions have to travel across this diffusion layer, and this movement is often the rate-controlling factor for electrodeposition. A similar, but reversed, procedure applies to the dissolution of metals at the anode. Ultrasonic agitation decreases the thickness of this diffusion layer, and thus assists both electrode reactions.

Concentration polarization is another factor which affects electrode reactions. This occurs due to the difference in concentration of metal ions between the bulk solution and that at the electrode surface. At the anode the concentration of metal ions increases at high current densities because their rate of dissolution is greater than the rate of diffusion away from the metal surface into the bulk solution. Hence the process becomes more difficult and the rate of dissolution tends to decrease. At the cathode the surface concentration of the ions tends to decrease because the rate of diffusion from the bulk of the solution is lower than the rate of discharge at the surface, thus causing the deposition rate to fall.

Ultrasound increases the movement of metal ions because it minimizes the difference in concentration around the electrodes, therefore reducing the effect of resistance polarization which contributes to the ohmic drop of the cell. Hence, ultrasound in the solution is beneficial because it allows the use of higher rates of deposition. These phenomena are more fully mentioned in the previous discussion on electroanalytical studies in Section 2.

Drake [59] has studied the electrodeposition of copper from an acidic sulfate ultrasonic bath. He measured the thickness of the diffusion layer, which was about 200 μm for the silent system, and obtained values of 34 μm with ultrasound of frequency of 1.2 MHz, and 3.4 μm with a frequency of 20 kHz. The corresponding changes in the values of the limiting-current density were from 8 to 50 and to 500 $A\,m^{-2}$, indicating a significant increase in the rate of deposition. These results show a dependence of limiting-current upon ultrasonic frequency, although absolute power levels were not given, which is different to that reported for small-scale electroanalytical experiments. However, it should be borne in mind that the experimental systems are quite different [31].

Aramaki et al. [60] confirmed that ultrasonic waves modified cathodic polarization, so that a higher plating rate could be achieved for gold. Walker et al. [52] have found similar results for zinc plated out from a sodium zincate bath. The zinc plated out at a much faster rate in the presence of ultrasound (i.e. 1950 $A\,m^{-2}$), when compared to the rate measured using a silent bath (250 $A\,m^{-2}$). Similar effects have also been reported for many other metals including chromium [61] and silver [62]; for nickel a 5- to 10-fold rate increase was observed [63].

An important feature in electrodeposition is the efficiency of the cathodic process, i.e. the percentage of the cathodic current which is used to deposit the metal rather than produce hydrogen. A higher value is important in giving a better coating because hydrogen evolution causes porosity within the metal.

Nowack and Habermehl [64] recorded a significant drop in cathodic efficiency, from 90% to 20%, as the current density in a silent system was changed from 100 to 200 $A\,m^{-2}$ for bright nickel. Ultrasound was highly beneficial and the efficiency remained very high (70%) at the much higher current density of 800 $A\,m^{-2}$. Similar improvements were also recorded for copper from acidic sulfate baths [65], and for zinc from a zincate solution [66].

The kinetics and mechanism of the electrodeposition of indium from acid electrolytes in the presence of an ultrasonic field was studied by Kadyrov and Rakhmatullaev [67] in the early 1980s. They studied the mechanism of the electrodeposition of In from different baths in an ultrasonic field in cells as a means of intensifying the process. They carried out experiments on baths containing various concentrations of Indium: (i) $In_2(SO_4)_3$ 30g/L, (ii) $In_2(SO_4)_3$ 70g/L, (iii) $In(NO_3)_3$ 30g/L, and (iv) $InCl_3$ 30 g/L at a pH 2.5 and a temperature of 20 °C. The ultrasonic vibrations were created by a magnetostrictive transducer with frequency of 17.5 kHz fed by a 800-W generator. They found that ultrasound greatly influenced the shape of the polarization curves and increased the deposition efficiency of the reaction for all systems.

Kamat and co-workers have looked at the electrogalvanizing of mild steel wire using noncyanide alkaline bath with ultrasonic agitation [68]. The aim of their work was to develop a suitable noncyanide alkaline Zn electroplating bath for the wire industry to meet the specific requirements for a high rate of plating, but still giving adherent and uniform coating. The effects of various parameters on the noncyanide alkaline bath, such as NaOH and ZnO concentration, temperature, and ultrasonic agitation were studied, and the best conditions for Zn electroplating of wire using such a bath were deduced. It was reported that with mechanical agitation alone a current density of 4–5 A/m^2 was achieved, giving an anodic current efficiency of 100–102% (this figure includes competing chemical dissolution) and a cathodic current efficiency of 97–99%. However, ultrasound at a frequency of 25 kHz and power 250 W gives a current density as high as 22–26 A/m^2 at a cathodic current efficiency of 94–96%, while the anodic current efficiency increases slightly to 104–105%. A comparison of the salient features of this noncyanide bath and the commonly used acidic sulfate bath showed that the former was superior with respect to throwing power and metal distribution, brightness, cathodic current efficiency, and limiting cathodic current density in the presence of the ultrasonic field.

Since ultrasonic agitation reduces the thickness of the double layer, it permits a faster plating rate and in turn may also change the composition of the deposit during plating. This can be particularly advantageous to increase the proportion of iron in nickel–iron alloys, giving a cheaper coating. In one deposit that was examined the concentration of iron increased from 3.5% in a silent system to 19.2% in the presence of ultrasound at a frequency of 24.8 KHz, and to 18.8% at a ultrasonic frequency of 37.9 KHz [69].

Prasad and co-workers have published a review dealing with the use of ultrasound in electroplating in which they compare the properties of deposits obtained from several still and ultrasonically agitated plating baths [70]. Work carried out by these workers into the use of ultrasonics during electrodeposition of several metals (Zn, Sn, Ni) has shown that in the presence of ultrasound smoother and harder deposits are obtained [71]. If a small grain size is produced, as during commercial plating, then according to the Hall–Petch relationship this should result in a increase in coating hardness. Ultrasonic agitation was originally considered to produce harder deposits because it formed a smaller grain size [72] and the deposits contained more material [73]. It is now generally accepted that these explanations are not entirely satisfactory because many exceptions have been observed, and the increase in hardness is now considered to be due to a combination of factors resulting from cavitation, such as the reduction in porosity [74,75]. Copper deposits produced from sulfate baths in the presence of ultrasound show a increase in hardness, but here the result may be explained by reduction in grain size [76,77].

The hardness of a coating is also dependent upon the composition of the plating bath; for example, the Vickers hardness values for copper from a still acid bath were

75 Hv, and 146 Hv in the presence of cyanide; these increased to 111 Hv and 185 Hv, respectively, during ultrasonic agitation of the system [78]. Similar results were obtained for nickel plating; the silent and ultrasonically agitated solutions gave 151 and 317 Hv values, respectively. When chloride is added to the plating bath the values obtained for the silent and ultrasonically agitated solutions were 193 and 305 Hv, respectively. For sulfate values of 151 and 407 Hv were obtained, and for fluoroborate values of 208 and 391 Hv were recorded.

Meyer and Nowack [79] reported a sonoelectrochemical increase in the hardness of nickel without loss of luster, and the transport of ions was enhanced with changes in the resistance polarization and rest-potential. Harder deposits were also found for nickel, but not for zinc, from an acidic sulfate bath [78]. However bright zinc from a sodium zincate electrolyte was harder under ultrasound (136 Hv, compared with 86 Hv), although the main advantages were an increased efficiency and a higher plating rate, so giving a smooth surface [66]. Harder (and brighter) chromium has also been plated from a modified, self-regulating, high-speed bath [80].

The increase in the brightness of metals electrodeposited in the presence of ultrasound is thought to be due to the shock waves that are generated by cavitation; these waves not only harden the surface and remove hydrogen bubbles, but also break off any perpendicular growths, such as dendrites. This in turn leads to the high surface quality and brightness which has been observed with copper electrodeposits [81].

The thickness of the electrodeposited metal can also be controlled by using ultrasound. An example of this is in the chromium electroplating of steel plates which are used as cathodes and electrolyzed in aqueous solutions containing CrO_3, or chromates and dichromates, to form an inner coating of Cr and an outer coating of chromate. However when the plates are treated in the same solution with ultrasound present it is possible to control the amount of chromate coating. Thus, a cold-rolled steel plate was electrolytically chromated from a silent aqueous solution giving a chrome coating of 13 mg/m^2 after 1 s electrolysis; in the presence of ultrasound a coating of 45 mg/m^2 is obtained [82] the same time period. Other workers have studied the electrodeposition of metal–polymer coatings during ultrasound treatment and observed similar results [83]. They managed to increase the coating during cathodic electrodeposition of $Cu(O_2CH)_2$ epoxy primer to a maximum of 36 or 45 g/m^2 by applying ultrasound for 15–25 s during electrolysis. An increase of 25% in the thickness of phosphate coatings, as well as an improvement in the quality of the coating, has been reported by Mikhoski and Pushev [84] in the presence of ultrasound. Ultrasonic vibration of a sulfuric acid bath used for anodizing aluminum has given thicker anodized films on aluminum. This improvement in anodizing was found to be more effective when an alternating current is used in the presence of ultrasound rather than a direct current [85].

Ultrasound not only influences the hardness of the electrodeposited material but can also greatly affect the porosity of the coating. This has been shown by work carried out by Chistyakov et al. [86] who studied the influence of ultrasound upon electrolytic nickel–cobalt coatings for diamond powders. Diamond powders, used for cutting edge wheels, showed greater wear resistance with a Ni–Co alloy coating than with Ni alone. The authors found that a Ni alloy containing 75% Co had the optimum properties, and when it is deposited in an ultrasonic field at a frequency of 880 kHz and an ultrasound intensity of 1.4×10^{-4} W/m^2, the coatings were less porous and harder.

Porosity is particularly important in thin cathodic coatings, such as gold contacts on copper used in the electronics industry, because corrosion will occur on the more reactive substrate and result in either perforation of the base, or the formation of corrosion products on the surface, which in turn have a detrimental effect.

The common cause of porosity is the formation of hydrogen bubbles on the cathode surface during plating. These bubbles obviously shield the plating surface from the substrate and prevent deposition on "local sites" below the bubble. The hydrogen, which is formed by the electrolysis of water, is also detrimental since it reduces the current efficiency. Inert particles on the surface or electrode fouling can have a similar shielding effect. Ultrasonic agitation of the solution assists in the removal of these bubbles and surface particles by microstreaming, thus enabling a more uniform deposition to occur. Grain coalescence and two-dimensional growth are also encouraged so denser deposits are formed. This improvement in quality, due to the reduction in porosity with ultrasound, has been reported for gold [59,87], chromium [60,88], and nickel [89,90]. A reduction in porosity in plated copper, together with a related increase in smoothness, has been reported by Walker and Benn [91].

Yamashita and co-workers studied the effects of ultrasound on the electrodeposition of Zn from $ZnBr_2$ solution [92]. They examined the effects of ultrasound on the cathodic polarization curves and the impedance characteristics of the system. From their results, it was shown that ultrasound increased the reaction rates of deposition and dissolution of Zn. However, ultrasound had no effect on the overpotential of the reaction. Electron microscopy studies also showed that uniform and fine crystals of Zn were obtained in the presence of ultrasound, thus increasing hardness of the coating.

Another study carried out by these authors [93] modeled the collapsing motion of a single bubble near an electrode surface, and equations for the motion of a spherical gas bubble were obtained. The jet speed and water hammer pressure during jet flow (liquid jet) were calculated, and when the jet speed was 120 m/s, the water hammer pressure was approximately 200 MPa upon the electrode surface. This pressure played an important part in the fineness of the crystal deposits. Mass transfer during the electrode reaction was by turbulent diffusion. The diffusion layer thickness was reduced to approximately 1/10th its size in the presence of the ultrasonic field. The baths contained the ions Cl^-, SO_4^{2-} , and Zn^{2+}. The ultrasonic frequency employed in the experiments was 40 kHz and it was seen that ultrasound considerably increased the deposition rate and current efficiency, as well as the smoothness and hardness of the deposit. Microscopy studies showed that the particle sizes of the deposits were finer, and X-ray diffraction showed that the texture coefficients for all the planes were affected. The kinetic parameters, such as exchange current densities, active overvoltage, and electrical double layer capacitance all increased in the presence of ultrasound, while the thickness of the diffusion layer decreased, all of which are of considerable interest in increasing the efficiency of the electrodeposition reactions.

Chiba et al. have also studied ultrasonic agitation effects on the electrodeposition of copper from a cupric–EDTA bath [94]. They found that the texture coefficients of the deposited films of Cu were greatly increased the {200} plane by ultrasonic agitation. Limiting-current densities and cathodic current efficiencies were both increased, while the grain size was reduced. The rate-determining step was not greatly affected by ultrasound, but the overall reaction rate was increased. These results are in good agreement with work carried out earlier by Smirnov and his co-workers [95]. They studied the effect of ultrasound on the electrolytic deposition of copper from industrial sulfate. The effective activation energies were calculated under various conditions of electrolysis and the effect of ultrasonic vibrations on the formation of cathodic deposits at various current densities was determined. They found that all coatings produced under the influence of ultrasound were smoother and harder than those produced in the absence of ultrasound.

Ratajewicz and co-workers have also studied the effect of a ultrasonic field on electrode polarization and activation energy in zinc electroplating [96]. The conclusion from their work was that the zinc electrochemical reduction reaction was influenced by the presence of organic additives, especially CN^- ions, and that in the presence of ultrasound the cathode fouling is reduced.

Ultrasound has been employed not only in the electrodeposition of wear-resistant coatings but has also been implemented in the manufacture of conducting materials, such as platinum black coated electrodes and wires. Platinum black has been electrochemically plated onto a conductive substrate (e.g., a Pt electrode) by positioning a counter electrode and the conductive substrate in a Pt(IV) plating solution. An electric current was passed for a predetermined period of time under ultrasonic agitation. The solution contained chloroplatinic acid (H_2PtCl_6). Pt wire electrodes coated with Pt black in this manner are more durable and have a higher measured electrochemical surface area than Pt wire electrodes coated with Pt black without the use of ultrasonic agitation [97].

Workers in Japan have shown that ultrasound can be used in electrochemical pickling and etching and smut removal in plating a beryllium–copper alloy [98]. The plating surface of Be–Cu alloys which are used commercially as electrical connectors, need to be pretreated by anodic treatment in an acid aqueous solution. These workers have shown that the smut may be effectively removed by application of ultrasound. This leads to a more resilient finish and a thickness of 0.2–3 μ coating being achieved more rapidly.

Ultrasound has also been successfully applied for surface treatment. A method for selective electroplating and electroetching was described in the early 1980s by

Drake [99] in which ultrasound is used to promote fast electron exchange at the electrode surface. The ultrasound agitation is provided by an ultrasonic transducer immersed in the bath. This method is particularly suitable for the selective plating of precious metals, e.g. Au on printed circuit boards.

Skomorokhov and workers have studied the effect of ultrasound upon electrolytic boronizing processes [100] and found that the sonicated system produced metal with a smoother finish which was also considerably harder.

While anodizing certain metals, such as aluminum, the color of the final metal is dependent on surface oxide formation and other factors. Ultrasound has been successfully utilized to produce different colored coatings. The colors of Al anodic coatings, which were produced by ac electrodeposition (4 min) in aqueous $CuSO_4$–H_2SO_4 mixture at 5 and 6.25 V, were 5YR8/2 and 5YR616, respectively in silent conditions, while the colors were 2.5YR616 and 7.5R5/10, respectively, when ultrasonic agitation was used. Also the color tone of the coatings obtained with ultrasonic agitation was better than that obtained without ultrasonic agitation. The effects of ultrasonic agitation on coloring Al anodic coatings in $CuSO_4$–$KMnO_4$ mixtures was also studied by changing the duration of ultrasonic agitation between 0–4 min, while keeping the ac electrolysis time to 4 min; it was found that both the uniformness of color and hardness of the anodic coatings were superior when the time of ultrasonic agitation was 2–3 minutes [101].

The coloring of anodized Al from ac current electrolysis in a bath containing $NiSO_4$–$CuSO_4$ or $CuSO_4$–$KMnO_4$ and additives HOAc, HNO_3, $H_2C_2O_4$, tartaric acid, citric acid, or H_3BO_3 revealed that the coloring is related more to the additive type and concentration rather than to the pH of the bath, and that the behavior of the anion of the additive contributes significantly. Thus for mixed coloring in a $NiSO_4$–$CuSO_4$ system, H_3BO_3 and HOAc are appropriate as additives, and it is necessary to maintain the Cu concentration low relative to the Ni concentration. For the $CuSO_4$–$KMnO_4$ system, HNO_3 and H_2SO_4 are appropriate as the acid additives, and that the application of ultrasound can achieve uniform coloring. Research is continuing in this area.

Electroplating generally employs very corrosive conditions creating an environmental problem that concerns the formation of corrosive mists above a bath surface due to competing hydrogen evolution processes. Several researchers have studied the effect of ultrasound on reducing some of these problems.

One promising area of investigation is the use of ultrasound to speed up the operating rates of existing aqueous electroplating systems which were not economically viable in the past. These systems can then become feasible and Fitch [102] has described the use of aqueous based compounds such as sodium salts (e.g. hydroxide, silicate, phosphate, carbonate, and borate) as an alternative to the more toxic materials commonly used. These inorganic compounds are inexpensive, readily available, and highly effective in the presence of ultrasound.

Ultrasound is also of interest in electroless deposition, which is not strictly the subject of this review. However, an example of this type of deposition has been published by Kathirgamanathan who studied ultrasonic-assisted electroless deposition of copper on to polymer substrates [103].

3.4 Electrochemical Corrosion/Dissolution, Erosion, and Passivation

It is well known that high-intensity ultrasound causes the formation of cavitation bubbles which in turn collapse in a manner that in the presence of a solid surface form high-velocity fluid microjets directed toward the surface. This microjetting, or intense fluid agitation at the surface, enhances transport of the electroanalyte, causes heat, and influences the behavior and integrity of surface films on metal electrodes. It also leads to the erosion or corrosion of the surface.

Work carried out by Bogoyavlenskii and co-workers in the late 1960s [104] showed the effect of ultrasound upon the passivation and repassivation of plating surfaces. Ultrasound at a frequency of 23 kHz applied during the anodizing of an Al alloy in 20% sulfuric acid electrolyte at current densities of 0.1–0.3 A/dm^2 yielded oscillograms which do not differ significantly from those obtained in the absence of the ultrasonic field. In a 25% Na_2CO_3 electrolyte, however, the presence of the ultrasonic field delays the passivation of the surface until higher current densities are reached, i.e. without ultrasound passivation, occurs at 0.3 A/dm^2, but not until 0.9 A/dm^{-2} in its presence. Insonation also results in the formation of a porous film with a finely divided structure.

Recently Perusich and Alkire [105] have proposed a mathematical model to determine the reaction and transport between liquid microjets and a reactive solid surface. Conditions were established under which oxide depassivation and repassivation occurs as a function of ultrasonic intensity, surface film thickness, and fluid microjet surface coverage. The model was based on the concept that cavitation induces sufficient momentum and mass transfer rates (water hammer pressures as described earlier) at a surface to create oxide film stresses leading to depassivation. The model was used to evaluate experimental data on the corrosion behavior of iron in sulfuric acid [106,107]. Focused ultrasound was used to investigate processes that influence depassivation and repassivation phenomena on pure and cast iron in 2N H_2SO_4 at two ultrasound frequencies and at power intensities of up to 7.8 kW/cm^2.

A curved piezoelectric transducer was used for the high frequency 1.58 MHz work since this created cavitation at a focal point with intensities of approximately 3.4 kW/cm^2. Low-frequency (20 kHz) ultrasound was produced with a commercial sonicator equipped with an exponential microhorn. At high focal intensities (>1.5 kW/cm^2), a single (100 ms) pulse of ultrasound produced depassivation; at low intensities continuous ultrasonic exposure was required. In all cases, the induced depassivation was followed by precipitation of a metal salt film upon the metal surface prior to the oxide film formation.

At high ultrasound intensities the time of passivation was affected significantly and repassivation was hindered completely. The researchers also found that the critical acoustic focal intensity required either for depassivation or to prevent repassivation varied exponentially with electrode potential. Once the oxide film had formed on the metal, the acoustic focal intensity needed to breakdown the film depended logarithmically on the time of passivation. The mass transfer coefficient for the electrode process was also found to be proportional to the acoustic focal intensity.

The effect of ultrasonic field on the polarization curves of Cu–Pb, and some brasses has been studied in chloride and sulfate solutions in the presence and absence of the respective metal ions [108]. The main effect of the ultrasound at low current densities is the acceleration of diffusion. The passivation current density in solutions free of the respective metal ions is considerably increased when ultrasound is applied. Stable passivity cannot be attained because of the periodic destruction of the salt film. The hydrogen evolution reaction is accelerated because of the destruction of the solvation shell. The oxygen depolarization reaction is also enhanced due to the increased diffusion. The rate of metal deposition is likewise increased by ultrasound. The steady-state potentials of reactions with anodic control are shifted in the negative direction when ultrasound is applied.

The role of potential in cavitation-erosion wear of metals was studied using steels 45 and 12Khl8N10T in aqueous citric acid solution and Na_2HPO_4 buffers(pH 6.5) [109]. Experiments were conducted using magnetorestrictive vibrators at a frequency of 22 kHz. Steel 45 has no passive zone while steel 12Kh18N1OT and alloy VT1-1 possess passivity. For steel 12Khl8NlOT the maximum wear was noted at –0.8 V and for alloy VT1-1 the minimum wear was at 0.4 V. In the case of steel 45 it was found that the reduction in potential to –1 V reduced the weight loss between 40 and 60% as compared to the original potential; on further reduction in potential, the reduction in weight increased. On shifting the potential into the anodic region, a uniform increase in the weight loss was observed for steel 45. Thus the potential affected chiefly the intensity of the corrosion process and its role during cavitation–erosion wear was predominant as compared to the role of surface phenomena.

Other workers have published improved procedures for inspecting both reinforced concrete and prestressed concrete structures with regard to determination of the embedded steel components [110]. A prototype ultrasonic procedure was developed to determine the condition of prestressed and pretensioned tendons in concrete. The application of electrochemical surface-mounted systems for estimating the rate of corrosion of reinforcing steel and other embedded steel components in large concrete structures was described using this technique.

A study was made of the complex effect of the special surface treatment of steel (40 kHz with formation of white films nonsusceptible to etching) with simultaneous use of inhibitors (bipyridine halides and inhibitor FMI-1) on its electrochemical characteristic in corrosive media. The white film on the tested steel (quenched from 850 °C in oil and tempered at 180 °C for 2 h) was produced by the following methods: (1) mechanical treatment by ultrasound, (2) friction-strengthening treatment, and (3) treatment by use of lasers.

The white film constitutes a specific state of the metal as regard to its structural and phase condition and as a result of pulse, forms mechanical and thermal reactions. This film consists of finely dispersed martensite, residual austenite, and in some cases fine carbides. Polarization measurements denoted that use of these inhibitors along with surface strengthening leads to increased overvoltage both of the anodic and cathodic reaction. In the case of the white films obtained by using ultrasound, the inhibition of the cathodic reaction is more pronounced, whereas for the laser method, the anodic reaction is more pronounced; as regard to the friction-strengthening treatment, the overvoltage of the cathodic and anodic process is about equal. Strengthening of the steel specimen surface by formation of white films leads to increased adsorption of inhibitors from acidic solutions and, thereby, to a substantial lowering of corrosive activity.

The dissolution of passivated steel surfaces in oxidizing media was accelerated by the use of ac at 50 Hz [111] . The systems studied were stainless steel Khl8N9T in a solution containing HNO_3 (10%) and NaF (2%) and stainless steel Khl8NlOT in a solution containing HNO_3 (8%) and HF (2%). It was observed that the time required to remove the oxide scale by chemical etching, for both stainless steel systems, was reduced significantly in the presence of ultrasound at 20 kHz frequency. For example, the time required for the removal of the oxide layer from stainless steel Khl8NlOT system was reduced from 30 min to 3.5 min when carried out in the presence of ultrasound.

3.5 Electropolishing, Electromachining, and Electrochemical Etching

Cavitation and the resulting surface-hardening effects of ultrasound due to the reduction in the thickness of the diffusion layer, has been utilized to improve performance in the finishing of machined components and other surface treatment techniques such as electro-polishing and pickling.

Electropolishing is a process that is very similar to anodizing. When a metal is subjected to a high anodic current in a suitable electrolyte solution dissolution may occur preferentially at certain raised points of the metal surface which leads to a mirror finish. Electropolishing baths generally have high acidity with added oxidizing agents such as chromic, nitric, and perchloric acids, and anodic current densities for electropolishing can be as high as 5 kA m^{-2}.

Another technique used in metal finishing is electrochemical machining, which employs anodic current densities of up to 5 MA m^{-2}. The principle of this technology is to advance a shaped tool, which serves as the cathode, towards the anodic workpiece. As the interelectrode separation narrows, the workpiece dissolves

$$M(s) \rightarrow ne^- + M^+(aq)$$

at an ever-faster rate until a millimeter-thick layer of aqueous solution separates the two electrodes. In this way the workpiece soon comes to adopt a shape that conforms to the shape of the tool, and a dissolution rate that matches the rate at which the tool is advanced. Typically, the working electrode is made of copper or steel and the tool behaves as an inert cathode, producing hydrogen.

$$nH_20(e) + ne^- \rightarrow 2H_2(g) + nOH^-(aq)$$

Since the electrodes are in close proximity in electrochemical machining, a major problem is that the ions produced can interact with one another, producing slurries of the metal hydroxide.

$$Mn^+ + nOH^-(aq) \rightarrow M(OH)_n(s)$$

Electrochemical grinding and electrochemical deburring are techniques related to electromachining in which excess metal is removed by a combination of mechanical abrasion and anodic dissolution.

An example of ultrasonic treatment of metals is in electrolytic polishing. This is an economical method of surface treatment for metal workpieces and in particular delicate products such as metal filters made from fine wire. These products can be finished more rapidly and economically by this method than by usual mechanical cleaning techniques. The condition of the electrolyte in this case plays an important part. Investigations on ultrasonically treated electrolyte solutions have shown that they enable workpieces to be polished in half the time required by conventional solutions. In addition it is possible to attain improved metal structures with these electrolytes as a result of their particularly aggressive action on the surface of the product [112].

In the mid-1970s Iliuteanu showed that electrolytic metal polishing is greatly accelerated with ultrasound [113]. The baths used for electropolishing of Cu and steel, consisted of H_3PO_4 and EtOH and $HC10_4$ + EtOH, respectively, and were previously irradiated for 12 min by ultrasound at a frequency of 1.2 MHz and a power intensity of 1.9 W/cm^2 to reduce their viscosity and increase conductivity. The 120 s polishing time needed when using untreated solutions was reduced to 60–70 s at 8 V and 250–140 mA/cm^2 for the presonicated samples of Cu. Similar results were obtained for steel.

The rate of electrochemical discharge machining was doubled by vigorous agitation with encircling jets of ultrasonically activated dielectric fluid. The machining rate did not decrease appreciably at machined depths up to 1/2 inch and was at a maximum at a frequency of 400 kHz with 17 W energy from each of the 4 piezoelectric transducers operated in-phase with one another and placed 6 inches from the machining interface. The transducers were mounted in flexible steel tubes which could be manually positioned around the electrical discharge tool to deliver the pumped streams of ultrasonically activated fluid to the machining interface. The tubes were fed from a pump chamber in which a magnetostrictive transducer was mounted to add a 0.1–50 kHz vibration to the fluid jets.

Okudaira [114] has studied the removal of sulfates from spent electromachining solutions by precipitation in the presence of ultrasound. It is quite common for sulfates to be formed during the electrolysis of Cr(VI) electromachining solution containing $Na_2S_2O_4$. The authors found that the sulfate contaminant can be removed from the system by precipitation, and that the rate of precipitation was aided significantly by the use of ultrasound.

Japanese researchers have developed a method for electromachining using a tubular cathode, either a alkaline or neutral salt solution, and employing ultrasound with the ultrasonic vibration being continuously supplied to prevent the precipitation of chiefly metal hydroxides within the cathode. Clogging of the cathode tube can thus be prevented [115].

Ultrasonic agitation was applied to the electrochemical etching of CR-39 neutron detectors at different temperatures of 40, 60, and 80 °C in NaOH of 2.5 and 5 mol L^{-1} concentration. The results are compared with conventional agitation. In 5 mol L^{-1} NaOH, a reduction of the minimum detectable dose was obtained by the use of ultrasonic agitation. The largest reduction, by a factor of 3, was found at 80 °C, and the lowest absolute value of the minimum detectable dose was achieved under these conditions [116].

Anodic dissolution of hard alloys has been enhanced by the application of ultrasound, apparently because of the increase in cavitation and the hydrodynamic pressure resulting in an increase in current density. The cyclic nature of the hydrodynamic pressure helps to remove passivating oxide films from the surface of the workpiece, thereby raising the process efficiency. This increase in current density resulting from the application of ultrasonic vibrations was most evident in hard alloys containing appreciable quantities of Ti and Ta carbides [117].

The effect of ultrasound on the rate of anodic dissolution of metals was studied by Karavainikov [118] in 1973. He found that the rate of dissolution of Fe in 10% HCl was slightly increased with ultrasonic vibration in the current density range of 0.08–0.4 A/m^2. The surface after dissolution under ultrasound had a uniform, fine-grained structure giving diffuse dispersion of light.

Vodyanov, a year earlier, had examined the effect of an ultrasonic field on the anodic dissolution of iron in sulphate solutions [119]. The dependence of an ultrasonic field (23 kHz) on the rate of anodic dissolution of Fe was investigated at pH 0.45–2.0 and at a SO_4^{2-} concentration of 0.1–1 N. The current versus time curves at controlled potential showed that the ultrasonic field increased anodic polarization.

Marshakov et al. studied the effect of an ultrasonic field on the anodic dissolution of a variety of metals [120]. Anodic polarization curves were measured in 0.1–0.5 N HCl, NaCl, H_2S0_4, and Na_2SO_4 solutions in an ultrasonic field of 20 kHz frequency. The effect of ultrasound was different for various metals due to differential effects on the rate-controlling process. Anodic dissolution of Fe was actually slowed down in the presence of ultrasound because the rate-controlling adsorption of anions at the iron surface was inhibited. At a cadmium anode, the energy of metal atoms in a lattice was increased in the presence of ultrasound and, therefore, the ionization of cadmium was accelerated. Anodic dissolution of copper and silver was found to be controlled by ion transport from the electrode. This process was accelerated by ultrasound and the rate of dissolution increased. In the presence of halide ions, insoluble passive layers on Cu and Ag were attacked by ultrasound, and the limiting-current in the passive region increased.

Similar results were obtained by Pan and Wan [121] who studied anodic dissolution of gold in concentrated potassium cyanide solution. The kinetics of anodic dissolution of Au in concentrated aqueous KCN were examined as a function of concentration, potential, and current efficiency. The authors deduced a mechanism for this process and showed the rate-determining step to be the reaction of CN^- with adsorbed Au(CN) to give$[Au(CN)_2]^-$. Again it was found that ultrasonic agitation at a frequency of 20 kHz increased the dissolution rate fivefold but did not change the mechanism. Potassium gold cyanide is an important reagent commercially due to its usage as a plating agent in the gold plating of metals.

Rajagopal et al. [122] studied the electrochemical preparation of potassium gold cyanide using both galvanostatic and potentiostatic techniques. From their results it can be concluded that potentiostatic dissolution method is at least 10-fold faster than the galvanostatic method. The rate of dissolution can be enhanced further by using ultrasonic stirring. The $KAu(CN)_2$ produced by this ultrasonic method is free from chloride and auric cyanide; however there is some loss of Au due to the disintegration of the anode.

A electrochemical method for measuring the amount of Cu leached from antifouling marine coatings has been described [123]. The authors used inverse voltammetry for determining the amount of Cu_2O leached from marine antifouling coatings using an automated rotary electrode. The results obtained were found to irreproducible owing to sorption of Cu to the surrounding tank material in which the experiment was carried out. The authors managed to eliminate this problem by accelerating the desorption of Cu from the surface.

Severdenko and his co-workers studied the effect of ultrasound on the dissolution rate and chemical activity of aluminum metal [124]. Al disks were subjected to a 20-kHz ultrasound field and then samples were cut from sites corresponding to the potential antinode or supersonic bias antinode of the wave. The electrochemical properties and the dissolution rate of the Al samples pretreated in ultrasound fields were compared with data for blank Al samples. Their results showed that samples cut from sites located in the potential antinode of the wave were more negative than the values inherent to untreated samples. This was due to the decrease of the thermodynamic stability of the metal as a result of the formation of microdefects, microcracks, etc. The dissolution rate of the ultrasound-treated samples cut from these sites in aqueous NaOH solutions was also enhanced. The reverse effect was observed with samples cut from supersonic bias antinode sites, i.e. electrode potentials shifted in the positive direction, dissolution rate in NaOH solutions decreased, and the overall increase of the thermodynamic stability of the metal was attributed by the authors to the redistribution under the effect of the ultrasound field of dislocations to energetically more stable configurations.

3.6 Semiconductor Systems Including Solar Cells

Japanese workers have published a paper regarding the study of photoelectrochemical reactions at a n-type polycrystalline zinc oxide electrode using photoacoustic detection [125]. They monitored *in situ* photoelectrochemical reactions at semiconductor electrodes using photoacoustic techniques.

Ottova et al. looked at two-compartment semiconductor-septum electrochemical photovoltaic cells with cadmium selenide and cadmium selenide telluride for water photolysis [126]. They used cells consisting of two chambers separated by a CdSe or CdSe/CdTe bipolar electrode. The bipolar electrodes were prepared by painting a CdSe slurry on a metal substrate or by ultrasound-aided electrodeposition from CdSe solution in $ZnCl_2$. The photoresponse (voltage and current output) and hydrogen yield from photo-induced electrolysis of H_2O in the dark chamber of the cell were evaluated as a function of CdSe preparation method. The ultrasound-aided deposition technique gave excellent coatings of CdSe.

3.7 Batteries

The use of ultrasound in battery technology is a growing area of research and scientists are continually developing techniques throughout the world which can enhance energy output and capacity of batteries. Ultrasound is widely used in battery studies, but mainly for some purpose during fabrication, and has been less-widely used during actual discharge or charge. Production uses include for example the preparation and compaction of powders in porous electrode formation, the cleaning of metal surfaces, and the pretreatment of electrolyte solutions. While it is not the purpose of this review to address systems which employ ultrasound outside the period of actual electrochemical activity, it is perhaps worth mentioning the effects of ultrasonic pretreatment of electrolyte solutions since this appears to produce significant benefits.

A study of the influence of ultrasound on the charging of lead–acid batteries [128] showed that there is a great improvement in the performance of these batteries caused by enhancement of ion transport. An increase of 10–22% in battery capacity is obtained in the presence of ultrasound, and even after deducting an estimated increase in capacity due to temperature changes caused by ultrasound, there is still an 8–14% increase in capacity for the battery. The results of the studies led to the conclusion that the improvement in performance of lead–acid batteries under the action of ultrasound can be attributed to the enhanced energy transfer and accelerated mass transport.

The voltages during discharge, and discharge capacities of Daniel–Jacobi electrochemical cells, containing 0.1N $CuS0_4$ and 0.1N $ZnS0_4$ as electrolytes, increased with sonication intensity from 0.6 W/cm^2 to 3.3 W/cm^2 of ultrasonic irradiation at 1 MHz. This is said to be due to the increase in the activities of the Cu^{2+} and Zn^{2+} ions in the solutions, with the increase in the activity of Cu^{2+} being greater than that of Zn^{2+}. The electrolytes were irradiated for 7–9 min intervals prior to the electrochemical measurements which were not performed under ultrasound [129].

Gavrila and co-workers [130–132] at the University of Bucharest, Romania, have extensively studied the effect of ultrasound on a number of battery systems and the reader is directed towards the work of this author. This work includes studies on charge and discharge rates and capacity of lead–acid accumulators. They found that an increase of 12–40% in the capacity of the lead accumulators was obtained by the action of ultrasound and that the number of discharge cycles was increased significantly. During the charging of the batteries to the charging limit where water electrolysis occurs the authors found that in the presence of ultrasound the charging time was increased by 1 h.

Ultrasound promotes the action of cavitation and streaming effects on diffusion processes and ion mobility in the electrolyte within the accumulator as well as increasing the active surface of the electrodes. This in turn causes an increase in the delivery capacity of the accumulator. However these phenomena are complex and are still under investigation.

Ultrasound has also been successfully employed in the preparation of other battery electrode materials. An example of this is the electrochemical impregnation of nickel hydroxide cathodes for batteries which was increased by 15% under ultrasonic irradiation. Active material content was 14.0 g/dm^3 under ultrasonic irradiation, and 12.0 g/dm^3 without. The active nickel hydroxide species formed by both electrochemical or chemical impregnation was not affected, but the deposition speed was higher and the grain size was smaller [133].

The effect of ultrasonic irradiation on the electrolytic production of manganese dioxide using platinum electrodes has been studied by Tasaka et al. [134]. The stripping and pulverization effects of ultrasonic irradiation on the electrolytic production of MnO_2 were investigated at various frequencies (28–2000 kHz) in a $MnSO_4$–H_2SO_4 system. At current densities greater than 1.5 A dm^{-2} and ultrasound frequencies of 28, 50, and 200 kHz, the cell voltage diminished, and the optimum stripping and pulverization effect was obtained at 28 and 50 kHz. At a current density (cd) of 3.0 A dm^{-2}, oxygen evolution took place which as expected diminished both the stripping effect and the current efficiency of the system.The available oxygen content of the MnO_2 produced was found to be independent of the ultrasonic frequency at current densities of both 1.0 and 3.0 A dm^{-2}, and the MnO_2 produced exhibited the same X-ray diffraction pattern as that deposited under mechanical stirring.

Other researchers have also looked at the electrochemical preparation of MnO_2 [135]. They describe a suspension process for producing MnO_2 for dry cells by electrolysis of an aqueous solution of $MnSO_4$ in which is dispersed or suspended a small quantity of Mn oxide (e.g. MnO_2, Mn_2O_3, and some Mn_3O_4 and MnO) particles. By using this procedure, the cd is increased; for example, in the usual procedure the cd maximum for C anodes is ~ 1.2 A/dm^2 and for Ti anodes 0.8 A/dm^2, while for Ti anodes in the suspension process a cd of 1.6 A/dm^2 was possible. The particles were kept in suspension by employing ultrasound at 19 kHz prior to electrolysis, i.e. a pretreatment protocol rather than a sonoelectrochemical one.

The effect of ultrasonic agitation on the electrolytic MnO_2 (EMD) deposition at a ferrite anode has been studied and its electrochemical activity determined by intrinsic polarization [136]. Ultrasonic agitation at 28–55 kHz produced a significant decrease in polarization and an increase in both current efficiency and pulverization at various current densities on the ferrite anode. The authors found that the performance of the ultrasonically produced MnO_2 excelled in the polarization for an alkaline medium.

Japanese researchers have brought out a patent [137] on the usage of ultrasound in the manufacture of various batteries and the reader is directed to this for further information. In general, the use of ultrasound in battery technology seems to offer great promise.

3.8 Other Inorganic Systems

Several processes for the electrolytic preparation of ferricyanides from ferrocyanides have been examined [138]. For example, a divided cell was employed with asbestos paper or rope wound over a polyvinyl chloride frame or microporous rubber or cation exchange membrane as a diaphragm, an anode cd of 0.2–10 A/dm^2, temperature of 10–70 °C with 2–10% alkali metal hydroxide as catholyte. A graphite or Cu anode was used under stationary or rotating conditions, or the cell was kept under the influence of ultrasound and a stainless steel cathode was used.

In the electrochemical oxidation of Fe^{2+} to Fe^{3+} (initial concentration approximately 1 g Fe^{2+}/L), a 1.8-fold higher rate, increased current efficiency from 12 to 31–45%, and lowered potential from 4.4–4.2 V were achieved in comparison with the process without ultrasound [139].

Work has also been carried out on the effect of an ultrasonic field on the electrochemical oxidation of ferrous ions to ferric ions [140]. When ultrasound was employed it was observed that the electrochemical oxidation of Fe^{2+} to Fe^{3+} was increased by a factor of 2–3. The current efficiency improves considerably since the ultrasonic field decreases the thickness of the diffusion layer and hence increases the limiting current.

The effect of various ultrasonic fields on the yield and rate of electrochemical processes in the oxidation of Fe^{2+} to Fe^{3+}, $Fe(CN)_6^{4-}$ to $Fe(Cn)_6^{3-}$, and Cr^{3+} to Cr^{4+} are also reported [141]. Percentage yields and current efficiencies for these reactions were studied at a cd of 0.25 A/mm^2 with and without ultrasound at frequencies of 15, 25, and 200 kHz. It was found that ultrasound always accelerated the process and increased current efficiencies dramatically. The authors found the optimum ultrasonic frequency to be 25 KHz, and also confirmed that ultrasound raised the limiting-current density considerably, causing a reduction of the diffusion layer thickness and therefore increasing the efficiency of the electrolytic reaction.

The kinetics of several well-known electrochemical reactions have been studied in the presence of an ultrasonic field by Altukhov et al. [142]. The anodic polarization curves of Ag, Cu, Fe, Cd, and Zn in various solutions of HCl and H_2SO_4 and their salts were measured in an ultrasonic field at various intensities. The effect of the ultrasonic field on the reaction kinetics was found to be dependent on the mechanism of metal anodic dissolution, especially on the effect of this field on the rate-determining step of the reaction. The results showed that the limiting factor of the anodic dissolving of Cu and Ag is the diffusion of reaction products, while in the case of Fe it is the desorption of anions of solution from the anode surface, and at Cd the limiting factor is the rate of destruction of the crystal lattice. Similar results were obtained by Elliot et al. [143] who studied reaction geometry in the oxidation and reduction of an alkaline silver electrode.

Kadyrov et al., who studied electrochemical hydrogen evolution on indium electrodes in the presence of an ultrasonic field [144], have also proposed a mechanism of ultrasound action on the cathodic reduction of indium [145]. Other workers have looked at the effect of ultrasound on the electroreduction of nickel and cobalt divalent ions catalyzed by ligands [146], obtaining similar results.

In another study, the application of a weak ultrasonic field (0.3 W/cm^2; 25 kHz) during the electrochemical oxidation of ferrocyanide ions on Pb anodes at 20 °C and at a fixed cd (2.5–15 mA/cm^2) markedly increased the reaction rate and the current, while the polarization was substantially decreased [146a]. The effects, which were most pronounced at the beginning of the electrolysis and at low current densities, were attributed to a considerable thinning of the diffusion layer on the anode in the presence of ultrasound.

Russian workers have looked at acoustic waves produced during the electrochemical oxidation of antimony [147], almost a reverse application of sonoelectrochemistry. Antimony was anodized in aqueous H_3BO_3 solutions galvanostatically (2.2×10^{-3} A/cm^2) and isothermally (292 K). The formation voltage increased to >200 V with time, which is characteristic of the valve metals. Acoustic waves were observed in this electrochemical oxidation with amplitudes that did not differ essentially from the very beginning of the oxidation. The energy of the acoustic wave had only one sharply distinct peak which coincided in time with the appearance of the electrochemical breakdown products.

The effect of ultrasound on the process of tellurium anodic dissolution in alkaline solutions was studied by the method of plotting polarization and galvanostatic curves [148]. Tests were made in NaOH solutions (concentrations of 0–20 g/L), subjected to the action of ultrasound at a frequency 17.5 kHz and using Te electrodeposited under ultrasound. The anodic polarization curves plotted without ultrasound and in its presence shifted with increased NaOH concentration towards negative values as a result of the increasing rate of Te anodic dissolution. The presence of ultrasound inhibited the process of Te anodic dissolution, probably due to the desorption of OH^- anions from the anode surface. This sonoelectrodeposited Te thus showed greater corrosion resistance in alkaline solution than that deposited by normal electrolysis. The same workers later described the reductive electrodeposition of Te from acid solutions containing species such as $Te(OH)_3$+ [149]. Here ultrasound mitigated the passivation phenomena leading to "good quality" Te electrodeposits.

Glassy carbon electrodes, irradiated with 20-kHz ultrasound from a 475-W generator in dioxane, exhibit enhanced heterogeneous electron transfer rates for aqueous redox probes [150]. When sonications are performed in water, however, no significant enhancement effects were observed. Several electroanalytical techniques with different time scales were employed along with SEM to characterize surfaces before and after ultrasonic modification in different solvents. Surface roughness does not change appreciably after brief sonication in dioxane, although a small amount of surface pitting occurs. These electrodes are demonstrated to remain active for up to 5 days and are be more prone to adsorb aromatic redox probes in aqueous media than mechanically polished electrodes. After sonication in water, the carbon surfaces are highly pitted and show evidence of an increase in the diffusion of electroactive surface oxides. Thus, the improvement in kinetics observed after sonication in dioxane is probably not associated with either increased microscopic electrode area or mediated electron transfer between surface oxides and solution analytes, but instead is likely to involve surface cleaning processes. This is an example of electrode pretreatment with ultrasound. Another such example—the roughening of platinum—has already been mentioned in Section 2 [41], and other examples may be found in, for example, the references of Zhang and Coury's paper [150].

Finally a novel application of ultrasound in cold nuclear fusion has been published. This whole area remains controversial, but a Japanese patent claims that sonication of a cell in which D_2O is electrolyzed at a palladium cathode causes an improvement in the "efficiency of cold nuclear fusion" [151]. Russian workers also report the "generation of nuclear-fusion products" during combined action of cavitation and electrolysis on the surface of titanium in deuterated electrolytes [151a]. And there has been recent speculation regarding the capability of ultrasound to drive fusion effects within a cavitating bubble [151b].

4. ORGANIC SONOELECTROCHEMISTRY

4.1 Introduction

This section concerns not only organic and organometallic electrosynthesis, but also electroinitiated polymerizations and other processes based on organic systems. Reference to sonoelectrochemiluminescence is made where appropriate. (This topic is further elaborated in Section 5.2.) All these aspects have been less widely studied under ultrasound than either electroanalysis or the electrochemistry of metal systems, with only a few sporadic reports prior to the 1980s. There has since been a gradual upsurge of interest, and this area has become an exciting topic of sonoelectrochemical research.

4.2 Organoselenium and Organotellurium

One of the first coherent series of reports upon ultrasonically enhanced electrosynthesis came in the mid-1980s from a French group who used the technique to explore routes to organoselenium and tellurium derivatives. Thus electroreductive synthesis of Se_2^{2-} and Se^{2-} dianions was enhanced by simultaneous irradiation from an ultrasonic cleaning bath [152]. Instead of employing a sacrificial cathode of elemental selenium, this procedure allowed the direct use of selenium powder with carbon cloth as a cathode. A further benefit was that this method also allowed production of the corresponding tellurium anions. These species could be employed *in situ* in aprotic solvents such as DMF, THF, and MeCN for the synthesis of selenides and tellurides by nucleophilic displacement from haloalkanes (Scheme 1). This system was extended to the electroreductive synthesis of a cyclopentadienyl titanium pentaselenide $[(C_5H_4R)_2TiSe_5]$ from selenium powder and $[(C_5H_4R)_2TiCl_2]$. When R = Me, the optimized process gave a 70% yield [153].

The same group also addressed the formation of unsymmetrical (di)aryl chalcogenides by electroreduction of the symmetrical diaryl chalcogenide at a carbon cloth cathode to give, for example, the $PhSe^-$ anion. Bromobenzonitrile was then added and electroreduction under ultrasound was continued, leading to 57% yield of $C_6H_5SeC_6H_4CN$ and a 42% yield of $C_6H_5TeC_6H_4CN$ with some of the corresponding symmetrical functionalized dichalcogenide as a side product [154]. This SRN1 mechanism was further exploited to produce a number of unsymmetrical phenylseleno benzophenones and their tellurium analogues [155], again by a two-step electroreductive process with both steps being performed under ultrasound. In the first procedure, the potential was made more negative after formation of the $PhSe^-$ anion, at which point haloketone was then added. Thus $PhCOC_6H_4SePh$ could be made in 49% yield, but production of the tellurium analogue was less efficient, giving only 17% isolated yield with symmetrical species such as $PhCOC_6H_4TeC_6H_4COPh$ as by-products.

These are complex reaction systems with a number of side reactions, and it was found that addition of weakly acidic species such as malononitrile and fluorene suppressed competitive solvent deprotonation and side reaction pathways so initiated. The systems were further sophisticated when a redox mediator such as azobenzene was added. Now the cathodic potential in the second step of the procedure could be substantially reduced until only the mediator remained elec-

$$E \xrightarrow{e} E^{2-},\ E_2^{2-} \xrightarrow{R\text{-}X} RER + REER$$

where E= Se, Te. X= Cl, Br.

Scheme 1.

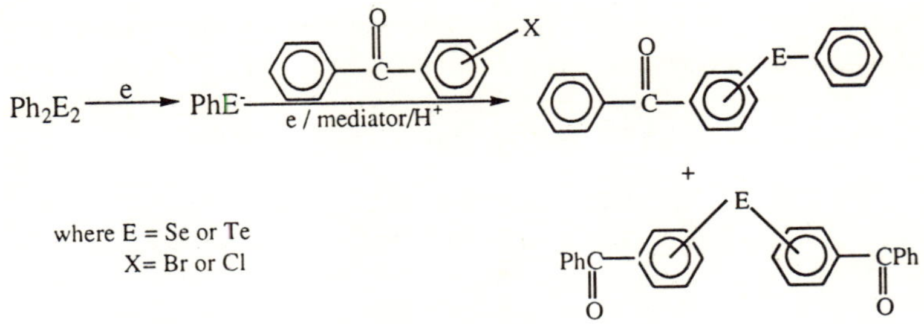

Scheme 2.

troactive. Thus the electroreductive generation of RSe^- or RTe^- comprised the first step; then the haloketone, the acidic species such as fluorene, and the azobenzene redox mediator were all added and the potential reduced for the second step. By this means, isolated yields of 86% $PhCOC_6H_4SePh$ (before crystallization) and 45% of the tellurium analogue are given. The procedure is outlined in Scheme 2. The role of ultrasound in such complex systems is not easy to ascertain, and interestingly in this paper, the isolated yield for the tellurium derivative under mechanical stirring is given as 44%, showing little difference from the sonoelectrochemical yield.

These reactions are of course different from those in ref. 155 where activation of the solid elemental chalcogenide is involved, and where particle-size breakdown and increased mass transport of solids to the electrode surface will themselves be useful benefits of ultrasonic irradiation.

The French group also reported the anodic cleavage of REER or RER (where E = chalcogenide) in MeCN, in contrast to the previous electroreductive systems, and a number of products were prepared where RE^+ had been trapped by appropriate nucleophiles [156].

4.3 Organosilicon and Organogermanium

A recent Japanese effort has involved the production of organosilanes and germanes as small molecules or polymeric systems by electroreduction of the appropriate halospecies at a reactive metal cathode in aprotic media. The same metal—Mg, Al, or sometimes Cu—is used as a sacrificial anode, and the cell is undivided. Thus with lithium perchlorate as the preferred electrolyte salt and in THF solvent, a dichlorosilane such as $PhMeSiCl_2$ gives a polysilane of $M_n \sim 3000$ in 22% yield [157]. This is in contrast to earlier work at Hg cathodes in divided cells, where Si–O-containing polymers and cyclotetrasilanes were obtained by other workers [158,159]. Simultaneous ultrasonic irradiation at 47 kHz in the Japanese procedure increased the yield of the quoted polysilane to 33%, although without obvious or consistent effect on M_n or polydispersivity in the given systems.

The Japanese work is the subject of a number of patent applications assigned to the Osaka Gas Company, for disilanes obtained from $R^1R^2R^3SiX$ for a variety of R groups [160] and also digermanes similarly [161]. The silane and polysilane process has been grouped into a European patent application [162] where the benefit is stated of switching the cell polarity at regular intervals as well as using ultrasound. In all the use of this technique to prepare these materials is said to be "safe and causes no pollution."

Other variations on this theme concern the electrosynthesis of germane polymers [163] or silane–germane copolymers [164] from dihalo derivatives. Alternatively, the use of trihaloderivatives ($RSiX_3$) allows the formation of network polysilanes [165]. In all these cases, the mechanism of ultrasonic influence is not elucidated and in general there are few specific details of the ultrasonic system, its geometry, and other relevant factors in English language digests of this work.

4.4 Electroinitiated Chain Polymerizations

Another coherent series of reports upon sonoelectrosynthesis, beginning in the mid-1980s, concerns the effect of ultrasound upon electroinitiated chain polymerizations. The authors examined the potentiostatic control of a number of copolymerizations and compared this with the more often used galvanostatic control technique. In such copolymerizations one monomer reacts at a different potential to the other, such that the copolymerization requires high oxidation potential to proceed appreciably. However, even higher potentials are often required to minimize blocking of the electrode surface by a thin film of polymer which impedes further electrochemical reaction; the authors rationalized their use of ultrasound in terms of its removal of this layer in a "sweeping clean" manner, thus allowing the true reactivity ratio of the monomers to operate throughout the electrolysis.

Figure 13 shows the effect of ultrasound upon the electrooxidatively initiated copolymerization of isoprene with α-methyl styrene in CH_2Cl_2/Bu_4NBF_4 at –30 °C using Pt electrodes in a divided cell [166]. Thus with ultrasound, copolymerization proceeds effectively at lower potentials, with some curvature in the dependence of composition upon potential, and the equal 50/50 copolymer was obtained at a "critical polymerization potential" of –2.6 V (vs. Ag^0/Ag^+), whereas without ultrasound the equal copolymer was obtained at 3.0 V (Figure 12a). Thus sonication can be said to alter the effective reactivity ratio of the monomers. Figure 12b demonstrates the overall conversion of the system. Here sonication produces a relatively consistent effect throughout the potential range employed, whereas although the silent system has a better conversion at lower potentials it decays at the higher potentials required to produce the equal copolymer. In addition, the molar mass of the copolymer decreases with increasing potential in the silent system, but this is less pronounced in the sonicated case. Overall, ultrasound produces a more uniform reaction system; the lower degree of conversion at the lower potentials in

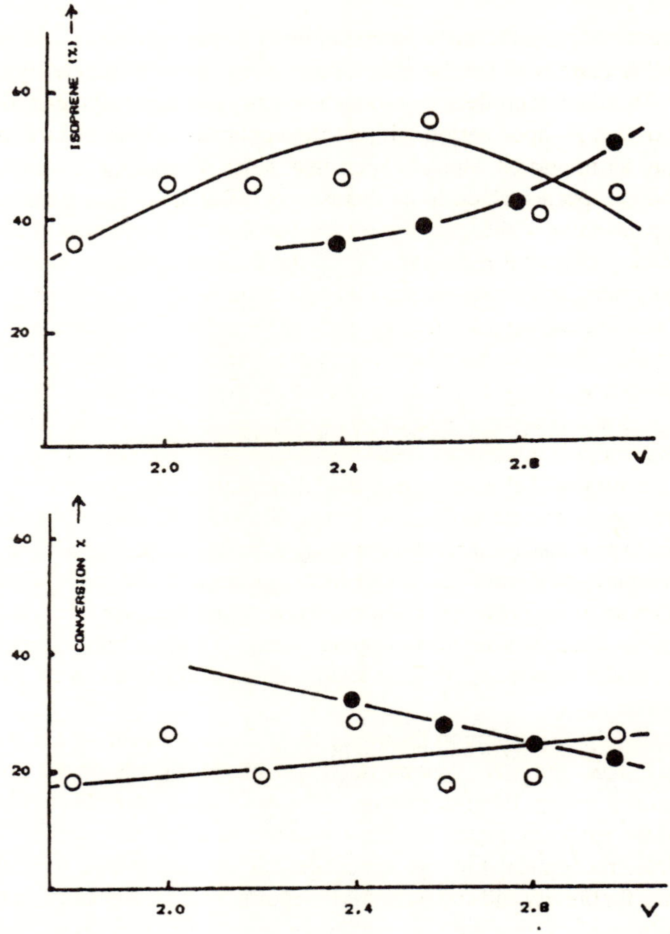

Figure 13. Electroinitiated copolymerization of α-methyl styrene and isoprene (● - non ultrasonic; ○ - ultrasonic). Top figure—isoprene content of copolymer; bottom figure—overall conversion (taken from ref. 166).

Figure 12b is not without precedent since simple mechanical stirring retards styrene polymerization [167].

Very similar effects were also found upon copolymer composition, total conversion, and RMM-control in the styrene–isoprene copolymer system [168] where the analogous traces in Figure 12 shifted to slightly more anodic values, with a better total conversion at high potentials under ultrasound. Ultrasound in both systems was provided by an Ultrasonic Cleaner at 25 kHz. Copolymer compositions were determined by infrared spectroscopy.

Also examined was the copolymerization of α-methylstyrene and 4-bromostyrene [169], again with similar effect, here using "a small Buehler-type" probe of 25 kHz. This did not produce any sonochemically induced polymerization of the monomer over a 24-hour period without the application of an electric potential, representing an important control experiment since ultrasound is well known to produce radical species which could themselves influence polymerization.

The main objective of these studies was to establish the usefulness of controlled-potential electrochemical methodology for the production of the electroinitiated chain copolymers, and it was found that both monomer reactivities (m_1 and m_2) increased with increasing potential, as might be expected. Their relative ratio in the copolymer also changed, but there was not a linear dependence either with or without ultrasound. The significance of this work is that care should be taken with the alternative and simpler electrochemical procedure of controlled-current electrolysis, especially if inhibiting effects operate, because alteration of the effective electrode potential will alter m_1 and m_2 and their ratio so that copolymer composition will change as the electrolysis proceeds. Ultrasound in the first instance was simply used to provide an enabling technology to prevent blocking of the electrodes so as to maintain the desired electrochemical parameters, and more subtle effects of irradiation upon the polymerization system were not discussed. In addition, and as is often the case in sonoelectroorganic studies to date, only a single set of ultrasonic conditions were applied, although a range of electrochemical parameters were addressed.

Incidentally, those wishing to follow up these studies should note that nowhere in refs. 167, 168, and 169 is given full experimental details of the preparative electrochemical set-up. Even the nature of the electrode material is not obviously stated, and the reader is invited to refer for details to an earlier paper [170] which concerns electrochemical but not sonochemical studies. While journal editors rightly strive for brevity and compactness in reports, a reader approaching this work from the ultrasonic viewpoint (i.e. unfamiliar with the history from the polymer side) will find the paucity of experimental detail in such seminal papers to be a significant omission.

The same authors do give more experimental details of the preparative electrochemical system used in a recent study of ultrasound upon the electroinitiated homopolymerization of butadiene sulfone (also known as 3-sulfolene) in $MeCN/Bu_4NBF_4$ at platinum [171], although they monitored polymerization kinetics in a reportedly novel way by cyclic voltammetry with a six-electrode system that they again describe elsewhere [172]. Interestingly, ultrasound in the preparative constant-potential reaction did not completely clear the electrode of a blocking film, although it did produce an improvement in the percentage-conversion versus time characteristics of the polymerization.

The final polymer can contain unbroken ring systems or linear units as shown in Scheme 3. By infrared spectroscopy the authors found more of the linear units, suggesting C–S bond cleavage to be the major process.

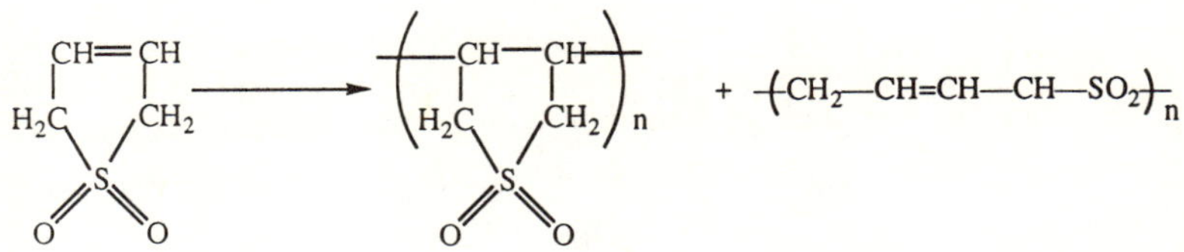

Scheme 3.

There is some variation in literature reports about the polymerization of this monomer, and by their electroinitiation method the authors claim "high yields at room temperature in reasonable polymerization times." The M_n of the polymer is ~5500, and the percentage conversion can be enhanced by increase of polymerization potential by increasing the temperature, or by using ultrasound.

4.5 Electrically Conducting Polymers

Electrically conducting polymers are quite different systems to the above electroinitiated chain polymerizations since they are formed by an unusual step-growth mechanism involving stoichiometric transfer of electrons. The polymers are obtained directly in a conductive polycationic form in which charge-compensating counter anions from the electrolyte system are intercalated into the polymer matrix [173]. Exact mechanistic details remain the subject of discussion, but Scheme 4, which shows polypyrrole formation is plausible. Polythiophene is similar where S replaces NH in the ring.

These conducting polymers are sometimes termed "organic metals," and their formation as thin-film coatings on an electrode involves nucleation phenomena; so perhaps by analogy with metal electrodeposition under ultrasound there have been studies to improve the quality of the deposited materials. Polypyrrole readily forms acceptable films under a wide variety of conditions [174], although there are subtle distinctions in behavior as a result of exact preparation procedure [175]. Ultrasound from 38 or 500 kHz baths does not appear to appreciably improve polypyrrole film formation, at least from MeCN electrolyte at the normal current density of 1 mAmp

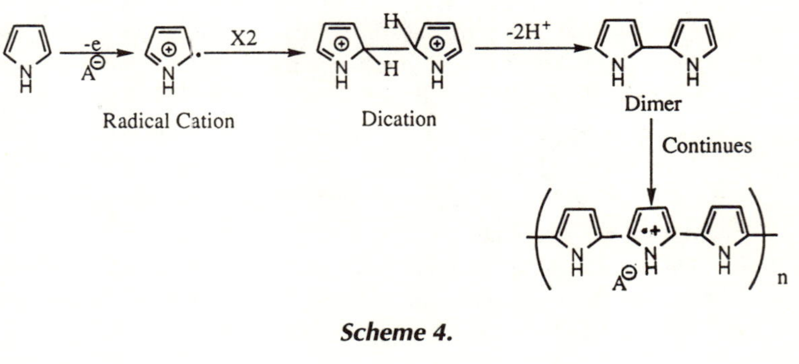

Scheme 4.

cm^{-2} [176], although this does not preclude useful effects of sonication in other conditions since the variation of behavior with preparation protocol is widely recognized as a source of the reputation for irreproducibility gained by these materials.

Polythiophene, however, is less tolerant of preparation conditions than polypyrrole. It also has less two-dimensional cohesion and cannot so easily be peeled from the electrode as free-standing films. Workers at Tokyo University have found that the quality of polythiophene films deposited on an electrode can be enhanced by ultrasound. By conventional methodology the films become brittle as the electrolysis current exceeds 5 mAmp cm^{-2}, but by using ultrasound from a 45 kHz cleaning bath, flexible and tough films (tensile modulus 3.2 GPa and strength 90 MPa) can be obtained even at high current density [177].

These workers have extended this work [178] and found that in nitrobenzene solvent without ultrasound an increase in current density (using the procedurally simpler galvanostatic methodology) produced an increase in effective electrode potential giving a low polymer yield and macroscopically inhomogeneous films. However, with ultrasound, increasing cd does not similarly influence the effective potential and there is a higher yield of better-quality film. The sonochemical benefits were especially marked at low temperature (5 °C), relatively low monomer concentration (0.1 M), and at high cd (up to 10 mAmp cm^{-2}). The best polymer conductivities obtained were 100–150 S cm^{-1}; and the sonoelectrochemical enhancement was attributed to effect upon the diffusion layer during the electropolymerization.

A different Japanese group (at Matsushita) have patented the use of ultrasound in the preparation of polypyrrole, polythiophene, or their derivatives for the manufacture of solid electrolyte capacitors. The polymer is electrodeposited on a dielectric film prepared on a valve metal such as aluminum or tantalum [179]. Use in capacitors is becoming a recognized application of conducting polymers, which otherwise have yet to make a substantial mark in the commercial world; it is significant that ultrasound contributes to this electrosynthetic system. However, given the sensitivity of conducting polymers to preparation procedures, and given the mechanistic complexities in these systems, care in the implementation of ultrasound is necessary to produce the best results.

4.6 Electroorganic Synthesis: Electrooxidations

In the preceding sections, ultrasound has been employed in electroorganic systems to facilitate an expected reaction, and this use also applies to many of the following synthetic systems. However, the possibility that ultrasound might also alter the course of the reaction had not really been addressed prior to studies of the Coventry group in the late 1980s. The reaction chosen was the electrooxidation of carboxylate anions (the Kolbé reaction), long known as one of the earliest discovered organic reactions and the subject of a wealth of empirical data, regularly

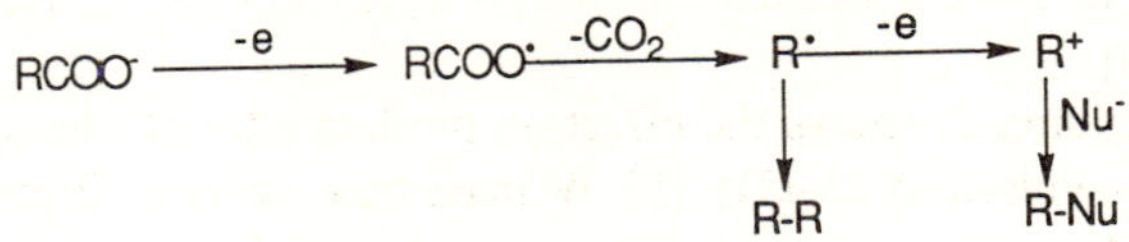

Scheme 5.

reviewed [10,180,181], but where there still remains mechanistic controversy [182]. Different reaction pathways exist under different kinetic regimes, and adsorption and other electrode phenomena are known to be important, all of which might be influenced to differing extents by ultrasound. Scheme 5 gives the usual and plausible scheme to account for the range of products obtained from electrooxidation of a typical carboxylate anion ($RCOO^-$). The general mechanisms break down into a pathway involving one electron per molecule of starting material, giving products from the radical intermediate, e.g. the dimer [R–R] (which is the actual Kolbé reaction), and a two-electron pathway per starting molecule, giving products from an intermediate cation (often called the Hofer–Moest Reaction).

Empirical rules have been elaborated to account for competition between these pathways, depending on electrolysis conditions [180]. The Coventry group chose to examine a system almost at balance where both pathways operate [183] in order to best identify any sonoelectrochemical effect on mechanism [184]. Table 4 shows product ratios (by glc) from the electrooxidation of partially neutralized cyclohexanecarboxylate in methanol at platinum, at a current density of 200 mAmp cm^{-2}. The first column shows a substantial amount (49%) of the dimer bicyclohexyl from the one-electron pathway, together with cyclohexylmethylether, cyclohexanol, and other products from the two-electron pathway (totaling ~30%). The methyl cyclohexanoate ester (17%) is considered to arise from acid-catalyzed chemical esterification of the starting material with methanol solvent, due to the quantity of protons produced around the anode; since at the high current densities needed, the parasitic

Table 4. Electrooxidation of Cyclohexanecarboxylate[a,b]

	Without Ultrasound	*With Ultrasound*
Bicyclohexyl	49.0	7.7
Cyclohexane	1.5	2.6
Cyclohexene	4.5	32.4
Methoxycyclohexane	24.9	34.3
Methyl cyclohexanoate	17.0	2.5
Cyclohexanol	2.1	6.8

[a]Relative product ratios by g.l.c. after the passage of 2.2 F mol^{-1}

[b]Average cell potential to maintain current density of 200 mA cm^{-2} is 8.3 V in the absence of ultrasound and 7.3 V in its presence.

discharge of the solvent methanol, which produces protons, remains a competitive reaction [180].

Table 4 (column 2) shows the effect on product ratio of ultrasonic irradiation from a Kerry Pulsatron 35-kHz (50 W maximum power) cleaning bath during electrolysis. Now there is only ~8% of the bicyclohexyl dimeric one-electron product, with ~41% of two-electron products from nucleophilic capture of the intermediate carbocation, and a striking enhancement to 32% of cyclohexene, with only <3% of cyclohexane. The preponderance of cyclohexene over cyclohexane shows its formation by proton loss from the carbocation intermediate, since free-radical routes to cyclohexene (i.e. hydrogen atom abstraction) also produce cyclohexane in equal if not greater amounts [180,183]. It is also noted in column 2 that the parasitic formation of methyl cyclohexanoate ester is lessened under ultrasound, perhaps suggesting enhanced adsorption of carboxylate with concomitant suppression of solvent discharge. This has precedent since ultrasound is thought to enhance adsorption phenomena in some dissolving-metal chemical reactions [185].

Other procedural benefits in the electrooxidation of cyclohexane carboxylate under ultrasound include a drop in overall cell voltage from 8.3 to 7.3 V needed to maintain the same cd for the galvanostatic system. The reaction approached completion in a shorter time-span despite the apparent switch to the two-electron process, suggesting diminution of parasitic processes.

Overall, ultrasound appears to favor the two-electron mechanism for the reaction, but the greatest effect of sonication upon product distribution was the substantial enhancement of alkene formation. Accordingly it was decided to examine a carboxylate electrooxidation system where there is no proton-loss pathway from the intermediate carbocation, namely using phenylacetate as a substrate.

Table 5 shows relative product ratios for phenylacetate in similar conditions to those used for cyclohexane carboxylate, but employing 100 mAmp cm^{-2} cd [186,187]. There are four columns in the table due to another complication: simple

Table 5. Electrooxidation of Phenylacetate[a,b]

	Without Ultrasound		*With Ultrasound*	
	No Pyridine	*13% Pyridine*	*No Pyridine*[c]	*13% Pyridine*
Bibenzyl	0	59.8	52.7	51.3
Toluene	0	0.4	3.1	1.5
Benzyl methyl ether	0	21.1	32.3	27.8
Methyl phenylethanoate	0	10.2	6.2	4.2

[a]Relative product ratios by g.l.c after the passage of 1.1 F mol^{-1}.

[b]Average cell potential to maintain current density of 100 mA cm^{-2} is 7.9 V in the absence of ultrasound and 6.6 V in its presence.

[c]A fine white powder precipitate was formed in this electrolysis.

electrolysis of the partially neutralized salt in methanol causes a very rapid increase in applied cell voltage due to the coating of the anode with a pale-colored material that causes the reaction to cease. This is well known [188], and it is necessary to add pyridine (up to 50% v/v) to keep the electrode clean, presumably by simply solubilizing the inhibiting layer. Table 5 (column 1) therefore shows 0% effective yields of products, while column 2 shows product ratios from a silent reaction in the presence of 13% (v/v) of pyridine. Here there is 60% of the Kolbé dimer bibenzyl, 21% of the two-electron ether product, and some 10% of the parasitic methyl ester. A major component of the remaining (8%) material is benzaldehyde, a persistent by-product of arylacetate electrooxidations whose exact mechanistic origin remains uncertain. The reactions were terminated after 1.1 Faraday mole^{-1} and were not run to completion.

Table 5 (column 3) shows the effect of ultrasound upon the product ratio from methanol/pyridine. There is now 53% bibenzyl, 32% methyl ether, and 6% methyl ester (with total 5% of other products), suggesting only a slight shift towards the two-electron products, but with an overall diminution of solvent discharge and side reactions. Phenyl acetate electrooxidation, however, is known to favor the one-electron route to bibenzyl in a wide range of conditions [188], and to be much less sensitive to mechanistic switches by manipulation of parameter than is cyclohexane carboxylate electrooxidation [180]. This trend remains even under ultrasound.

Table 5 (column 4) shows the product ratios under ultrasound in the absence of pyridine. Overall, there is the same trend with sonication, namely a slight shift from one-electron towards two-electron pathway, although here there seems to be a higher yield of benzaldehyde-derived by-products.

The most significant factor is that there is no evidence of electrode fouling, and the reaction maintains cd at a steady and lower voltage. In addition, a fine white powder was formed during the electrolysis, which was isolated by filtration (yield 14% by weight). This appears to be a polymer of moderate RMM containing aromatic rings as well as two types of methylene groups (from NMR) and an aliphatic ester carbonyl (1735 cm^{-1} in the IR). This suggests a structure incorporating $-C_6H_4-CH_2-$ units and $-C_6H_4CH_2COO-$ units. It may be supposed that enhanced mass transport under ultrasound and the abrasion effect near the electrode surface has swept the inhibiting species into solution, thus keeping the electrode clean. There is much less of this powder under ultrasound in the presence of pyridine, suggesting that it is indeed solubility factors that demand the use of the cosolvent in silent conditions. It is very unusual to obtain products of carboxylate electrooxidation that originate from the acyloxy radical. Previous work has involved the trapping of this intermediate by intramolecular cyclization before it has time to decarboxylate (Scheme 6) [189,190], but here a linear product involving units from the first-formed radical has been isolated after intermolecular reaction.

The obviation of pyridine under ultrasound represents a significant procedural enhancement since its presence considerably hampers work-up, and also has environmental implications. On an industrial scale the pyridine would have to be

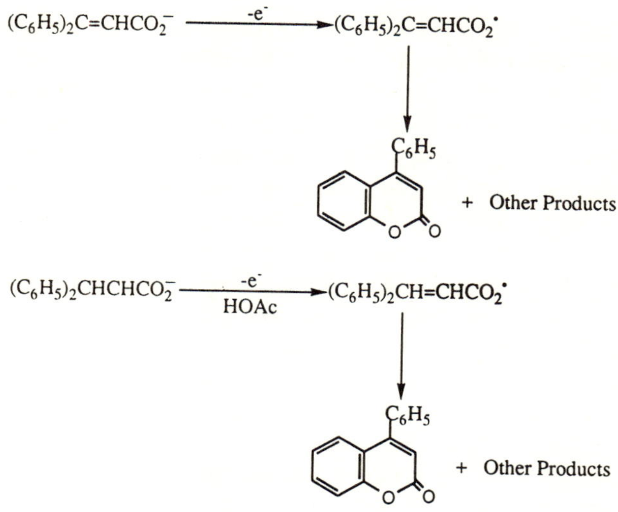

Scheme 6.

recycled on economic grounds. These results therefore have implications for electrolyses that are affected by electrode fouling problems. A further feature is again the lowering of the applied cell voltage from 7.9 V to 6.6 V under ultrasound, representing an energy saving.

Shortly after our publications on the sonoelectrochemical oxidation of phenylacetate [186,187], a parallel study was reported by Japanese workers [191] who employed crossed Kolbé electrolyses of phenylacetates and succinates, variously deuterated, to produce deuterated derivatives of 4-phenylbutyric acids. In control experiments to produce deuterated bibenzyls from phenylacetate without succinate present, they obtained 11% of dimer with pyridine present. Under normal conditions, this rose to 47% yield of and 41% of dimer under ultrasound from a cleaning bath, although here the reaction time is stated to be reduced threefold. However, it is ambiguous whether this shortened reaction-time benefit also applies to the reaction with pyridine but without ultrasound. The authors state that ultrasound "helps to keep the electrode surface clean" and it would seem that in their conditions, which employ aqueous solution instead of methanol, the electrode is not completely "switched off" by the insulating film under normal conditions. The authors did not examine the system with both pyridine present and ultrasound, but the observed yield drop from 47 to 41% might suggest the same trend towards the two-electron pathway under ultrasound, although other products were not identified and quantified.

These workers also extended the cross-Kolbe system to palmitic acids, although the significance of ultrasound was not detailed in this short paper [192].

The Coventry group has also examined the behavior of *p*-chlorophenyl acetate electrooxidation under ultrasound [187]. This substrate is known to markedly favor the two-electron mechanism [180], showing that the choice of reaction pathway is more dependent on substrate nature than upon manipulation of electrolysis parameters. A further feature of this system is the appreciable yield of *p*-chlorobenzaldehyde-derived products. This is shown in Table 6 where it can be seen that sonication produces little change in relative product ratio, although there is an increase in total yield after ether extraction. Thus 46% by weight of mixed product is obtained (unreacted acid is not recovered by this procedure) with ultrasound, but only 23% by weight from the silent reaction. This represents increased reaction efficiency since the same quantity of charge was passed in each case.

An unexpected advantage of sonication was found here. There was no formation of polymeric coating on the anode, tending to confirm that the free *para*-position of unsubstituted phenylacetate is necessary for production of the inhibitory species; instead the cathode became coated in a black deposit as the electrolysis proceeded. In order to maintain the cd as high as 100 mAmp cm^{-2} in methanol, it is necessary to have the working and counter electrodes close together in an undivided cell to minimize resistance losses; it is thus possible to envisage a reductive cleavage process occurring on the protonated acid molecule (the salt is only partially neutralized) with loss of chloride ion to yield species that could polymerize on the cathode instead of the normal cathodic reaction of hydrogen evolution. Fouling on either electrode will increase cell resistance and energy loss, hence ultrasound here offers a benefit to the system by suppressing coating of the counterelectrode.

The origin of ultrasonic effect upon carboxylate electrooxidation is not straightforward to establish in view of the complex mechanism of the reaction with different kinetic regimes, the loss of carbon dioxide, and also the role of adsorption

Table 6. Product Distribution from Electrochemical Oxidation of p-Chlorophenylacetate[a]

	Relative Product Distribution[b]	
	Silent[c]	*Sonicated*[d]
p-Chlorobenzyl methyl ether	71.5	69.3
Methyl *p*-Chlorophenylacetate	3.4	2.5
p-Chlorobenzaldehyde	9.1	4.2
p-Chlorobenzyl alcohol	3.5	9.0
p-Benzaldehyde acetal	10.8	11.2
Others	1.7	3.8

[a]Average potential difference required for current density of 100 mA cm^{-2} is 5.63 V in the absence and 4.83, in the presence of ultrasound.

[b]Relative peak area % after passage of 1.1 Faraday mol^{-1}.

[c]Total weight yield after extraction–23%.

[d]Total weight yield after extraction–46%.

and other electrode phenomena. It has also been suggested that the second electron transfer to give the carbocation need not occur in bulk solution, although this requires methoxyradicals from solvent discharge or other species to act as redox mediators in the solution phase [182].

It may be that ultrasonic enhancement of mass transport sweeps the intermediate radicals that have escaped the electrode back to the electrode surface where they are further oxidized, although this would depend upon radical lifetimes. Direct enhancement of the second electron transfer to give the cation while species remain in first contact with the electrode would be complicated by the decarboxylation step after the first electron transfer, and also by an observation of weak electrochemiluminescence (ECL) from the electrolysis cell in phenylacetate electrooxidation (see Section 5.2) [215]. This is enhanced by ultrasound, as are a number of other ECL systems. This suggests that at least a proportion of the reaction pathway involves benzyl radicals which escape the electrode, although the sonoelectrochemiluminescence reaction conditions of low carboxylate concentration, low current density, and presence of electrolyte salt are different to those for the preparative electrolyses.

Ultrasound, of course, can provide appreciable microscopic heating effects, e.g. via cavitation phenomena. Thus to minimize possible contribution from bulk heating effects in carboxylate electrooxidation, the cells were cooled during sonoelectrolysis, but to examine other possible temperature effects of ultrasound the Coventry group performed electrolyses at 60 °C and found that shifts in product distribution so caused were not of the order observed under sonication at lower temperature. The group also confirmed the stability of reactants and products under ultrasound without application of an electric potential. This does not preclude microscopic heating effects being the cause of some sonoelectrochemical phenomena since the "hot spot" theory predicts large-temperature gradients associated with cavitation.

The Coventry group also examined the use of higher ultrasonic frequencies (500 and 800 kHz) [193]. Here the trend in product distribution from cyclohexane carboxylate electrooxidation at platinum in methanol was found to be the same as under sonication in the 20 to 40 kHz frequency range. However, the yields are better and there remains the lessened cell voltage requirements and other benefits. There also seems to be fewer of the numerous low-yield methoxylated species and other side products. This is evidenced visually by the clarity of the electrolyte during reaction and the absence of buildup of a brown coloration during electrolysis. It would seem that higher frequencies are more suitable for the insonation of these reaction systems, although this is not straightforward to explain since cavitation phenomena, effects of acoustic impedance, cell geometry implications, and other factors all change with frequency.

Cavitation is substantially more difficult to induce at higher frequencies, and since these experiments were performed at the same ultrasonic power (measured calorimetrically), suggests that the observed sonoelectrochemical benefits are not cavitational in origin. However, it may be simply an effect of the shorter wavelength at the higher frequency. Thus in the 20–60 kHz frequency region, the half-wave-length is of the order of centimeters so that only a limited number of nodes and antinodes span the dimensions of the electrolysis cell used; in the 500–800 kHz region, however, the distance between nodes and antinodes is only in the order of millimeters. If, as seems likely from electroanalytical studies, certain sonoelectrochemical benefits originate from nodes or positive regions of the wave, but antinodes or negative regions of the wave do not interfere, then simplistically an increase in the number of nodes within the active region of the cell will produce more uniform and effective sonoelectrochemical phenomena.

4.7 Electroorganic Syntheses: Electroreductions

One of the potentially most useful preparative-scale electroreduction reactions involves the transformation of aryl nitrocompounds to amino derivatives, but in practice this is a complicated system in which there are a number of competing reactions including, for example, hydrodimerization to hydrazo compounds [11,194]. An early sonoelectroreductive paper concerns a variant of the nitroreduction system in which nitrobenzene is converted directly into *p*-amino phenol on a 50 g scale by reduction at 50 mAmp cm^{-2} in a catholyte of 30% aqueous H_2SO_4 containing some 0.5% of $SnCl_2$ and 0.01% of the nonionic surfactant poly(ethyleneglycol)nonyl phenyl ether [198]. Ultrasound is applied at 2.375 kW power (full cell dimensions are not given, thus precluding assessment of power density), electrolysis is continued for 16–30 h, and the cell temperature rises to 80–90 °C. A simple workup yields some 70% yield of product. The anolyte in the divided cell contains only a more dilute 8% solution of H_2SO_4. Despite the multipathway nature of nitroelectroreductions, there appears to be no further reports of sonoelectrochemical studies on this type of reaction.

Recent work from the group at the Tokyo Institute of Technology has been specifically directed at ultrasonic control of product selectivity in electroreductions.

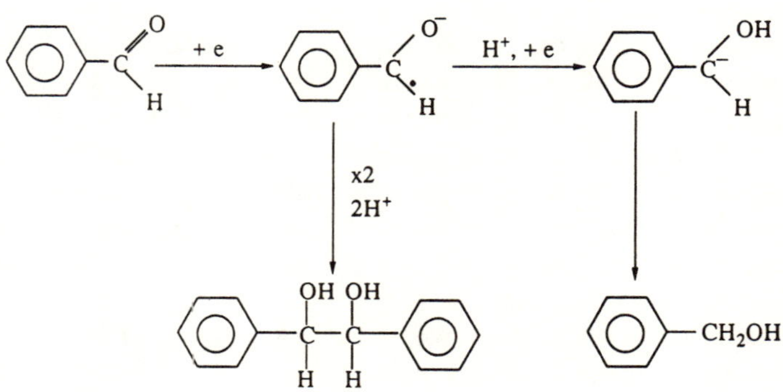

Scheme 7.

Thus the electroreduction of benzaldehyde [196] can lead to either the hydrodimer in a one-electron per substrate molecule process, or to the benzyl alcohol in a two-electron process, as indicated in Scheme 7.

Using a lead cathode in dilute methanolic sulfuric acid at constant current of 20 mAmp cm^{-2}, the benzyl alcohol was the major product from an unstirred solution, while mechanical stirring reversed the position to favor the hydrodimer. However, ultrasonic irradiation from a cleaning bath (100 W, 36 kHz) so strongly favored the hydrodimer that the alcohol was barely evident. The effect varied somewhat with the position of the cell in the ultrasonic bath, and increased in magnitude with ultrasonic power, but throughout all electrolyses there was no appreciable change in the stereochemistry (dl/*meso* ratio) of the benzoin hydrodimer.

For benzaldehyde there was always some 30% yield of unidentified products, whether the reaction was sonicated or not. These side products are thought to be hydroquinoid dimers. *p*-Methyl benzaldehyde, which also shows the substantial sonoelectrochemical switch to the hydrodimer, has a blocked *para*-position and is less affected by side reactions (results are summed up in Table 7). Similar effects were obtained by ultrasonic irradiation from a horn probe (20 kHz), and by employing a range of different silent and sonicated conditions (powers, sources, cell geometries) and comparing product ratios and limiting current densities from

Table 7. Electroreductions of Benzaldehyde, *p*-Methylbenzaldehyde, Dimethylmaleate, and Benzyl Bromide[a]

System	*Product Ratio 1-Electron/2-Electron*	*Current Efficiency (%)*
Benzaldehyde		
still solution	0.6	36
mechanical stirring	2.5	63
ultrasound(bath)	3.4	65
***p*-Methyl benzaldehyde**		
still solution	0.0	34
mechanical stirring	0.1	74
ultrasound(bath)	>100	77
Dimethylmaleate		
still solution	0.0	66
mechanical stirring	0.3	93
ultrasound(horn)	0.4	96
Benzyl bromide		
still solution	0.0	40
mechanical stirring	0.3	40
ultrasound(horn)	0.9	40

[a]From reference 199 for details see text.

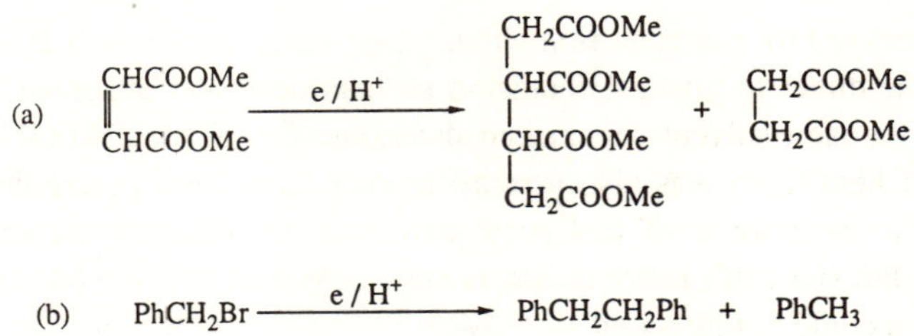

Scheme 8. Electroreductive pathways of (**a**) dimethylmalate and (**b**) benzyl bromide. (From ref. 196).

voltammetric data. The authors then used their previously reported method for estimation of mass transport coefficients [197] to conclude that ultrasound was acting simply by agitation and not by cavitation.

In the same paper, it was noted that less striking but still significant switches towards the one-electron products occurred in other sonoelectrochemical reductions, including dimethylmaleate at a lead cathode in an aqueous mixed-phosphate buffer, and benzyl bromide at a lead cathode in methanolic tetraethylammonium bromide solution, both shown in Scheme 8.

The reduction of benzoic acid at a lead cathode in aqueous sulfuric/citric acids does not give a one-electron hydrodimer, but instead yields the two-electron products benzaldehyde and the four-electron product benzyl alcohol. Here ultrasound produces some switch towards the two-electron products; thus in all cases studied the authors found that ultrasound favored the process involving the smaller number of electrons per molecule. This is opposite to the sonoelectrochemical effect seen in carboxylate electrooxidation [184,186,187] where the process involving the greater number of electrons was favored by ultrasound, and shows that in the present state-of-the-art generalizations are inappropriate. The nature of the electrochemical system is an important consideration in the establishment of sonoelectrochemical phenomena.

The group at Tokyo Institute of Technology have also examined the electroreduction of methyl halides at a reactive tin cathode [198]. This is a different type of electrochemical system in which the cathode is a reactive metal. A surface intermediate is formed on the electrode which reacts to give either a three-electron dimeric distannane or a four-electron tetramethyl stannane, as shown in Scheme 9.

For reduction of CH_3I in DMF/Bu_4NClO_4 at room temperature galvanostatically at 10 mAmp cm^{-2}, an unstirred reaction gave a 1:10 ratio of distannane to stannane,

$$CH_3I \xrightarrow[\text{Sn Cathode}]{e} (CH_3)_3SnSn(CH_3)_3 + (CH_3)_4Sn$$

Scheme 9.

slightly improved by mechanical stirring, while ultrasound from a 20-kHz probe increased the dimer to give a 1:5 ratio. A similar trend was observed for methyl bromide in the same solvent system, and also in an aqueous MeCN/Et4NBr solvent system, but here there was also a small amount of a solid precipitate formed. Interestingly the amount of this precipitate was significantly increased under ultrasound, but since this material was not characterized, the role of ultrasound in its formation could not be stated.

In their first paper [196] the authors quote enhanced mass transfer across the electrode interface as the origin of the sonoelectrochemical trend towards products from the lesser amount of electrons per substrate molecule, but here the involvement of surface species on the reactive electrode complicates such an explanation.

Unusual compounds have been produced by sonoelectrochemical techniques. Dichlorodimesitylsilane produced tetramesityldisilene when reduced under controlled-potential conditions in the presence of ultrasound (–3.2 V vs. Ag/Ag+) at yields greater than 90% [196a]. This species, which contains a double bond between the silicon atoms, was isolated in >90% purity and reacted further with trapping agents.

A group at Wesleyan University, Middletown, Connecticut, have examined reductive silylation of geminal dihalides [199]. Here ultrasound was originally introduced only because a new design of cell was too constricted to be stirred magnetically [200]. The reaction employs a stainless steel cathode, a sacrificial magnesium anode, and occurs in two steps with stereochemical implications, although there are significant competitive pathways (Scheme 10)

Ultrasound substantially enhances current efficiency and stereoselectivity over silent electrolyses, but results were somewhat erratic which was attributed to the need for care in positioning the electrolysis cell in the Branson ultrasonic cleaning bath. Again, the observation refers to the need to address geometric considerations when dealing with soundwaves. The authors suggested that enhancement could result from improved "single electron transfer" (SET) as defined by Luche [185].

The situation became complicated when a divided cell was employed. Although unsuitable for effective preparative electrolyses because of resistance losses, this control experiment yielded an unexpected reduced product, an α-halosilane, from the anode (oxidizing) compartment containing the sacrificial magnesium electrode. This suggested the involvement of a sonochemically induced reaction, without

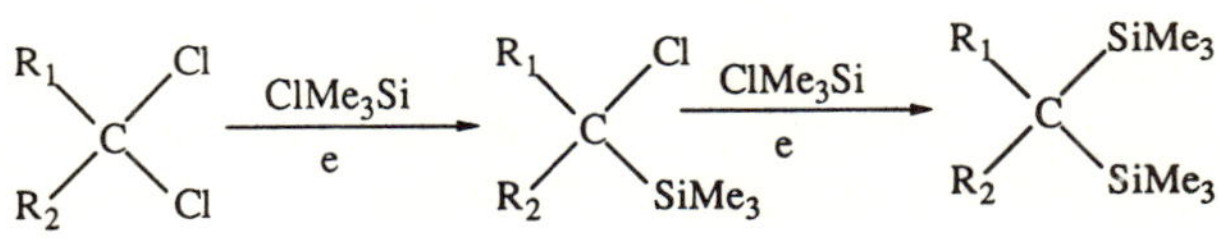

Scheme 10.

electrolysis, and therefore complicated the interpretation of the sonoelectrochemical events in the normal undivided cell.

The possibility that ultrasound can be involved chemically in an intended electrochemical system is not without precedent at Wesleyan University. In earlier work involving the electroreduction of α,α'-dibromo ketones at a mercury cathode, ultrasound was employed just for stirring [201,202]. It was then realized that although the dihaloketone was stable to mercury (without electrolysis) over weeks in silent conditions, the ultrasonic irradiation, which tended to produce a range of finely divided mercury droplets above the pool, induced sonochemical reduction on the metal [203]. The authors later deduced experimental conditions using either electrochemistry or chemical reaction on ultrasonically dispersed mercury that could select from the range of possible products [204].

There is thus a need for caution in the assignment of sonoelectrochemical phenomena. However, since the eventual aim of synthetic chemists is to produce new or improved routes to target molecules, it is not necessarily important in the first instance to establish mechanistic details. The fact is that ultrasonic irradiation has been shown to produce a number of benefits in electrosynthetic oxidations and reductions, and may act before, during, or after the fundamental electrochemical processes.

5. OTHER ELECTROCHEMICAL SYSTEMS

5.1 Introduction

The distinction in previous sections of electroanalysis, inorganic electrochemistry (particularly metal systems), and electroorganic synthesis leaves out a number of other electrochemical systems. Ultrasound has been applied to many of these, to interesting effect, and this section concerns a number of such systems. There is, of course, overlap in any attempt at compartmentalization, and here some studies on batteries, electrochemiluminescence, and micellar systems could be considered as contributing to electroanalysis, while other multiphase electrolyses might be considered as electrosynthesis. In addition, most multiphase electrolysis is directed to the destruction of haloorganics and is aimed at waste treatment. There are also "one-off" applications of ultrasound in electrochemistry, which are collected at the end of this section.

5.2 Sonoelectrochemiluminescence

It is well known that when acoustic waves are generated in aqueous and other media luminescence can be observed [205]. This type of luminescence is known as sonoluminescence and has been attributed to the process of cavitation during which bubbles are formed. The collapse of these cavitation bubbles during the compression portion of the sound wave represents a very high energy output which results in the formation of light, i.e. sonoluminescence [206]. Insonation also produces radical species, e.g. hydroxyl radicals from water, which may contribute to luminescence mechanisms. Sonoluminescence has been extensively studied since the 1960s and there are several excellent reviews on the subject. However scientists recently have studied the effect of ultrasound upon electrochemiluminescent systems.

Electrochemiluminescence (ECL) is defined as the production of light (luminescence) from an electrochemical system undergoing electrolysis. However, the emission of light from a electrolysis cell is a relatively uncommon phenomenon. The available electrode potential range of roughly –3.0 V vs. NHE (which causes benzene reduction) to +3.0 V (which causes benzene oxidation) certainly provides sufficient energetics within the span to achieve intermediates in suitable excited states, but there are also kinetic limitations to be overcome. Thus reaction events of a very short duration are required to prevent thermal relaxation by bond vibration or other mechanical reorganization of energy, or some other nonradiative relaxation phenomenon [207].

However, ECL is known in a relatively restricted number of systems, and is a useful phenomenon for the identification of intermediates and the establishment of mechanisms and in analytical systems—e.g. in immunoassays in which an ECL-active moiety is attached to an antibody, and the change in ECL is monitored when this combines with its appropriate antigen.

A potentially useful ECL system for immunoassays is ruthenium bipyridine [208]. This has a number of advantages: it can be made to work by a number of different procedures, including the use aqueous media, important in biological systems; and depending on the conditions it can be somewhat insensitive to oxygen, unlike the majority of ECL-systems where oxygen is either a component in the mechanism or else interferes with radical species, either way requiring fine control of oxygen availability with procedural consequences.

However, the ruthenium bipyridine ECL system has a number of drawbacks. Diffusion constraints mean that greater brightness occurs at an exposed edge of a planar electrode, and there is often patchiness over the electrode surface. These factors require the use of small electrode areas with low emitted-light intensities for collection. In addition, mechanisms involving the oxidation of Ru $(bpy)_3^{2+}$ to Ru $(bpy)_3^{3+}$ suffer from gradual fouling at the anode, requiring some form of cleaning procedure with continued usage.

The known effects of ultrasound to enhance diffusion near an electrode and to minimize electrode fouling by preventing buildup of coatings on the electrode surface prompted the Coventry group to examine this ECL system under ultrasonic irradiation [209,210]. Here ECL originates from the excited state of Ru $(bpy)_3^{2+}$, which can be attained by three main routes:

1. The "reductive–oxidation" route involves the formation of $Ru(bpy)_3^{+}$,+ by cathodic reduction. This then reacts with an oxidizing species either present

originally in the solution or else simultaneously electrogenerated. However, the reduced Ru^+ species is highly reactive and aqueous solution cannot be used.

2. An alternative procedure is to oxidize at the anode to $Ru(bpy)_3^{3+}$, which then meets a reducing agent in solution. This oxido–reduction method is most useful since an appropriate solvent is water containing oxalate as reducing agent.
3. Finally the "regenerative" route involves the electrochemical generation of both Ru^+ at the cathode and Ru^{3+} at the anode. However, these species do not need to diffuse across the cell to meet, since the working electrode can be switched rapidly from –ve to +ve potentials (which need not be the same) to create diffusion waves of both species that react as they escape the electrode.

The Coventry group examined the oxidoreductive and regenerative systems, employing a 1-cm^2 platinum foil electrode. This is too large for conventional ECL use and the light emission is visibly restricted mostly to the edges of the flag electrode in silent conditions. Insonation from a 40-kHz probe system immediately produces a brighter and uniform emission across the electrode surface, evident to the naked eye. Data from a photomultiplier shown in Figure 14 for the oxido–reduction system is aqueous oxalate; note also the weak sonoluminescence obtained in the absence of an electrode potential.

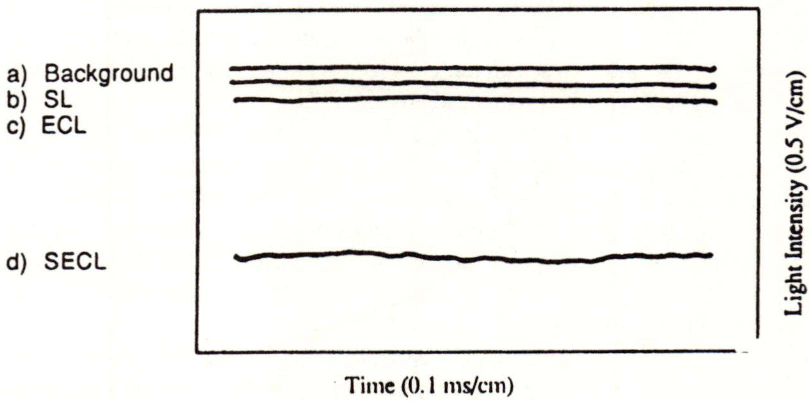

Figure 14. Sonoelectrochemiluminescence of tris-bipyridine ruthenium (II) dichloride ($[Ru(bpy)_3]Cl_2$) at platinum in aqueous oxalate solution. Potentiostatic control +1.2 V (SCE). (**a**) Background zero line. (**b**) Sonoluminescence from 40-kHz probe, no applied potential. (**c**) Electrochemiluminescence at +1.2 V (SCE), silent. (**d**) Sonoelectrochemiluminescence at 1.2 V (SCE) and ultrasound (40-kHz probe). (Taken from reference 209).

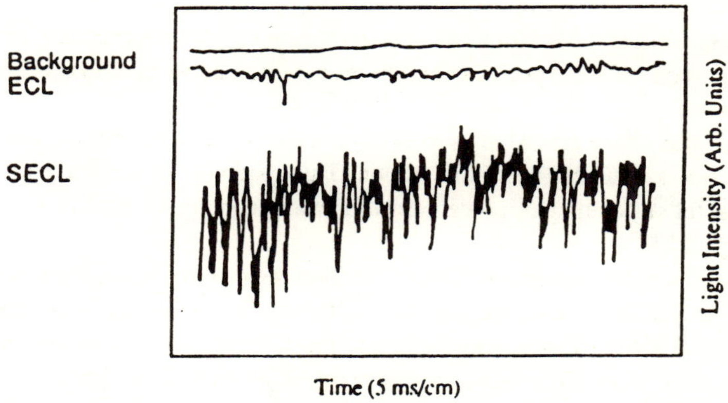

Figure 15. Sonoelectrochemiluminescence of tris-bipyridine ruthenium (II) dichloride ([Ru(bpy)3]Cl2) at platinum in acetonitrile using "regenerative" procedure. Background zero line. Sonoluminescence: Insonation from 40-kHz probe, no applied potential. Electrochemiluminescence: potentiostatic control +1.2 V (SCE) and –1.9 V at 1 Hz duty cycle. Sonoelectrochemiluminescence using 40-kHz probe.

Figure 15 shows the sonoelectrochemical enhancement of the regenerative procedure for $Ru(bpy)_3$ ECL. The noise in the trace is due to the pulsing technique.

The diminution of electrode fouling in oxido–reductive mode is shown in Figure 16. Here in the silent system there is a substantial drop in light emission with repeated use, whereas with ultrasound there is no loss of emission over the same period.

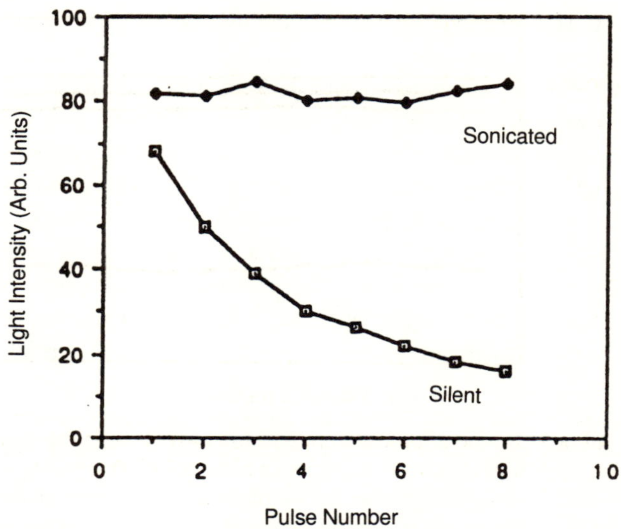

Figure 16. Diminution of electrode fouling ([Ru(bpy)3]Cl2) in aqueous oxalate at +1.2V (SCE) at platinum foil.

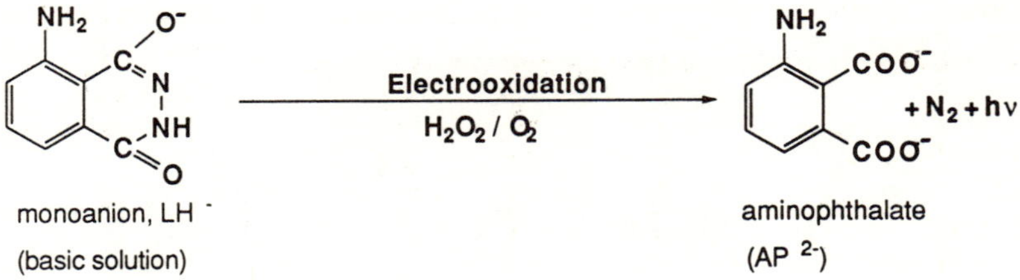

Scheme 11.

Other ECL systems were also examined, and sonoelectrochemical enhancements were observed for luminol. This is perhaps the most well-known luminescence system in which electrooxidized luminol (3-aminophthalhydrazide) is converted by reaction with peroxide or a derived species into the excited state of aminophthalate from which emission originates [211] (Scheme 11).

The luminol system has been more thoroughly studied by the group at Glasgow Caledonian University who use the sono-ECL enhancement of luminol for the detection of very low concentrations of residual hydrogen peroxide in cleaning solutions for contact lenses [212].

Zhivnov et al. have obtained similar increase in the intensity of ECL under ultrasound from a 9,10-diphenylanthracene system [213]. The ECL intensity of 9,10-diphenylanthracene in DMF solutions under sonication was increased by factors of 20–30 depending on the concentration of the solutions, ultrasound power, and frequency. The same authors also studied the acoustic field effect on ECL of several organic molecules in solutions [214] and obtained similar results for the following ECL molecules: 1,5-diphenyl-3-styrylpyrazoline, rubrene, 9,10-diphenylanthracene, 9,10-dimethylanthracene, and perylene.

An intriguing example of a new ECL system originated from the studies of the Coventry group on carboxylate electroxidation [215]. Low levels of light emission were observed from electrooxidation of phenylacetate and a number of ring-substituted derivatives in methanol and particularly in acetonitrile solution. Again ultrasound enhanced ECL-emission. The intermediacy of benzyl radicals was suggested by the effect of a radical quencher, by the interference of oxygen when present, and by the influence of substituent groups. Thus both electron-withdrawing groups and electron-donating groups at the *ortho* and *para* positions diminished light output, and any substituent anywhere produced less light than the meta-chloro derivative, which was similar to the unsubstituted compound as shown in Table 8. This unusual pattern of reactivity with directing group is known in the reactions of benzyl radicals [216,217].

As might be expected, neither acetate, cyclohexane carboxylate, or 4-phenyl butyrate cause light emission in this system. In the latter case the benzene group is

Table 8. Electrochemiluminescence (ECL) from Aryl Acetate Electrooxidation[a]

ECL Intensities from Different Alkyl and Aryl–Alkyl Carboxylates

	PAA	*3PPA*	*4PBA*	*Acetate*	Cyclohexanoate
Current Density (mA cm^{-1})	10 mA	10 mA	10 mA	100 mA	100 mA
ECL	5 V	5 V	0.0 V	0.0 V	0.0 V
SL(20 kHz)	1 V	2 V	0.1 V	0.1 V	0.2 V
SECL(20 kHz)	15 V	15 V	2 V	0.2 V	0.5 V

Positional Effect of Substituent Groups on Phenylacetate ECL Intensity

	PAA	*2ClPAA*	*3ClPAA*	*4ClPAA*
Current Density (mA cm^{-1})	10 mA	10 mA	10 mA	10 mA
ECL	5 V	1.2 V	15 V	4.3 V
SL(20 kHz)	1 V	0.4 V	1 V	0.6 V
SECL(20 kHz)	12.5 V	5 V	15 V	11.8 V
	PAA	*2MeOPAA*	*3MeOPAA*	*4MeOPAA*
Current Density (mA cm^{-1})	10 mA	10 mA	10 mA	10 mA
ECL	5.3 V	1.0 V	2.5 V	0.8 V
SL(20 kHz)	1 V	1 V	1 V	1 V
SECL(20 kHz)	15 V	2.0 V	3.5 V	1.5 V
	PAA	*$2NO_2PAA$*	*$3NO_2PAA$*	*$4NO_2PAA$*
Current Density (mA cm^{-1})	20 mA	20 mA	20 mA	20 mA
ECL	8.7 V	1.0 V	0.8 V	0.2 V
SL(20 kHz)	1 V	1 V	1 V	1 V
SECL(20 kHz)	12.3 V	1.0 V	1.2 V	1 V

[a]Conditions for all systems 0.005 M acid in acetonitrile; 50% neutralized with tetramethylammonium hydroxide in methanol at platinum electrode. ECL measured as output voltage from photomultiplier tube (15V maximum).

three Sp^3-carbons away from the radical site, but perhaps surprisingly 3-phenyl propionate also gives substantial light emission. Formally the radical site is two Sp^3-carbons away from the aromatic ring, but processes can be proposed which could produce a benzyl radical species, although all these introduce mechanistic complication. It is known that preparative electrooxidations of 3-phenyl propionate produce 1-phenyl products from nucleophilic capture in the two-electron pathway, attributed to rearrangement of the 2-phenyl ethyl carbocation to the 1-phenyl isomer. Here ECL involves radical species, but it should be noted that preparative conditions of high carboxylate concentration, which also provides the ionicity, are quite different from ECL experiments where very low carboxylate concentrations are employed in the presence of an inert electrolyte salt to bolster conductivity. However, ECL clearly gives a handle for mechanistic studies, and the influence of ultrasound provides yet another parameter for elucidation.

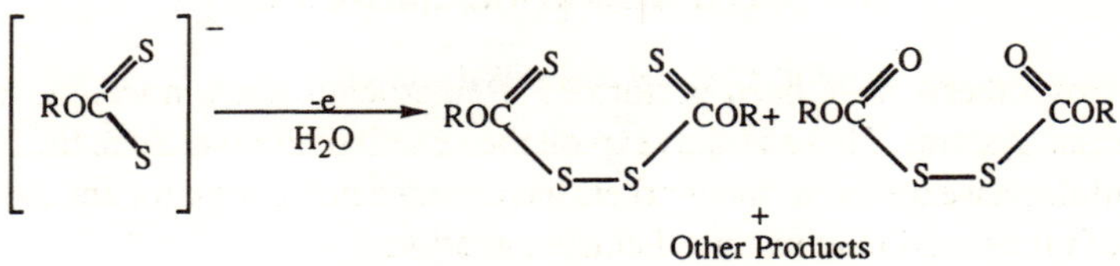

Scheme 12.

5.3 Flotation in Mineral Processing

A series of reports from the early 1980s onward concerning the effect of ultrasound upon ostensibly electroorganic synthetic systems involved Russian work on xanthate electrooxidation. The purpose was to assist froth flotation in mineral processing where xanthates are the most widely used "thiols" in the sulfydryl anionic-type of collection system [218]. The mechanisms of flotation are not completely understood; adsorption phenomena and polarity effects are involved and for xanthates the formation and involvement of dixanthogen, xanthic acid and other species has been proposed. The group at Alma–Ata therefore studied the xanthate–dixanthogen system and in an early paper [219] found that combined ultrasound and electrooxidation increased the amount of dixanthogen (see Scheme 12) and gave a uniform emulsion with 80% of droplets below 5 μ diameter. The recoveries of galena, sphalerite and chalcopyrite were increased by use of this prepared flotation medium [220].

The reports continued, involving other areas and identifying, among the sonolectrochemical products from butyl xanthate, the compounds butyl monothio carbonate disulfide $(BuO_2C)_2S_2$ and bis(butyl xanthogen) [221], leading to the description of an apparatus for the activation of flotation reagents [222]. This work was reported in limited-issue Russian-language journals; there was also a brief review in a Russian book [223], but like much of the considerable body of Russian work on sonochemistry this was not readily accessible or comprehensible to scientists in the West. Thus none of this data is mentioned in recent English-language textbooks on mineral processing [218].

There is, however, a recent paper in English from a Bulgarian group on the sonoelectrochemical oxidation of sodium butyl xanthate [224]. These workers confirm the benefits of sonoelectrochemistry, but make the striking observation that in these xanthate flotation systems they obtain similar effects under 50-Hz irradiation (i.e. infrasound, within human hearing) to those at 20 kHz (ultrasound). Irradiation power details are not easy to compare from the report, but the result has considerable implications for sonoelectrochemical significance of cavitational phenomena, and further confirms that studies across the widest of frequency ranges are necessary for mechanistic elucidation.

5.4 Electrolysis in Multiphase Media

Electrosynthesis have been performed in microemulsions, micelles, and other multiphase systems. These studies exploit the benefit of ultrasound in the formation of the multiphase medium, and workers have tended not to look for any more subtle effect of ultrasound than just mechanical agitation.

The major thrust of this work has, in particular, concerned the electrolytic dechlorination of polychlorinated biphenyls and similar species for environmental control of pollutants. However, there is a wider potential for this technology in the use of aqueous solvent systems for syntheses involving otherwise water-immiscible organic compounds. Water is of course the cheapest, most widely available and environmentally friendly solvent and with modern concerns over ecology and the search for 'clean technologies' there is considerable opportunity for this particular application of ultrasound in electrochemistry.

A major effort into electrolytic dechlorination technology has been achieved by a group at the University of Connecticut. This team has produced a series of papers since 1984 examining electroreductive dehalogenations of substrates ranging from simple halobiphenyls, using mercury-pool cathodes irradiated by an ultrasonic cleaning bath [225], through to DDT itself [1,1-bis(4-chlorophenyl)-2, 2, 2-trichloroethane]. Thus, at carbon felt cathodes using irradiation from an ultrasonic horn both aromatic and aliphatic chlorines were removed [226]. In this latter instance cobalt bipyridine species were employed as redox mediators, and other mediators examined include zinc phthalocyanine and nickel tetrasulphonato phthalocyanine [227]. In all cases, surfactants such as didocyldimethyl ammonium bromide (DDAB) were employed; sonication was considered simply to enhance mass transport, being used to produce, for example, bicontinuous microemulsions and thence to maintain these during subsequent electrolysis, often with duty cycles involving appreciable off-times to avoid overheating from the ultrasound. The electrochemical methodology avoids the use of quantities of metal reductants, such as nickel, which are themselves an environmentally damaging species. This field of electrocatalysis in microemulsions and media of controlled microstructure has been thoroughly reviewed recently by Rusling [228–230].

An alternative strategy is to use anodic oxidation to destroy halocompounds in an indirect process [231]. Thus 1,2 dibromoethane was degraded to bromide ions, carbon dioxide (trapped as carbonate salt), and water by use of barium peroxide in the presence of a surfactant such as dodecyl trimethyl ammonium chloride. Ultrasound was provided by a Branson cell disruptor, employed in the preparation of the multiphase electrolyte. During electrolysis mechanical stirring was employed. At a graphite anode using either potentiostatic control or the experimentally simpler galvanostatic control up to 89% destruction of the substrate was observed. The reaction is pH-sensitive and the semioptimized system operated at pH 5. The key step is thought to be oxidation of barium peroxide to the superoxide, which then causes the desired decomposition since hydrogen peroxide is not an effective replacement for the barium compound. The authors had previously used the system to decompose carbon tetrachloride to chloride and carbon dioxide (as barium carbonate) [232].

Ultrasound was also used for the dispersion of a surfactant pyrrole, prior to electrooxidation to the conducting polymer [233]. An amphiphilic (pyrrolylalkyl) ammonium monomer dispersion was used to coat the electrode surface with monomer, subsequently electropolymerized to thin films using an aqueous electrolyte for this step. Ultrasound has also been used to assist impregnation of pyrrole monomer into, for example, a conventional polymer matrix prior to polymerization to yield a composite of the conducting and conventional polymers, but is also a pretreatment effect of ultrasound rather than a sonoelectrochemical one [234].

The amphiphilic pyrrole is a fairly specialized application of ultrasound for the formation of a multiphase electrolyte system, but a more general case of ultrasonic assistance in both electrolyte formation and subsequent electrolysis is given in a recent Japanese patent by Permelec Electrode [235] in which benefit is claimed by use of a hydrophobic electrode to electrolyze poor water-soluble materials under ultrasound an aqueous emulsion.

5.5 Waste Treatment

A novel area of ultrasound technology is in the control and breakdown of pollutants in the environment. Over the past few years there has been extensive research into this field. Electrochemical breakdown of pollutants has been reported by several groups around the world; for example, Comninellis et al. have studied the electrochemical breakdown of various phenol contaminants using SnO_2 [236]. The electrochemical treatment of aromatic polyhydrocarbons in waste and portable waters have also been reported by several authors [237–239]. Other workers have looked at the electrochemical oxidation of organic pollutants carried out under conditions of simultaneous oxygen evolution using various electrode materials [240–242].

Ultrasound recently has been studied as a new and novel technique in the removal and breakdown of both chemical and biological pollutants, and there are several articles in the literature dealing with this subject [243,244]. Over the past few years there has been considerable interest shown in combining the two techniques of ultrasound and electrochemistry for the treatment of a variety of pollutants, both biological and chemical. Some of the most recent and important papers in this field are reviewed below.

The removal of PhOH from industrial effluents by electrochemical oxidation was speeded up by passing ultrasound waves across the flow of ions between the electrolyzer electrodes [245]. Thus, over 80% oxidation of PhOH to maleic acid was achieved when ultrasound (25 kHz, 104 W/m2) was applied to a solution containing 100 g PhOH/L and 2 g NaCI/L in an electrolyzer. Without ultrasound, only ~ 50% of the PhOH was oxidized under the same conditions.

Metal ions in wastewater are recovered as powdered metals by chelation and electrolysis in an ultrasonic cell to prevent metal precipitation on an electrode. The metal ions are concentrated by chelating with a resin, eluted to give a concentrate, and then electrolyzed. The electrolysis cell had a cylindrical electrode, and ultrasonic transducers were submerged in the electrolyte. Cu powder (average size 5 μ) of 99.86% purity was recovered by using UR-30 chelating agent [246].

Wastewaters were also treated in a circulating system consisting of an electrolyzer; a reactor containing a heat exchanger system, a partition, an ultrasound generator, and a UV irradiation source; valves; and a connecting piping system with an inlet and an outlet. Wastewater is subjected to electrolysis during which hydrogen is generated on the cathode and chlorine is generated on the anode—the chlorine produced hydrolyses to HCl and HOCl. In the reactor, the contaminants (e.g. NaCN) are decomposed. If the electrolysis is done at 60 °C and treatment in the reactor is done at 30 °C, efficiency of the electrical current in the process is ~80% in the presence of ultrasound. The electrochemical treatment is especially suitable for wastewaters containing cyanides and phenols [247].

Japanese researchers [248] have designed an apparatus for the electrochemical reduction of carbon dioxide employing ultrasound. The apparatus comprises an electrolytic cell containing CO_2-dissolved electrolytic solution, a porous Pt-group metal anode, a proton-conductive solid electrolyte having a porous metal cathode used as a catalyst for the electrochemical reduction of CO_2 on one side and a second anode on the other side facing oppositely to the cathode, and an ultrasonic vibrator. CO_2 can be reduced effectively for a long time.

A new electrochemical system to hydrogenate coal with active hydrogen generated from water has been developed by Mhiyake [249]. The system consists of electron transfer from a Pb cathode to a powdered Ni suspension via a Cr^{3+}/Cr^{2+} mediator dissolved in an aqueous electrolytic solution. Active H produced on the Ni powder, which has a large surface area, hydrogenates coal chemically by contacting coal particles. The electrolysis was carried out under ultrasonic irradiation to prevent adhesion of coal-derived material on the Pb cathode and to agitate the system efficiently. The current efficiency attained by the system was 11.2% at 144 °C; 13.3 H atoms per 100 C atoms were added by electrolysis with 17,280 C atoms. Analyses of the hydrogenated coal indicate that cleavage of ether bonds and hydrogenation of aromatic rings result from the reaction.

Ultrasound has been successfully employed for the removal of radioactive "cruds" by Japanese workers [250]. Radioactive crud deposited on the various apparatus of a nuclear power plant was removed by electropolishing followed by ultrasonic cleaning. Their method exhibited excellent removal efficiency, and was found to be especially useful for removing deposits on the inside wall of coolant conduits. Similar results were obtained by Gauchon et al. [251] who studied the decontamination of German BWR nuclear reactors by chemical, electrochemical , and water-jet methods employing ultrasound. Their results showed that simultaneous electrolysis and ultrasound gave the most efficient method for decontamination.

Sandstede et al. have looked at electrochemical *in situ* processes for removal of mercury and other heavy metals from soils, sludges, and waters [252]. Heavy metals are removed from contaminated material. The solutions tested were saturated with oxidants such as $KMnO_4$, H_2O_2, and O_3, etc., and surfactants and an electrical charge was passed through the solution. Liquid mercury formed the cathode and in turn forms amalgams with the heavy metals; these are removed and the amalgam layer is thermally regenerated. The anodes used are catalyzed titanium, and the soils prior to this treatment were sonicated to loosen them and enhance metal removal.

In general, the benefit of ultrasound upon metal electrodeposition (discussed in length in Section 3) offers promise for the electrochemical removal of metals from solution, although it should be recognized that the systems are somewhat different. Thus, commercial metal electrodeposition requires high concentrations of metal ions in solution while wastewater treatment involves very dilute concentrations of metal ions. However studies are in hand to define the scope of sonication for metal–ion removal [253], and this area offers great promise.

6. CONCLUSIONS

Ultrasonic irradiation produces a number of significant benefits in a wide range of electrochemical systems. Thus in electroanalysis it provides another time-dependent variable to be used for mechanistic elucidation, and which further extends the range of hydrodynamic regimes available to the modern electroanalyst. The technique also provides a probe into the fundamental physicochemical principles of electrolyte solutions, electrode phenomena, and associated processes.

Insonation of an electrosynthetic reaction can produce altered product ratios, greater efficiencies, lessened cell power requirements, and a diminution of detrimental electrode fouling. In electrodeposition, ultrasound alters the properties of the product coating, be it a metal deposited, a semiconductor, a polymer, or some other electrogenerated material. Sonication also affects corrosion and electrode dissolution, and is useful, for example, in systems employing sacrificial electrodes.

Ultrasound influences multiphase systems such as the production of microemulsions. It is useful in electrosynthesis involving immiscible materials—this effect has been particularly exploited for several applications in environmental science. Ultrasound can also enhance electrochemiluminescence systems, and has been applied to many other aspects of electrochemistry, including the as yet unexplained benefits of pre-treating electrolyte solutions. It has even been proposed to enhance 'electrochemical cold-fusion'.

The best established effect of ultrasound in electrochemistry is the diminution of the diffusion layer and the enhanced limiting currents so produced. This is, per se, of benefit towards sensitivity improvement in electrochemical sensors, and is also the origin of many sonoelectrochemical phenomena. Ultrasound also affects electrode surfaces, which has been exploited as a pretreatment protocol, and has a beneficial effect during electrolysis.

However, there is some discrepancy between reports of observed phenomena, e.g. effects on electrode roughening or surface abrasion, and it is fair to say that, at present, this is one of a number of aspects of sonoelectrochemical mechanisms that remain open to discussion. In particular the role of cavitation is not explained with certainty. Cavitation is the most widely described consequence of ultrasonic irradiation, producing large localized temperature and pressure gradients, and streaming phenomena, but the apparent frequency independence (at constant power) of both limiting-current in electroanalysis and product distributions in electrosynthesis is inconsistent with cavitational differences in these frequency regimes. However, the situation is complicated, because change of power (at constant frequency), which affects cavitation, does affect limiting-currents in electroanalysis.

Therefore, as might be expected, there may be a number of sonoelectrochemical effects that originate from different combinations of sonochemical and electrochemical phenomena. An assumption that cavitation is the root cause of a particular observation may need to be justified; it is certainly desirable whenever possible to obtain an independent assessment of cavitational phenomena in any particular experimental system.

Thus at the time of this writing, it is recognized that there remain outstanding issues concerning the quantitative definition of ultrasonic effects in different media. It may turn out by reverse logic that sonoelectrochemical studies provide a means to gain a better understanding of cavitation. In any event, the practicing electrochemist needs to consult the sonophysicist for mutual benefit. However, it is clear from all this that ultrasound can modify electrochemical behavior, and this can be of benefit across the field of electrochemistry.

Those interested in pursuing sonoelectrochemistry should note that ultrasound is sometimes employed in electrochemical systems without this fact being obvious from the title and abstracts of a published paper. Often, in these reports, the purpose for the ultrasound and the significance of its use are not discussed in detail. One such example is, "Competitive Electrochemical Synthesis of Polydimethylsilane without Solvent," where sonication was found to increase the yield, although without explanation [254]. The reader is advised that there may be more such reports in the literature, and the reviewers apologize for inadvertent omission of work in the field.

Sonoelectrochemistry is not a new concept. Pioneering sonoelectroanalytical experiments were performed quite some time ago; while for electrosynthesis a reactor offering "high-speed coulometry" using ultrasonic stirring was described in 1963 [255]; but the topic will now benefit from a thorough study exploiting modern developments in ultrasonic and electrochemical methodologies. It is hoped that, for example, greater use will be made in bulk electrosynthesis, perhaps to redress what is often perceived by workers in the field as the disappointing commercial exploitation of large-scale electrosynthesis [256]. A methodology to widen the viable applications of electrosynthesis is widely sought.

In this regard, assessment of the energy savings due to lessened cell power requirements in a sonoelectrosynthetic system is often countered by mention of the energy required to input ultrasound. Clearly the overall energy balance of the system requires both to be taken into account. However, it should be remembered that ultrasound need not be provided by piezoelectric transducers or magnetostriction. The phenomenon of cavitation was first elaborated from erosion of ship's propellers with use, and the whistle effect is an effective means of producing vibration. Most scaled-up electrosyntheses incorporate flow systems that require pumping of an electrolyte, and in 1981 a Japanese patent application [257] concerned the use of appropriate baffles to produce 20 kHz ultrasound through pumping of the solution. It may therefore be that commercial exploitation of sonoelectrochemistry is not necessarily disadvantaged. What is necessary for this is that electrochemical engineers talk to ultrasonic engineers and that conclusions about sonoelectrochemical viability should not be drawn until all options are clear to both parties.

With all the modern methodology now available, this is an exciting time for this fascinating subject.

REFERENCES AND NOTES

[1] Southampton Electrochemistry Group. *Instrumental Methods in Electrochemistry*. Ellis Horwood, 1985.

[2] Kissinger, P.T. and Heineman, W.R. (eds.) *Laboratory Techniques in Electroanalytical Chemistry*. Marcel Dekker, 1984.

[3] Bard, A.J. (ed.). *Monographs in Electroanalytical Chemistry and Electrochemistry*.

[4] Bockris, J. O'M. and Reddy, A.K.M. *Modern Electrochemistry*. Plenum, Rosetta, 1970, Vols. 1 & 2.

[5] Christensen, P.A. and Hamnett, A. *Techniques and Mechanisms in Electrochemistry*. Blackie Academic & Professional 1994.

[6] Brett, C.M.D. and Oliviera-Brett, A.M. *Electrochemistry, Principles, Methods and Applications*. Oxford University Press, 1993.

[7] Bryn, H.D. *Introduction to Electrochemistry*. Macmillan Physical Science Series, 1993.

[8] Pletcher, D. *A First Course in Electrode Processes*. The Electrochemical Consultancy, 1991.

[9] For example, Bockris, J.O.M., Conway, B.E., and White, R.E. (eds.). *Modern Aspects of Electrochemistry*. Plenum.

[10] Fry, A.J. *Synthetic Organic Electrochemistry*. Harper and Row, 1972.

[11] Baizer, M.M. and Lund, H. (eds.). *Organic Electrochemistry*, Third Edition. Marcel Dekker, 1991.

[12] Kyriacou, D. *Modern Electroorganic Chemistry*. Springer-Verlag, 1994.

[13] Weinberg, N.L. (ed.). *Techniques of Electroorganic Synthesis*. Wiley.

[14] Pletcher, D. and Walsh, F. *Industrial Electrochemistry*, Second Edition. Chapman & Hall, 1990.

[15] Linden, D. (ed.). *Handbook of Batteries and Fuel Cells*. McGraw-Hill, 1984.

[16] Jones, D.A. *Principles and Prevention of Corrosion*. MacMillan, 1991.

[17] Albery, W.J. *Electrode Kinetics*. Clarendon Press, 1975.

[18] Compton, R.G., Wellington, R.G., Dobson, P.G., and Leigh, P.A. *J. Electroanal. Chem.*, 370 (1994) 129 and references therein.

[19] Brett, C.M.A. and Oliviera-Brett, A.M. In C.H. Bramford and R.G. Compton (eds.), *Comprehensive Chemical Kinetics*. Elsevier, 1986.

[20] Yeager, E. and Hovorka, F. *J. Acoust. Soc. America*, 25 (1953) 443.
[21] Barrett-Gultepe, M.A., Gultepe, M.E., and Yeager, E. Final Report on Tasks NR-051-162 and NR 384-305, Order No. AD-A113616 (1982).
[22] Zana, R. and Yeager, E. In J. O'M. Bockris, B.E. Conway, and R.E. White (eds.), *Modern Aspects of Electrochemistry*. Plenum, 1979, Vol. 14, Chapter 1.
[23] Moriguchi, N. *J. Chem. Soc. Japan*, 55 (1934) 34.
[24] Cataldo, F. *J. Electroanal. Chem.*, 332 (1992) 325.
[25] Walton, D.J., Burke, L.D., and Murphy, M., *Electrochimica Acta*, in press.
[26] Kowalska, E. and Mizera, J. *Ultrasonics*, 9 (1971) 81.
[27] Walker, R. In T.J. Mason (ed.), *Advances in Sonochemistry*. JAI Press, 1993, Vol. 3.
[28] Walker, R. and Holt, N. S. *Surf. Tech*, 22 (1984) 165.
[29] Bezzulov, A.L., Nikulin, V.N., and Abutalipova, L.N. *Tezisy Dok. Vses. Soveshch Elektrochim 5th*, 2 (1974) 281.
[30] Kaplin, A.A., Biamin, V.A., and Stas, I.E. *Zh. Anal. Khim.*, 43 (1988) 1157.
[31] Walton, D.J., Phull, S.S., Chyla, A., Lorimer, J.P., Mason, T.J., Burke, L.D., Murphy, M., Compton, R.G., Eklund, J.C., and Page, S.D. *J. Applied Electrochem.* 25 (1995) 1083.
[32] Klima, J., Bernard, C., and Degrand, C. *J. Electroanal. Chem.*, 367 (1994) 297.
[33] Compton, R.G., Eklund, J.C., Page, S.D., and Rebbit, T.O. *J.C.S. Dalton Trans.*, (1995) 389.
[34] Hagan, C.R.S. and Coury, Jr., L.A. *Anal. Chem.*, 66 (1994) 399.
[35] Compton, R.G., Eklund, J.C., and Page, S.D. *J. Phys. Chem.*, 99 (1995) 4211.
[36] Huck, H. *Ber Bunsenges Phys. Chem.*, 91 (1987) 648.
[37] Berlan, J.A., Faid, F.A., Contamine, F., Delmas, H., and Ratsiula, B. Presentation to Fourth Meeting of the European Sonochemistry Group, Blankenberge, Belgium, September 1994.
[38] Wang, J. *Talanta*, 28 (1981) 369.
[39] Dewald, H.D. and Peterson, B.A. *Anal. Chem.*, 60 (1990) 779.
[40] Fleischmann, M. and Pons, S. *Ultramicroelectrodes*. Datatech, Morganton, NC, 1987.
[41] Compton, R.G., Eklund, J.C., Page, S.D., Sander, G.H.W., and Booth, J. *J. Phys. Chem.*, 98 (1994) 12410.
[42] Compton, R.G. Address to European Network Meeting on Ultrasonic Modification of Electroorganic Processes, Coimbra, Portugal, February 1995.
[43] Yegnaraman, V. and Bharathi, S. *Bulletin of Electrochemistry*, 8 (1992) 84.
[44] Walsh, F. *A First Course in Electrochemical Engineering*. Electrochemical Consultancy, Romsey, 1993.
[45] Bockris, J. O'M. and Khan, S.U.M. *Surface Electrochemistry*. Plenum, 1993.
[46] Lorimer, J.P., Walton, D.J., Phull, S.S., and Pollet, B. *Electrochimica Acta*, in press.
[47] Dutta, S.K. and Sinha, A.P. *Indian Chem. Eng.*, 23 (1981) 19.
[48] See for example, Mason, T.J. and Lorimer, J.P. (eds.). *Sonochemistry*, Ellis Horwood, 1988 and references therein.
[49] Neppiras, A.E. *Phys. Rep. Sect. Phys. Lett.*, 61 (1980) 159.
[50] Reisse, J., Francois, H., Vandercammen, J., Fabre, O., Kirsch-de-Mesmaeker, A., Maerschalk, C., and Delplancke, J.L. *Electrochim. Acta,* 39 (1993) 37.
[51] Fuchs, F.J. In M. Murphy (ed.), *Metal Finish 61st Guide Book Directory Issue*. Elsevier, New York, 1993.
[52] Walker, R. *Plating Surf. Finish*, 72 (1985) 63.
[53] Novotny, M., Barvy, A., and Laky, N.P. *Povrchove Upravy*, 21(1) (1981) 4–6.
[54] Emst, K. *Ind-Anz.*, 112 (1990) 12.
[55] Savioli, I. *Tecnol. Filo*, 6 (1988) 26.
[56] McDonald, D.J. *Wire Ind.*, 55 (1988) 78.
[57] Burstein, E. *Products Finishing*, 53 (1989) 60.
[58] Walker, R. *Intemat. Met. Rev.*, 19 (1974) 1.
[59] Drake, M.P. *Trans. Inst. Met. Finish*, 58 (1980) 67.

[60] Aramaki, Y., Yamashita, T., and Aramaki, Y.T. *J. Surface Finish Soc. Japn.*, 40 (1989) 701.
[61] Ciovica, D. *Cercetari Metalurgice*, 20 (1979) 487.
[62] Wei, T.Y., Wang, Y.Y., and Wan, C.C. *Plating Surf. Finish*, 66 (1979) 47.
[63] Logvinenko, L.P. and Rozdaibeda, L.V. *Tekhnol. Organ. Proizvod.*, 3 (1984) 58.
[64] Nowack, N. and Habermmehl, K. *Metalloberflache*, 41 (1981) 67.
[65] Walker, R. and Duncan, S.J. *Surface Technol.*, 23 (1984) 301.
[66] Walker, R. and Holt, N.S. *Plating Surf. Finish*, 68 (1981) 44.
[67] Kadyrov, R.K. and Rakhmatullaev, N.G. *Uzb. Khim. Zh.*, 2 (1981) 6.
[68] Kamat, J.T., Kinnerkar, P.S., and Roy, D.L. *Trans. Indian Inst. Met.*, 28(2) (1975) 173.
[69] Walker, R. and Halagan, S.A. *Plating Surf. Finish*, 72 (1985) 68, 144.
[70] Prasad, P.B.S.N.V., Vasudevan, R., Seshadri, S.K. *Trans. Indian. Met.*, 46(4) (1993) 247–52.
[71] Prasad, P.B.S.N.V., Vasudevan, R., and Seshadri, S.K. *Indian J. Eng. Mater. Sci.*, 1(3) (1994) 178.
[72] Kochergin, S.M. and Vyaselva, G. Ya. *Electrodeposition of Metals in Ultrasonic Fields*. Consultants Bureau, New York, 1966, p. 19.
[73] Muller, F. and Huss, K. *Helv. Chim. Acta*, 33 (1950) 217.
[74] Walker, C.T. and Walker, R. *Nature*, 244 (1973) 141.
[75] Walker, R. and Walker, C.T. *Nature*, 250 (1974) 410.
[76] Walker, R. and Benn, R.C. *Plating*, 58 (1971) 476.
[77] Walker, R. and Benn, R.C. *Electrochim. Acta*, 16 (1971) 1081.
[78] Walker, R. 12th Seminar on Electrochemistry, Proceedings, Central Electrochemical Research Institute, Karaikudi, India, 1972, pp. 385–412.
[79] Meyer, D. and Novack, N. *Metalloberflache*, 38 (1984) 524.
[80] Namgoong, E. and Chun, J.S. *Thin Solid Films*, 120 (1984) 153.
[81] Walker, R. and Walker, C.T. *Ultrasonics*, 13 (1975) 79–82.
[82] Kamata, M., Higuchi, S., and Tsukamoto, Y. *Jpn. Kokai Tokkyo Koho.*, 6 (1979).
[83] Ljlberg, Z.R., Koniashvili, S.A., Ninoshvili, B.I., and Nadel, L.N. *Lakokras. Mater. Ikh Primen.*, 4 (1984) 30.
[84] Mikhoski, M. and Pushev, G. *Tekh. Mis'l.*, 15 (1978) 93.
[85] Anon, J. *Alum. Finish Soc. Kinki*, (1985) 12.
[86] Chistyakov, E.M., Ouda, T.M., Pugach, E.A., and Bykova, M.I. *Sint. Almazy*, 3 (1977) 38.
[87] Anon. *Gold Bull.*, 12 (1979) 73.
[88] Moreva, N.P., Pokoev, V.A., and Strelkov, G.A. *Fig. Strukt. Sveist. Tverd. Tel.*, 3 (1979) 81.
[89] Walker, R. and Clements, J.F. *J. Met. Finish*, 16 (1970) 100.
[90] Walker, R. *Trans. Inst. Met. Finish*, 53 (1975) 40.
[91] Walker, R. and Benn. R.C. *Plating*, 58 (1971) 476.
[92] Yamashita, T. and Chiba, A. *Hyomen Gijutsu*, 40 (1989) 492.
[93] Chiba, A. and Yamashita, T. Interfinish 92, Int. Congr. Surf. Finish., Vol. 1, 8493. Assoc. Bras. Trat. Superficie: Sao Paulo, Brazil, 1992.
[94] Chiba, A. and Wu, W.C. *Plat. Surf. Finish.*, 79 (1992) 62.
[95] Smirnov, Yu. R., Khavskii, N.N., Zainutdinov, F.M., Panov, N.A., and Katerova, N.A. From ref. *Zh., Met.* (1972), Abstr. No. 5-368.
[96] Ratajewicz, Z., Sawa, J., and Saneluta, C. *Politech. Lubelaka Lublin*, *Przem. Chem.*, 70 (1991) 480.
[97] Marrese, A. U.S. Patent 47 50 977 (1988).
[98] Muneo, O. Japan Kokai Tokkyo Koho, JP 86 198 468 (1993).
[99] Drake, M.P. U.K. Patent Appl., 4 GB 2111530 A1 830 706 (1983).
[100] Skomorokhov, O.I., Buinov, M.P., Korotkov, V.O., Pasechnik, S.Y., and Zhurov, V.I. *Zashch. Pokrytiya Met.*, 8, (1974) 43.
[101] Ezawa, H., Kawaguchi, S., and Suzuki, S. *Arumin. Hyomin Shori Kenkyu Hokoku.*, 86 (1974) 9.
[102] Fitch, A. *Met. Finish*, 86 (1988) 69.
[103] Kathirgamanathan, P. *Polymer*, 35 (1994) 430.

[104] Bogoyavlenskii, A.F. and Kochergina, V.A. *Khim. Khim. Tekhnol.*, 12(4) (1969) 461.
[105] Perusich, S.A. and Alkire, R.C. *J. Electrochem. Soc.*, 138(3) (1991) 700.
[106] Alkire, R.C. and Perusich, S.C. *Science*, 266 (1991) 1121.
[107] Perusich, S.A. and Alkire, R.C. *J. Electrochem. Soc.*, 138(3) (1991) 708.
[108] Altukhov, V.K. *Sb. Nauch. Rab. Aspir. Univ.*, No. 4 85 (1968).
[109] Belyi, V.I. and Nekoz, A.I. *Treniya Iznashivaniya*, 14 (1978) 64.
[110] Gareth, J., Hladky, K., Gaydecki, P., and Dawson, J. ASTM Spec. Tech. Publ., STP 1137 (Corrosion Forms and Control for Infrastructure), 1992, pp. 246–257.
[111] Fokin, M.N., Tugarinova, M.V., and Mazurina, I.I. *Elektron. Obrab. Mater.*, 6 (1972) 82.
[112] Gavrila, I. *Wire World Int.*, 24(3) (1982) 96.
[113] Iliuteanu, N. *Constr. Masini*, 26(2) (1974) 74.
[114] Okudaira, Y. Japan Kokai, JP 7 626 184 (1981).
[115] Harada, Y. Japan. Kokai, JP 772 781 (1978).
[116] Pitt, E., Scharmann, A., Scheib, S., and Werner, B. *Radiat. Prot. Dosimet.*, 22 (1988) 259.
[117] Chaika, G.V. *Elecktron. Orab. Mater.*, 6 (1970) 23.
[118] Karavainikov, V.N. *Akust. Ul'trazvukovaya Tekh.*, No. 8 (1973) 30.
[119] Vodyanov, Yu. M., Marshakov, I.K., and Altukhov, V.K. *Elektrokhimiya*, 8(6) (1972) 896.
[120] Marshakov, I.K. and Altukhov, V.K. *Voronezh. Elektrokhimiya*, 5(6) (1969) 658.
[121] Pan, T.P. and Wan, C.C. *J. Chem. Technol. Biotechnol.*, 29(7) (1979) 427.
[122] Rajagopal, I. and Rajagopalan, S.R. *India Bull. Mater. Sci.*, 6(2) (1984) 165.
[123] Shcherbakov, V.M. and Mnatsakanov, S.S. *Mater. Ikh Primen.*, 3 (1984) 40.
[124] Severdenko, V.P., Petrenko, S.I., Petrenko, V.V., and Gurskii, L.I. *Vestsi Akad. Navuk Belarus. SSR, Ser. Fiz.-Tekh. Navuk*, 1 (1974) 43.
[125] Masuda, H., Morishita, S., Fujishima, A., and Honda, K. *J. Electroanal. Chem. Interfacial Electrochem.*, 121 (1991) 363.
[126] Ottova, A., Cerhàta, D., Otto, M., and Tien, H.T. Inst. Biotechnol., Slovak Tech. Univ. Bratislava 81237, Czech. Adv. Hydrogen Energy, 8 (Hydrogen Energy Prog. 8, Vol. 2), 825-32 52-2 (Electrochemical, Radiational, and Thermal Energy Technology) 1990.
[128] Lin, X., Jia, Z., and Yangfen, X. *Proc. I.C.A. 14 Beijing*, 2 (1992).
[129] Mihu, V.P. and Gavrilla, I. *An. Univ. Bucuresti*, 18 43 (1969) CP6.
[130] Gavrila, I. *Ultrasonics*, 16:4 (1978) 156.
[131] Mihu, V.P. and Gavrila, I. *An. Univ. Bucuresti, Fiz.* 21 (1972) 45.
[132] Gavrila, I. *Maschinenmarkt*, 89:26 (1983) 501.
[133] Chiba, A., Tani, T., and Ouchi, Y. *J. Mater. Sci. Lett.*, 12(9) (1993) 620.
[134] Tasaka, A., Miyasaka, A., Yamashita, M., and Kubokawa, M. *Denki Kagaku Oyobi Kogyo Butsuri Kagaku*, 45(1) (1977) 32-36.
[135] Anon. *Japan Belg.* (1981) 15.
[136] Yamashita, M., Takemura, H., and Kubokawa, M. Manganese Dioxide Symp., Proc. 2nd, Meeting Date 1980, 278 Edited by: Schumm, B., Jr., Joseph H.M., and Kozawa A., I.C., 1981. Kowalskei, E. *Mizera J., Pol.* 2 (1973).
[137] Furukawa Battery Co., Japan Kokai Tokkyo Koho, JP 8 494 367 A2.
[138] Council of Scientific and Industrial Research (India), Report No. 137245, 1975.
[139] Kowalska, E. and Mizera, J. *Przem. Chem.*, 48(10) (1969) 601.
[140] Kowalska, E. and Mizera, J. *Przem. Chem.*, 48(4) (1969) 210.
[141] Kowalska, E. and Mizera, J. *Ultrasonics*, 9(2) (1971) 81.
[142] Altukhov, V.K. and Marshakov, I.K. Nov. Metody Issled. Korroz. Metal., pp. 183–188. Edited by: Rozenfel'd, r. L. Nauka, Moscow, 1973.
[143] Eliot, B.A. and Blackham, A.U. NASA Contract. Rep., NASA-CR-96279, 1968.
[144] Kadyrov, R.K. and Rakhmatullaev, N.G. *Tashk. Elektrokhimiya*, 21(12) (1985) 1666–1667.
[145] Rakhmatullaev, N.G. and Kadyrov, R.K. *Uzb. Khim. Zh.*, (2) (1984) 6.
[146] Ruvinskii, O.E. and Vyskubova, N.K. *Krasnodar. Elektrokhimiya*, 22(1) (1986) 130.

[146a]Kowalska, E. and Mizera, J. *Przem. Chem.*, 48(10) (1969) 601–604.
[147] Girginov, A., Lilov, E., Ikonopisov, S., Vodenicharov, Kh., and Lozev, M. *Bolg. Akad. Nauk*, 43(2) (1990) 53.
[148] Rasulov, K.R. and Kadyrov, R.K. *UZb. Khim Zh.*, 17 (1973) 21.
[149] Rasulov, K.R. and Kadyrov, R.K. *Elektrokhimiya*, 14 (1978) 636.
[150] Zhang, H. and Coury, L. *Anal. Chem.*, 65 (11) (1993) 1552.
[151] Miyanaga, S. Japan Kokai Tokkyo Koho, 4 (1991).
[151a]Lipson, A.G. et al. *Zh. Tekh. Fiz.*, 63(7) (1993) 187.
[151b]Pool, R. *Science*, 266 (1994) 1804.
[152] Gautheron, B., Tainturier, G., and Degrand, C. *J. Am. Chem. Soc.*, 197 (1985) 5579.
[153] Gautheron, B., Tainturier, G., and Degrand, C. *Organometallics*, 5 (1986) 942.
[154] Degrand, C. *J. Chem. Soc., Chem. Comm.*, (1986) 1113.
[155] Degrand, C., Prest, R., and Compagnon, P.L. *J. Org. Chem.*, 52 (1987) 5229.
[156] Degrand, C., Prest, R., and Nour, M. *Phosphorus Sulfur*, 38 (1987) 201.
[157] Shono, T., Kashimura, S., Ishifune, M., and Nishida, R. *J. Chem. Soc., Chem. Commun.*, (1990) 1160.
[158] Hengge, E. and Firgo, H. *J. Organomet. Chem.*, 212 (1981) 166.
[159] Hengge, E. and Latscher, G. *Monatsh. Chem.*, 109 (1978) 1217.
[160] Japan Kokai Tokkyo Koho, JP 0 3264 683 (1991). Shono, T., Kashiwamura S., and Nishida, P. Assigned to Osaka Gas Co.
[161] Japan Kokai Tokkyo Koho, JP 05 306 483 (1993). Shono, T., Kashiwamura, S., Nishida, P., and Murase, H. Assigned to Osaka Gas Co.
[162] Eur. Pat. App. EP 446 578 (1991). Shono, T., Kashimura, S., Nishida, R., and Kawasaki, S. Assigned to Osaka Gas Co.
[163] Japan Kokai Tokkyo Koho, JP 0 5306 340 (1993). Shono, T., Kashiwamura, S., Nishida, R., and Murase, H. Assigned to T. Shono/Osaka Gas Co.
[164] Japan Kokai Tokkyo Koho JP 0 5306 342 (1993). Shono, T., Kashiwamura, S., Nishida, R., and Murase, H. Assigned to T. Shono/Osaka Gas Co.
[165] 3 patents, Japan Kokai Tokkyo Koho, UP 05 230 317/8 plus PCT Int. Appl. WO 9 306 152 (all 1993) Shono, T., Kashiwamura, S., Nishida, R., and Kawasaki, S. Assigned to Shono, T./Osaka Gas Co.
[166] Akbulut, U., Toppare, L., and Yurttas, B. *Polymer*, 27 (1986) 803.
[167] Akbulut, U., Birke, R.L., and Fernandez, J.E. *J. Polym. Sci. Polym. Chem.*, 13 (1975) 133.
[168] Akbulut, U., Toppare, L., and Yurttas, B. *Brit. Polym. J.*, 18 (1986) 273.
[169] Eren, S., Toppare, L., and Akbulut, U. *Polym. Commun.*, 28 (1987) 36.
[170] Akbulut, U., Eren, S., and Toppare, L.K. *Macromol, J., Sci. Chem.*, A21 (1984) 335.
[171] Aybar, S.P., Hacioglu, B., and Akbulut, U. *J. Polym. Sci. Part A Polymer Chemistry*, 29 (1991) 1971.
[172] Yigit, S., Kisakurek, D., Turker, L., Toppare, L., and Akbulut, U. *Polymer*, 30 (1989) 348.
[173] Skotheim, T.A. (ed.). *The Handbook of Conducting Polymers*. Marcel Dekker, New York, 1986, Vols. 1 & 2.
[174] Walton, D.J. In L.S. Miller and J.B. Mullin (eds.), *Electronic Materials—From Silicon to Organics*. Plenum Press, 1991.
[175] Walton, D.J., Hall, C.E., and Chyla, A. *The Analyst*, 117 (1992) 1305.
[176] Walton, D.J., unpublished results.
[177] Osawa, S., Ito, M., Tanaka, K., and Kuwano, J. *Synthetic Metals*, 18 (1987) 145.
[178] Osawa, S., Ito, M., Tanaka, K., and Kuwano, J. *J. Polym. Sci. Part B Polymer Physics*, 30 (1992) 19.
[179] Japan Kokai Tokkyo Koho JP 03 280 519 (1991). Fukuyama, M., Kojima, T., Tsuchiya, S., and Kudo, Y. Assigned Matsushita Electric Industrial Co.
[180] Eberson, L. and Utley, J.H.P. In M.M. Baizer and H. Lund (eds.), *Organic Electrochemistry* (2nd Ed), Marcel Dekker, 1982.

[181] Torii, S. *Electro-Organic Syntheses Part 1 Oxidations*. Kodansha, 1985.

[182] Vassiliev, Y.B. and Grinberg, V.A. *J. Electroanal. Chem.*, 308 (1991) 1 and earlier papers in this series.

[183] Hawkes, G.E., Utley, J.H.P., and Yates, G.B. *J. Chem. Soc. Perkin Trans 2* (1976) 1709.

[184] Walton, D.J., Chyla, A., Lorimer, J.P., Mason, T.J., and Smith, G. *J. Chem. Soc., Chem. Commun.*, (1989) 603.

[185] Luche, S.L., Einhorn, C., and Einhorn, J. *Tet. Lett.*, 31 (1990) 4125.

[186] Walton, D.J., Chyla, A., Lorimer, J.P., and Mason, T.J. *Ultrasonics 'Conference Edition*, November 1989, p. 1241.

[187] Walton, D.J., Chyla, A., Lorimer, J.P., and Mason, T.J. *Synth. Commun.*, 20 (1990) 1843.

[188] Coleman, J.P., Lines, R., Utley, J.H.P., and Weedon, B.C.L. *J. Chem. Soc., Perkins Trans. 2* (1974) 1064.

[189] Koehl, Jr., W.J. *J. Org. Chem.*, 32 (1967) 614.

[190] Bonner, W.A. and Mango, F.D. *J. Org. Chem.*, 29 (1964) 430.

[191] Tashiro, M., Tsuzuki, H., Goto, H., Ogashasa, S., and Mataka, S. *Journal of Labelled Compounds and Radiopharmaceuticals,* XXIX (1991) 475.

[192] Tashiro, M., Tsuzuki, H., Goto, H., and Makata, S. *Chemistry Express*, 6 (1991) 403.

[193] Walton, D.J., Phull, S.S., Mason, T.J., and Lorimer, J.P., manuscript in preparation.

[194] Torii, S. *Electroorganic Syntheses, Part 2 Electroreductions*. Kodansha, 1986.

[195] Gigi, S., Paucescu, V., and Kurth, S. Romanian Patent RO 72 382 (1980).

[196] Ono, Y., Nishiki, Y., and Nonaka T. *Chem. Lett.*, (1994) 1623.

[197] Cheng, P.C., Nonaka, T., and Chou, T.C. *Bull. Chem. Soc. Japn.*, 64 (1991) 1911.

[198] Abobe, M., Matsuda, K., and Nonaka, T. *Denki Kagaku Oyobi Kogyo Butsuri Kagaku*, 62 (1994) 1298–9.

[199] Fry, A.J., Touster, J., Sirisoma, N.U., and Raimundo, B. In Little, D.R. and Weinberg, N.L. (eds.), *Electroorganic Synthesis* (Manuel M Baizer Memorial Symposium) Marcel Dekker, 1991, pp. 99.

[200] Touster, J. Ph.D. Thesis. Electrochemical and Sonochemical Reductive Silylations of Geminal Dihalides (Wesleyan University 1991).

[201] Fry, A.J. and Bujanauskas, P. *J. Org. Chem.*, 43 (1978) 3157.

[202] Fry, A.J. and O'Dea, J. *J. Org. Chem.*, 40 (1975) 3625.

[203] Fry, A.J. and Herr, D. *Tetrahedron Lett.,* (1978) 1721.

[204] Fry, A.J. and Lefor, A.T. *J. Org. Chem.,* 44 (1979) 1270.

[205] Frenzel, H. and Schultes, H. *Z. Phys. Chem.*, B27 (1934) 421.

[206] Crum, L.A. *Physics Today*, (1994) 22.

[207] Faulkner, L.R. and Bard, A.J. In W.R. Ware (ed.), *Creation and Detection of the Excited State*. Muriel Dekker, New York, 1976.

[208] Phull, S.S. The development and Evaluation of an Electrochemically Generated Chemiluminescent Immunoassay System. Ph.D. Thesis, University of Warwick (1990) and references therein.

[209] Walton, D.J., Phull, S.S., Lorimer, J.P., and Mason, T.J. *Ultrasonics*, 30 (1992) 186.

[210] Walton, D.J., Phull, S.S., Lorimer, J.P., and Mason, T.J. *Electrochimica Acta*, 38 (1993) 307.

[211] Vitt, J.E. and Johnson, D.C. *J. Electrochem. Soc.*, 138 (1991) 1637.

[212] Burgess, A. et al. Presentation to 4th Meeting of European Sonochemistry Group, Blankenburg, Belgium 1994.

[213] Zhivnov, V.A., Rumyantsev, I.Y., Tomin, V.I., Dezhkunov, N.V., and Prokhorenko, P.P. *Ser. Fiz.-Tekh. Navuk*, 3, (1976) 86.

[214] Zhivnov, V.A., Rumyantsev, I.Y., and Tomin, V.I. *Beloruss, Spectrosc. Lett.*, 10(9) (1977) 763.

[215] Walton, D.J., Phull, S.S., Colton, C., Richards, P., Chyla, A., Javed, T., Clarke, L., Lorimer, J.P., and Mason, T.J. *Ultrasonics Sonochemistry*, 1 (1994) S23.

[216] Dincturk, S. and Jackson, R. *J. Chem. Soc., Perkins Trans 2* (1981) 1127.

[217] Fisher, T.H., Dersheim, S.M., and Crewitt, M.I. *J. Dry Chem.*, 55 (1991) 1040.

[218] Willis, B.A. (ed.). *Mineral Processing Technology*, Fifth Edition. Pergamon, 1992.

[219] Zholshibekova, M.R., Ischenko, V.V., Baishulakov, A.A., and Malakhov, Y.A. *Komplenskn Ispol'z Miner. Syr'ya*, 2 (1983) 7.
[220] Chanturiya, V.A., Zholshibekova, M.R., and Malakhov, Y.V. *Elektron Obrab Mater.*, 2 (1983) 76.
[221] Chanturiya, V.A., Dimitrieva, L.L., Zholshibekova, M.R., and Malakhov, Y.V. *Vestn. Akad. Nauk Kaz SSR*, 9 (1984) 47.
[222] Shautenov, M.R. and Malakhov, Y.V. *Kompleksn Ispol'z Miner. Syr'ya.*, 3 (1988) 24.
[223] Chanturiya, V.A. and Dimitrieva, L.L. In A.M. Gol'man and L.L. Dimitrieva (eds.), *Flotatsionnye Reagenty*. Nauka, Moscow, 1986.
[224] Tsonkov, T., Paulapanski, M., and Djendova, S. *Oxid. Commun.*, 17 (1994) 321.
[225] Connors, T.F. and Rusling, J.F. *Chemosphere*, 13 (1984) 415.
[226] Schweizer, S., Rusling, J.F., and Huang, G. *Chemosphere*, 28 (1994) 961.
[227] Zhang, S. and Rusling, J.F. *Environ. Sci. Technol.*, 27 (1993) 1375.
[228] Rusling, J.F. *Acc. Chem. Res.*, 24 (1991) 75–81.
[229] Rusling, J.F. In A.J. Bard (ed.), *Electroanalytical Chemistry*. Marcel Dekker, New York, 1994, Vol 19, pp. 1–88.
[230] Rusling, J.F. In B.E. Conway and J.O'M Bockris (eds.), *Modern Aspects of Electrochemistry*. Plenum Press, New York, 1994, No 26, pp. 49–104.
[231] Franklin, T.C., Darlington, J., and Solouki, T. *J. Electrochem. Soc.*, 138 (1991) 747.
[232] Franklin, C., Darlington, J., and Adeniyl, W.K. *J. Electrochem. Soc.*, 137 (1990) 2124.
[233] Coche-Guerente, L., Deronzier, A., Galland, B., Martet, J., Labbe, P., Reverly, G., Chevaliei, Y., and Amhron, J. *Langmuir*, 10 (1994) 602.
[234] Kathirgamanathan, P., Souter, A.M., and Baluch, D. *J. Appl. Electrochem.*, 24 (1994) 283.
[235] Japan Kokai Tokkyo Koho, JP 06 093 483. Nonaka, S., Kunngi, Y., Tei, Y., Watanabe N., assigned to Permelec Electrode.
[236] Plattner, E. and Comninellis, C. In S. Stucki (ed.), *Process Technologies for Water Treatment*. Plenum, New York, 1988.
[237] Murphy, O.J., Hitchens, G.D., Kaba, L., and Verostrko, C.E. *Wat. Res.*, 26 (1992) 443.
[238] Stucki, S., Kotz, R., Carcer, B., and Suter, W. *J. Appl. Electrochem.*, 21 (1991) 99.
[239] Parsons, R. and Vander-Noot, T. *J. Electroanal. Chem.*, 257 (1988) 9.
[240] Comninellis, C. and Plattner, E. *Chimia*, 42 (1988) 250.
[241] Comninellis, C. and Pulgarin, C. *J. Appl. Electrochem.*, 21 (1991) 1403.
[242] Lim, C.L. and Byung, H. *J. Kor. Chem. Soc.*, 35 (1991) 762.
[243] Low, C.M.R. *Synth. Lett.*, (2) (1991) 123.
[244] Fillion, H., Refouvelet, B., and Luche, J.L. *Synth. Comm.*, 19 (1989) 3343.
[245] Mizera, J. *Waste Treatment and Disposal.*, Pol. 2 (1982).
[246] Nakaji, Y. and Oishi, J. Japan Kokai Tokkyo Koho, JP 7 890 705 (1986).
[247] Dylewski, R. and Pisarska, B. *Waste Treatment and Disposal*, Pol 8 (1988).
[248] Anon. Japan Kokai Tokkyo Koho, JP 9 175 434 (1992).
[249] Mhiyake, M., Hamaguchi, M., and Nomura, M. *Energy Fuels*, 31 (1989) 362.
[250] Urata, M. and Kamiya, K. Japan Kokai Tokkyo Koho, JP 7 887 069 (1980).
[251] Gauchon, J.P., Mordenti, P., Bezia, C., Fuentes, P., Kervegant, Y., Munoz, C., and Pierlas, C. Comm. Eur. Communities, [Rep.] EUR 10043, 1985.
[252] Sandstede, G., Koehling, A., and Schoenbucher, Ger. Offen., DE 4210950 A1 931007 DE 92-4210950 920402 (1993).
[253] Pollet, B., Lorimer, J.P., Phull, S.S., and Walton, D.J., manuscript in preparation.
[254] Bordeau, M., Biran, C., Leger-Lambert, M.P., and Dunogues, J. *J. Chem. Soc., Chem. Commun.*, (1991) 1476.
[255] Bard, A.J. *Anal. Chem.*, 35 (1963) 1125.
[256] Walton, D.J. *Manufacturing Chemist*, (1981) 44.
[257] Suzuki, C. Japan Kokai Tokkyo Koho, JP 56 047 600 (1981).

INDEX

Volume 2, 1991, 322 pp. $109.50
ISBN 1-55938-267-8

CONTENTS: Sonochemical Initiation of Polymerization, *P. Kruss*. Free Radical Generation by Ultrasound in Acqueous Solutions of Volatile and Non-Volatile Solutes, *P. Reisz*. Ultrasonic Studies of Polymeric Solids and Solutions, *R.A. Pethrick*. The Action of Ultrasound on Solidifying Metals, *O.V. Abramov*. Ultrasound in Chemical Processing, *N. Senapat*. Ultrasonic Organic Synthesis Involving Non-Metal Solids, *T. Ando*. The Influence of Ultrasound on Oscillating Reaction, *M.A. Margulis*. The Manipulation of Particles in an Acoustic Field, *C.J. Schram*.

Volume 3, 1993, 292 pp. $109.50
ISBN 1-55938-476-X

CONTENTS: Preface. Ultrasound and Colloid Science: The Early Years, *David Gilling*. Contributions to Various Aspects of Cavitation Chemistry, *Arnheim Henglein*. Sonochemistry: from Experiment to Theoretical Considerations, *Jean-Louis Luche*. Ultrasonic Agitation in Metal Finishing, *Robert Walker*. Ultrasonic Atomization, *Andrew Morgan*. Reations of Organosilicon Compounds in Acoustic Fields, *O.I. Zinovev and M.A. Margulis*. Use of Ultrasound in the Identification of Biological Molecules, *John A. Evans*. Initiation and Catalysis of Oxidation Proceses in an Acoustic Field, *Evgen M. Mokry and Volodymyr L. Starchevsky*. Index.

FACULTY/PROFESSIONAL discounts are available in the U.S. and Canada at a rate of 40% off the list price when prepaid by personal check or credit card and ordered directly from the publisher.

J A I P R E S S